Advanced Systems
Development Management

Wiley Series on Systems Engineering and Analysis
Harold Chestnut, Editor

Chestnut

Systems Engineering Tools

Hahn and Shapiro

Statistical Models in Engineering

Chestnut

Systems Engineering Methods

Rudwick

*Systems Analysis for Effective Planning:
Principles and Cases*

Wilson and Wilson

From Idea to Working Model

Rodgers

Introduction to System Safety Engineering

Reitman

Computer Simulation Applications

Miles

Systems Concepts

Chase

Management of Systems Engineering

Weinberg

An Introduction to General Systems Thinking

Dorny

A Vector Space Approach to Models and Optimization

Warfield

Societal Systems: Planning, Policy, and Complexity

Gibson

Designing the New City: A Systemic Approach

Coutinho

Advanced Systems Development Management

House and McLeod

Large-Scale Models for Policy Evaluation

Advanced Systems Development Management

John de S. Coutinho

A WILEY-INTERSCIENCE PUBLICATION

JOHN WILEY & SONS, New York • London • Sydney • Toronto

Library of Congress Cataloging in Publication Data:

Coutinho, John de S.
   Advanced systems development management.

   (Wiley Series on Systems Engineering and Analysis)
   "A Wiley-Interscience publication."
   Bibliography: p.
   Includes index.
   1. Systems Engineering. I. Title

TA168.C68   620.'7   76-30531
ISBN 0-471-01487-7

Printed in the United States of America

10 9 8 7 6 5 4 3 2 1

SYSTEMS ENGINEERING AND ANALYSIS SERIES

In a society which is producing more people, more materials, more things, and more information than ever before, systems engineering is indispensable in meeting the challenge of complexity. This series of books is an attempt to bring together in a complementary as well as unified fashion the many specialties of the subject, such as modeling and simulation, computing, control, probability and statistics, optimization, reliability, and economics, and to emphasize the interrelationships among them.

The aim is to make the series as comprehensive as possible without dwelling on the myriad details of each specialty and at the same time to provide a broad basic framework on which to build these details. The design of these books will be fundamental in nature to meet the needs of students, engineers, and managers and to insure they remain of lasting interest and importance.

Preface

There are many books and college courses on management. Graduates of such courses are sought by all kinds of commercial and governmental enterprises covering a wide range of activities: manufacturing, transportation, distribution, financing, regulation, and so forth. The stress is on management principles, not on the enterprise being managed. In general, the success of most of our enterprises supports this premise.

There is one exception: the management of the development of advanced systems projects, that is, large projects that incorporate significant advances in new technology. The concept of advanced systems was born in the military shortly after World War II when it became necessary to integrate advanced electronics, computers, power operated flight control systems, jet engines, airframes, "integrated logistic support," specially trained air crews and maintenance personnel, all into an efficient, "cost-effective," workable "system." This "systems engineering" technology, which has come to include sophisticated hardware simulators with man in the loop, was developed in bits and pieces over time by the military and eventually evolved at an unprecedented rate to the level of sophistication that made possible the success of Apollo. It constitutes a national treasure of unimaginable significance. This book is written to help preserve this treasure.

The growth of the advanced systems industry after World War II soon had a major impact on the nation's economy and the United States budget. However, the basic nature of the relationship between government and industry, and the "nonmarket" characteristics of the process were somewhat of a mystery to responsible U.S. decision making authorities. A significant new process was evolving. To provide an unbiased insight into this new process, independent of governmental and industrial participants, the Harvard School of Business Administration, with the support of the Ford Foundation, conducted a series of studies to analyze the new process (1, 2). These investigations did a great deal to provide our national leadership with a better understanding of the advanced systems development process. However, the studies were limited mainly to economic, policy, and legislative considerations, and did not attempt to cover the management mechanisms described in this book.

The special techniques required for the successful development of an advanced system, the "how-to-do-it-techniques," are still known to and understood by only a few individuals, who have been personally associated with large military or space development projects and have learned the techniques on the job. This situation was acceptable as long as advanced systems development was limited to military and space programs. However, the new technologies are finding their way into applications that are impacting on and changing the lives of every citizen. A much more widespread knowledge is now needed of these special management techniques. The fire at the nuclear plant at Brown's Ferry (115), which almost resulted in a widespread catastrophe, provides an example of what can happen when the management principles described in this book are ignored.

In contrast to normal management procedures, it is essential that the manager of an advanced systems development project be thoroughly grounded in the new technology being incorporated into the new system. His knowledge must be basic and comprehensive, going beyond the mere capability of using the right words. It is not enough for him to know that one can see through glass because "glass is transparent." He must know why he and his development team either do or do not understand why one can see through glass. In the management of advanced systems development, what managers don't know can hurt them by leading to wrong decisions. Unfortunately, most managers do not realize the limitations of standard management concepts when applied to the development of advanced systems, which readily explains why so many advanced system projects fail and why they represent such a great waste of our national resources.

This book discusses the development cycle, beginning with the recognition of the need for the new system, its technical definition, source selection, contracting, development, qualification, acceptance, operational testing, and release for production, as applicable to the type of system. In covering such a broad scope, only the essential principles of each phase are outlined.

Normally no one individual involved in systems development is active in all activities described in this book. An individual immersed in a specific phase of the advanced systems development process normally functions at a much deeper level than discussed here. However, the book should help him to relate his own responsibilities to associated activities comprising the entire process, and give him a better understanding of the requirements associated with his own functions.

For those who are unacquainted with the advanced systems development process, this book provides an overview that will enable them to understand the relationships of the various phases of advanced systems development, and furnishes them with a guide to the process that must be followed if their developments are to be successful. I believe that many individuals need this information as our concept of the world changes from one where man can exploit the resources of the universe with impunity, to one where we have come to recognize that our resources are limited and that our survival depends on maintaining a balance between highly interdependent ecological systems. We must create a new treaty with nature, in which we can live in harmony with each other and responsibly preserve our environmental inheritance.

I belong to a community that normally does not write books: that of design engineers who learn by doing. When designers speak, normally they are not understood except by peers who have served a similar long practical apprenticeship. However, in an extension of my design activities, I have been exposed to so many extraordinary experiences that I feel an obligation to share them with others. I participated in the initial organization of the U.S. Navy's BuWeps—Industry Material Reliability Advisory Board (BIMRAB), later renamed NAVAIR Systems Effectiveness Advisory Board (NASEAB), and served on the Board for the 12 years of its existence, as well as on its Administrative Committee and as an ad hoc member of its technical committees (144). This experience brought me into close personal contact with a cross-section of top naval and industrial management. Much of the material in this book is a result of this experience.

With a background in stress analysis and structural airframe design, I participated in the painful introduction of power operated flight control systems. I recognized the need to assure the integrity of highly complex power operated flight control systems in the same way that structural integrity is assured by stress analysis, margins of safety, and demonstration tests. This led to my organizing the first reliability and maintainability control group within a design department, and preparation of the first formal quantitative design trade-off studies involving reliability, maintainability, and other pertinent design parameters (117). In cooperation with the Navy's Bureau of Naval Weapons (BuWeps), I helped to develop the concept of the failure mode and effect analysis, which first appeared as a requirement in MIL-F-18372 (35), and prepared the first such analyses for approval by the Navy (116, 145). I prepared many reliability and maintainability proposals for aircraft and spacecraft projects, including those for the Orbiting Astronomical Observatory (OAO) and the Lunar Module of Apollo (LM), and served as Director of Reliability for the Lunar Module while the project was first being organized, then as Special Assistant to the Vice President-Program Director/LM. In this capacity, I became acquainted with the operation of most of the major subcontractors and suppliers associated with the Apollo program.

The initial formulation of this book was developed under the guidance of Dr. Edgar Roessger, Director of the Institute for Flight Operations and Air Transportation at the Technical University of Berlin, and of Professor Olaf Peters, thesis advisor, while I was associated with the Institute as a doctoral candidate. Since the members of the Institute were not involved in the U.S. procurement system, they were able to review the material from an unbiased academic viewpoint.

The concept of rigorous structural design control and performance assurance was developed in the military in the 1940s and early 1950s during a period of intense creative technical progress. One of the men to emerge as an outstanding leader at that time was Ralph L. Creel of BuWeps, who received the 1971 Spirit of St. Louis Medal from the American Society of Mechanical Engineers in recognition of his achievements. Mr. Creel has meticulously reviewed this manuscript and made numerous valuable suggestions. This book could not have been written without his active assistance.

Mr. Otto Fedor of Kennedy Space Center has spent considerable time reviewing the manuscript. He has used parts of the text in his graduate class on reliability management at the Florida Institute of Technology.

Others who have reviewed the manuscript and offered valuable suggestions are John Riordan, Department of Defense (now retired); Robert Bussler, Naval Air Systems Command; Lewis Jones, National Science Foundation; Clyde Meade, U.S. Army Management Engineering Training Agency; and George Breen, Grumman Aerospace Corporation.

This book is the sole responsibility of the author and should not be construed to represent official or coordinated Department of Defense or Army policy, philosophy, or viewpoint.

JOHN DE S. COUTINHO

Aberdeen, Maryland
March 1977

Contents

1. Introduction 1

    1.1   Impact of Technology on Society 3
    1.2   Systems Assurance 6
    1.3   Hardware or Engineering Models 11
    1.4   Software Models 16
    1.5   Management Structures 17

2. Systems Procurement 18

    2.1   Systems Quality 21
    2.2   Customer Versus Contractor Responsibilities 23
    2.3   Dominance of Schedule Considerations 30
    2.4   Correlation of Requirements and Test Results 31
    2.5   Chapter Summary 34

3. The Systems Development and Demonstration Cycle 35

    3.1   Concept Formulation 36
    3.2   Request for Proposal 42
    3.3   Development and Qualification 44
    3.4   Customer Operational Tests 48
    3.5   Chapter Summary 51

4. Contracting 52

    4.1   Design-to-Performance Procedures 54
    4.2   Design-to-Cost Procedures 59

4.3	Contractual Incentives	61
4.4	Competitive Developments	63
4.5	Scoring Methodology	73
4.6	Life Cycle Cost	81
4.7	Reliability and Availability	85
4.8	System Performance	87
4.9	Program Management	90
4.10	Performance Growth Potential	97
4.11	Chapter Summary	100

5. Technical Requirements **101**

5.1	Contractual Requirements	101
5.2	Extending the Technical State of the Art	106
5.3	Operability and the State of the Art	109
5.4	Feasibility	110
5.5	The Reality of Hazard	111
5.6	Reliability Requirements	113
5.7	Producibility	114
5.8	Subcontracting	117
5.9	Contractor Motivation	118
5.10	Maintainability	119
5.11	Logistics and Support	121
5.12	Safety Requirements	127
5.13	Standards and Commonality	135
5.14	Chapter Summary	138

6. Systems Engineering **139**

6.1	Scope	141
6.2	Historical Origins of Systems Engineering	146
6.3	Mainstream Functions in Systems Development	150
6.4	Subsystem Definition	152
6.5	Component Development Procedures	155
6.6	Illustration—Weight and Cost Control Procedures	161
6.7	Design Data Requirements	164
6.8	Subsystems Integration Testing	166

6.9   Simulation Testing 168
6.10   Qualification Testing............................. 177
6.11   Acceptance Testing 180
6.12   Chapter Summary 182

7. Design Assurance 188

7.1   Limitation of the Experimental Method 188
7.2   The Need for Reliability Control 193
7.3   Technical Causes of Unreliability 195
7.4   Reliability as a Probability 198
7.5   Statistical Treatment of Reliability 201
7.6   Engineering Techniques of Design Assurance 202
7.7   Effectiveness of Current Design Assurance Techniques 226
7.8   Chapter Summary 230

8. Organization of the Technical Development Staff 233

8.1   Aerospace Industry Objectives 234
8.2   Stability of the Technical Staff 236
8.3   Management of the Technical Development Staff   238
8.4   Responsibilities of the Technical Development Staff . 241
8.5   Project Responsibilities 247
8.6   Management of a Functional Unit 254
8.7   Evaluation of Project and Functional Leadership 256
8.8   Customer In-House Technical Capability 259
8.9   Chapter Summary 263

9. Systems Development Management 265

9.1   Management Responsibilities 265
9.2   Theory X Versus Theory Y 267
9.3   Policy Manuals 269
9.4   Planning, Scheduling, and Control 271
9.5   Planning... 273
9.6   Scheduling 286
9.7   Project Control 289
9.8   Monitoring Technical Quality..................... 292

9.9   Subcontractor Monitoring 292
9.10 Computer Applications 299
9.11 Quality of Paperwork............................. 304
9.12 Quality of Engineering........................... 307
9.13 Configuration Control............................ 310
9.14 Acceptance Procedures 315
9.15 Risk Management 316
9.16 Management Effectiveness 324
9.17 Chapter Summary 324

10. Principles of Quality Assurance 327

10.1   Definition of a Quality Assurance Program 330
10.2   Initial Quality Planning 335
10.3   Quality Assurance of a Function 339
10.4   Control of Engineering Documents................. 339
10.5   Control of Measuring Devices 344
10.6   Control of Personnel Performance 347
10.7   Control of Purchases 348
10.8   Subcontractor Quality Programs 351
10.9   Government Source Inspection 356
10.10 Receiving Inspection 359
10.11 Government Furnished Material 363
10.12 Process Controls 365
10.13 Completed-Item and Final Inspection 370
10.14 Inspection Status and Cost Monitoring 370
10.15 Chapter Summary 373

11. Operational Testing and Product Improvement 374

11.1   Customer Engineering Tests 376
11.2   Customer Field Testing 380
11.3   Reliability Growth Monitoring 382
11.4   Product Improvement 387
11.5   Chapter Summary 396

12. Professional Responsibilities **398**

12.1 Professional Orientation of Engineers 398
12.2 The Challenge of Our Times 400
12.3 Technical Advancements by the Aerospace Industry 401
12.4 The Customer-Contractor-Regulator Roles 405
12.5 Extensions of the State of the Art 405
12.6 The Goal—Systems Assurance 406

Index **419**

Advanced Systems
Development Management

1

Introduction

This book is concerned with large, complex, high cost systems that make use of advanced technology and that are generally produced in small quantities or are one of a kind.

Advanced technology systems, be they airways, nuclear power plants, food processing and distribution systems, or space systems, provide the heart of our technical progress; but along with their benefits, each one creates new specific hazards to the health and safety of the community. The pollution created by the horse traffic in New York City reached intolerable proportions around the turn of the century, and the automobile was welcomed as a saviour that not only provided then inconceivable personal mobility, but also eliminated a most obnoxious health and safety hazard. Today, the automobile takes its turn as a leading offender in creating pollution and endangering the health and safety of the community.

Hazards of New Technology

The impact of new technology on health and safety has become so great, and the details involved are so technically complicated, that new specialized government agencies, such as the Food and Drug Administration, the Environmental Protection Agency (EPA), and the Occupational Safety and Health Administration (OSHA), have been organized for the purpose of protecting the public. For example, the protection of workers, the public, and the environment against the hazards associated with the processing of asbestos and other toxic industrial materials has become a matter of serious public concern. These problems are compounded by the lack of data because of the length of time required for testing, as when cancer occurs 20 years after exposure, or when the exposure of female workers in the very early stages of pregnancy to otherwise safe levels of contamination may cause deformation in future generations (194).

These problems are so widespread that new political philosophies such as consumerism are becoming major national issues. A majority of the proposed bills before the U.S. Congress involve technological matters.

One of the basic common laws of the land, *caveat emptor* (let the buyer beware), a law with a tradition of thousands of years, is being changed to make the designer and manufacturer responsible for the safety of his product.

The number of liability suits involving product safety is increasing exponentially, and comprehensive court decisions are being made that are creating a new regulated operating environment for the design engineer (195). The environment in which our technical progress has been accomplished to date is changing in a very basic way (146, 147). Design engineers, who shape all technical progress and who have been autonomous within their respective domains, will have to learn to operate in a more controlled environment and to give formal consideration to safety, reliability, and social factors, with someone representing the public interest looking over their shoulders, watching their every move. They will have to document and place in a retrievable file the technical justification for every design decision, so that their companies can defend themselves in court years after the particular design engineer has departed from the scene. The cost of these new procedures will be considerable; they will be added to the price of products and may have a profound effect on productivity, on our business structure and economy, and possibly on our standard of living.

Systems Assurance Techniques

There are management approaches to systems development that provide reasonable assurance that new systems will be relatively safe and will operate as intended. The procedure requires the establishment of an empirical data base and the construction of applicable models, and consists of establishing formal requirements, monitoring the design process to assure that the system will comply with the requirements, and demonstrating that the system conforms with the requirements. This book records these proven systems development management techniques and makes them available for the solution of the many sociotechnical challenges that our society will be facing during the next decades.

The procedures in this book consist of guidelines that will normally increase the chances of success of a program manager who is developing a new advanced system. If a program manager does not follow these applicable procedures, his chances of success will not be very high; on the other hand, use of the procedures will not automatically guarantee success. The applicable procedures must be applied with understanding, intelligence and imagination.

Management Acceptance of Systems Assurance Procedures

Acceptance of these assurance management procedures did not come easily in the industries where they are practiced; in every case they were adopted only as the result of accidents or threat of imminent disaster. Some other industries that to date have escaped the public safety and liability spotlight, have examined these procedures and rejected them as too costly; they have, however, not developed any better alternative solutions. Some major aerospace firms, which believe that they honestly have tried to apply these techniques for over 20 years, doubt that the benefits have been worth the substantial investment; however, at the present time they, too, have nothing better to offer. Furthermore, in all probability, some of the techniques described in this book, when applied to civilian systems, will be challenged before the courts and will be subjected to modifications with the envolvement of new design and development philosophies that conform to new court decisions. However, some of the models used in this book, such as the ASME Pressure Vessel and Boiler Code (196), have already survived courtroom tests. At this time, this book represents a base line for an interesting development that will profoundly influence our future capability to exploit advanced technology for the common good and that will shape the foundations of our developing civilization.

1.1 IMPACT OF TECHNOLOGY ON SOCIETY

The role of technology in molding modern society is most interesting and dramatic (3). The history of just one major engineering discipline, transportation, reflects the entire story of civilization (31). For prehistoric man, the dawn of civilization began when he discovered that rivers provided bonds of trade and friendship between the villages and tribes that prospered along their friendly shores. With the domestication of the horse and the invention of the wheel and of the stirrup, foot paths became highways and trade routes; the benefits of civilization spread inland from the banks of the rivers. One development led to another and soon rivers reached out beyond their natural banks as men learned to build canals. By such engineering achievements, which eventually included the railroad, the steamship, and the telegraph cable, at the beginning of this century civilization of one sort or another embraced about one third of the population of the earth.

Looking at the other side of the coin, however, two thirds of the population of the earth lived in isolation. Men lived their lives as their fathers had done before them from time immemorial, without any idea that the world reached further than the limits of their hunting grounds. Such men do not possess the concept of progress and do not dream of the future. Their myths and legends are concerned with the glories of the far past. As individuals, they can only aspire and hope that their lives will be as good as that of their fathers. This state of affairs, as old as the history of man on earth, has suddenly undergone a significant and dramatic change during this century.

Within our generation, the airplane has opened up access to all inhabited points on our globe. Today no significant group of people exists anywhere in the world who has not had some contact with twentieth century civilization. The impact of this development, the awakening of emerging nations embracing two thirds of the earth's population, promises to keep the world in a state of ferment for centuries to come. Conflicts arise as new nations attempt to modify their traditional values to make them compatible with those portions of western civilization that they decide to absorb. No serious statesman pretends to know where this will all end (118).

The impact of the airplane on the course of human history carries an important lesson for engineer and statesman. This impact was not foreseen or intended by the inventors of the airplane. Flying developed as a sport. Early airplane enthusiasts did not conceive that aircraft would ever have any significant commercial value. Much less did they envision that aerospace technology would some day develop weapons that could destroy cities and nations.

1.1.1 The Age of Power

In the past it generally has not been possible for the creators of new technical devices to foresee, even in their wildest dreams, the full impact of their inventions on society. Among the earliest boyhood recollections of the author is being awakened each morning by the klippety-klopp of the milkman's horse. Milkman and horse made a fine working team; the horse pulling the wagon and stopping at exactly the right place in front of each house, while the milkman walked along and up to each house with his deliveries. Alas, the friendly horse has now been replaced by the much more efficient milk delivery truck and supermarket. It is so much cheaper to generate horsepower by means of an internal combustion engine, that an entirely new milk delivery system is evolving. Even the tradition-bound U.S. Army Cavalry has replaced all of its beautiful, brave horses with mechanical powered devices. Those of us who are saddened by the obsolescence of the horse, acknowledge the technical superiority of the internal combustion engine, but wonder why, with such powerful technology available, we are forced to sacrifice so much of the charm we have come to cherish. Perhaps this vague sense of loss of beauty underlies our discomfort with the hectic tempo of life that seems to follow in the wake of technical strides.

These examples are only a small part of the total picture. Just as the horse has become obsolete as a source of power, so has the human slave laborer. Human muscle cannot compete with power tools. Human slavery has become as obsolete as has the horse, and for the same reasons: economics. Greek and Roman civilizations had the highest concepts of human dignity; but their state of technology allowed the upper classes to enjoy the good life only with the use of slavery. The achievements of statesmen and politicians are limited by the possibilities inherent in the technology of their times.

1.1.2 The Systems Approach

With the continuing application of our advanced technology, we foresee still more tremendous growth in our country. However, unless some present-day engineering practices are modified, life in the future may degrade rather than improve.

Today, for example, a community needs no longer tolerate inherited methods of engineering or modes of transportation that may be inadequate or undesirable from a social viewpoint, such as one river or one railroad. We have the resources and capabilities to provide whatever engineering works or modes of transportation are most socially desirable and humanly effective in any given situation.

Inherent in this new systems approach is the prior identification of the true needs of the community as a basis for developing an "optimum" solution, as well as the better definition of the social responsibility for the adverse effects associated with the introduction of technical advances. To establish the design requirements for his hardware, the engineer of the future will have to dig deeply into social relationships never before considered to be his concern. This is the systems approach in its broader sense, and it involves not only the engineer, but also the humanist and the politician, the sociologist, lawyer, economist, psychologist, geographer, historian, and experts in other disciplines.

Adjusting to a New World

So far within this twentieth century, more technological progress has been made than in all previous centuries since the beginning of time. A new technical world has been created with possibilities for the welfare of mankind that exceed anything ever imagined by men before. In contemplating these new opportunities, it is important to realize that all of our traditional institutions, political, legal, social, religious, educational, were founded and configured well before the beginning of this century. All of these institutions are based on concepts or models that today are no longer completely valid. All are geared to solving the problems of previous centuries. For example, of today's pressing problems, such as the welfare state, job security, welfare and the war on poverty, unionization, civil rights, social justice, women's liberation, urban renewal, mass transportation, environmental control, medicare, global defense, labor strikes, civil service strikes, guaranteed annual wage, and many more, none are mentioned in the Constitution of the United States. This situation, which impacts in a basic way on every aspect of human activity, is a new one; something like this, simultaneously embracing all of mankind, has never existed before in the history of the world, and there is little guidance that we can take from the past. In former times there have been local revolutions; today all of the world is in turmoil. All at once all of the basic institutions in our society face the fundamental problem of having to rethink and redefine their roles, their objectives, and their responsibilities in a new world offering new and tremendous possibilities, based on the technological advances of the past half century (148).

Developing a New Design Engineering Outlook

Although civilization has developed as a function of technical progress, the equation describing the process has remained a secret. Not all results of technical progress have been good, indicating that many engineering designs have not included adequate consideration of a balance of human needs.

In the 1930s we went through a phase when the technical community was accused of being antisocial because its machines were putting men out of work. Perhaps we are now approaching another period of public disfavor because we stand accused of creating safety and pollution hazards. In order to cope with these problems, the design engineer, in addition to his technical capabilities, must develop a thorough understanding of the impact of modern technology on society, and he must assume a new sense of responsibility for the safety and reliability of his systems.

1.2   SYSTEMS ASSURANCE

The industrial revolution of the nineteenth century provided men with the means for creating new systems based on the application of the scientific insights developed during the previous century. The steam engine provided a new source of power. Mechanical engineers rose to the challenge of providing every house in American cities with gas for illumination and cooking. Civil engineers laced the country with railroads, building enormous bridges to span our great rivers. Electrical engineers provided our commercial centers with telegraph service, and later made sure that every family had its own telephone. Mining, metallurgical, and chemical engineers discovered new sources of materials and developed new processes for mass producing them in the required quantities. Mankind acquired large "systems" such as factories, locomotives, and transoceanic liners. *Systems performance generally was the result of whatever could be achieved by the assembly of available components.*

Since World War II, however, the word "system" has taken on a new meaning. The development of control technology coupled with the invention of electronics and of new materials and power sources has made it possible to assemble specialized components into new advanced systems whose performance is increased by orders of magnitude. A new management approach has evolved: *system performance requirements are established in advance* to satisfy a defined need, and new specialized advanced technology system elements are invented, designed, and integrated on schedule so that the system will attain the required performance on the promised delivery date. A new methodology of formal systems assurance has been established to monitor the design, testing, and integration process *to assure that the various phases of component design and system development are carried out so that the required system performance will be attained.*

1.2.1   Advanced Systems Development

Human experience with the successful acquisition of very large new advanced systems exists only since World War II, and then only in three related areas: the development of major military systems, the application of atomic energy, and the space program. The applicable development procedures and techniques were evolved in bits and pieces over time, initially by the military services.

The impact of advanced technology on modern systems has resulted in dramatic increases in effectiveness, but only at equally dramatic increases in cost. Investments totaling billions of dollars are required. The design, development, and deployment of these new systems consumes considerable time, up to a decade. The risks associated with such systems-development projects are high. The men responsible for the expenditures of such large sums of public funds must have reasonable assurance that the projects to which they are dedicated will turn out to be safe and successful.

The engineering-management techniques for the development, acquisition, and assurance (including safety) of such new systems, which will not come into existence until after many years of intensive development, are different from the procedures for the procurement of existing commercial products, services, or systems. These are offered for sale on a competitive basis in the public market. To "acquire" a new technically advanced system, it is not possible to go into a market, point to an existing item and say: "I want one like that one on such and such a date."

What, then, are the procedures for the acquisition of new technically advanced systems, and who are the experts in the application of these principles and techniques?

The creation of advanced, safe, and reliable systems requires the utilization of large numbers of inventive engineers. Throughout the history of civilization, there have been organizations of large teams of men, such as those who built the pyramids or formed the great armies of World War II. But the many thousands of men in these organizations were not acting as creative individuals; they followed orders in executing the well worked-out details of a master plan, created and controlled by a small group of leaders. Most workers could have been replaced by machines, had these existed with the capability of performing the desired functions. In contrast, the development of new technically advanced systems is based on the application of specialized adaptations of advances in the "state of the art" at all system levels, requiring the coordinated contributions of teams of thousands of motivated, professional, and creative technical specialists. Such individuals do not respond to the same controls and motivational forces as do noncreative workers in very large organizations. The engineering management of the development of new systems represents a completely different and new problem, *one which has never before existed in the history of the world*. It is a challenge that must be met if our society is to continue to profit in full measure from the advances of modern science.

1.2.2   The Apollo Program

The largest single peacetime systems-development project so far has been the Apollo program, which had the objective of landing men on the surface of the moon and returning them safely to earth under unprecedented and imaginative full worldwide television coverage. As never before in the history of a nation in peacetime, men organized and managed, with the active support of the military, the required special skills and resources wherever they could be mustered throughout our vast continent and throughout the world, in a great team effort to satisfy the requirements of an inspiring national goal. The cost of the Apollo program was originally estimated at $30 billion; some eight years later it achieved its initial success at a cost of approximately $24 billion.

This example is all the more interesting because the Apollo Program was administered by a newly organized developer, the National Aeronautics and Space Administration (NASA). The necessary special management technology was transferred to NASA from the then existing military/industrial complex. This transfer made possible the spectacular success of the Apollo program which exceeded all initial expectations. The maintenance of a predetermined launch schedule for even one moon flight was not considered likely only three years before the first Apollo flight. Yet the initial Apollo moon flights proceeded within seconds of a schedule established many months in advance. The many contractors, who normally compete fiercely with each other, all worked together for the common goal as if they all belonged to the same volunteer fire company. The dedication of the individuals involved reached new heights of personal involvement (119). Although tremendous sums were involved, it is noteworthy that no one became really rich as is common practice in other industries. The conquest of the moon has required the highest degree of organizational and management talent; yet little of this was directed toward personal self-enrichment. No other industry involving similar sums of money can lay claim to such personal dedication; oil, railroads, steel, motion pictures, automobiles, all have produced their fair share of millionaires. The Apollo Program constitutes the single achievement in the second half of this century in which all Americans can be intensely proud; it is an achievement in which millions of people all over the world rejoice with us.

1.2.3 Future Applications

The systems engineering-management and assurance know-how that makes possible the success of programs such as Apollo constitutes a national treasure which should be widely studied and understood. It should be made available to other systems development projects, where the professional application of this know-how will determine the success or failure of the project, with corresponding impact on the national security, welfare, and economy.

At present these critical systems engineering-management techniques are known to, and understood by, a very small group of individuals who all have been personally associated with large military or space development projects and have learned the techniques "on the job." Lack of widespread understanding of these procedures was not important as long as defense systems were small. Today, however, expenditures for advanced military systems constitute a significant percentage of the national budget and affect the economy and welfare of the country in many ways. Furthermore, advanced technology is finding its way into almost all phases of modern life: transportation, communications, manufacturing, distribution, energy systems, housing, and so on. It, therefore, becomes more important that the fundamentals of systems procurement and assurance be more widely known and critiqued, among government policy makers as well as in the academic and industrial communities.

Importance of Advanced Systems Management

The management of the development-engineering process incorporating advanced technology is an aspect of a systems-development project that spells success or failure. It is emphasized that if those responsible for an advanced systems-engineering project do not master at least the level of technical material included in this book, there is little likelihood that they will be able to execute effectively a large advanced-systems development project. There is no shortcut to the hard work needed to develop a mastery of the technical aspects involved. The history of defense systems procurement is full of expensive, but unsuccessful, projects that bear witness that, even in the defense industry, the applicable and necessary assurance procedures are not universally understood or followed.

In the future our society will face national systems-engineering problems of similar magnitude and technical difficulty as those associated with Apollo. It will be difficult, however, to recreate the level of enthusiasm and professional dedication associated with the conquest of the moon. Yet future design requirements for new systems will call for greater efficiency and cost effectiveness than was achieved on the Apollo Program. To satisfy these requirements for greater efficiency it will be necessary to develop more effective procedures than those used on the Apollo Program.

In the attempt to highlight and clarify this particular problem, this book generally avoids the discussion of management or organizational structures which would only confuse the problems discussed. Even with this limitation, the scope of this book is very broad; as such it avoids in-depth discussion of individual techniques. These are already adequately recorded in individual textbooks, such as the other volumes of the Wiley Series on Systems Engineering and Analysis. This book supplements the series by presenting the management fundamentals and the overall framework into which individual techniques must be fitted harmoniously and *presents an overview of the scope of advanced systems development management activities*. In particular, the systems engineering methods in

Chestnut (4) are reviewed here from the management and applications viewpoint.

1.3 HARDWARE OR ENGINEERING MODELS

The traditional engineering procedure for analyzing past experiences and recorded data, so that they can be used by others, is to describe the experience in terms of generalized equations, and with these equations to construct the "models" that are the tools of the professional planner (5). The systems designer, planning for and solving the problems of the future, is dependent on the models that express the experiences of the past in applicable, generalized terms.

Hardware development is based on predictive deterministic mathematical models (formulas, procedures) that identify the hardware characteristics (variables) that must be controlled, and that describe the relationships between the variables. Although these models are always simplifications of the real world, they provide the professional tools for predicting hardware performance and for designing tests that can demonstrate with a high degree of confidence that the hardware complies with requirements. The control over all critical variables is *absolute;* if control over any single required hardware characteristic is found to be deficient, that is, the component will not operate as required, the engineers involved cannot rest until the problem is solved. Extensive laboratories are available to generate data to develop and verify hardware models; to determine the characteristics and properties of materials, subsystems, and systems; and for developing methods for demonstrating compliance with requirements.

The ideal procedures presented in this book seldom are applied in pure form, they always are modified to suit the particular circumstances. However, it should be possible to arrive at more effective solutions to individual problems if the basic principles are thoroughly understood and used as guidelines.

Equipment Complexity

As equipment and systems become larger, more complex, and more sophisticated, the associated predictive models will include a greater number of variables, and some of them may be very difficult to measure and control. An adequate data base may be lacking. The time available for design, the characteristics of available instrumentation, test procedures, and mathematical techniques limit the number and complexity of the variables that can be considered and processed. To be of practical use in the time-restricted, schedule-oriented, dynamic design process, models for sophisticated equipment are always oversimplified in a trade-off between usefulness versus accuracy.

In the application of models it is always necessary to remember that *engineering is the art of skillful approximations.* A rough estimate of the right variable is much more significant than a precise measurement of the wrong parameter. To some extent, if the design engineer is aware of or recognizes a necessary requirement and makes what he believes to be proper provisions, even though he cannot perform a quantitative analysis, the device may perform satisfactorily. It is primarily requirements, interfaces, and considerations that are unknown or that have been ignored or neglected that cause subsequent trouble.

However, "designer awareness" of the importance of a particular characteristic is by itself usually not adequate to produce a satisfactory advanced system design; in most cases control by reference to quantitative models is required.

1.3.1 Hardware Design Control

Complex models for sophisticated equipment and systems are evolved over time from simpler models in parallel with the development of more advanced equipment out of the experience with simpler equipment. In the process of developing more sophisticated models, one of the key tasks is the recognition, selection, and definition of the new variables and parameters required in the improved model to describe the increased performance of the more advanced equipment with the required accuracy, and the creation of an expanded data base.

This refinement of design control by the addition of new variables follows a set pattern. The approach can be best illustrated with a simple example, namely the introduction of formal weight control in the aircraft industry in the early 1940s.

Need for Weight Control in Aircraft Design

Early aircraft designers were very aware of the importance of weight. They went to great extremes to build their hardware as light as available technology would permit. Nevertheless, they felt no need for formal quantitative weight control procedures. The limiting factor in aircraft design was the performance of the power plant. Most quantitative analysis in the early days of aviation was concerned with making maximum use of the available power.

As propulsion plants improved and aircraft grew in size, there arrived a time when propulsion was no longer the limiting factor. Other parameters gained in significance, among them weight. The need for paved runways and the strength of the pavement, for example, became a limiting factor. On naval aircraft carriers the strength of the landing decks and the capacity of the catapults and arresting gear established limits on take-off and landing weights.

With increasing performance, there appeared an increasing number of factors that previously had been neglected without impairing operational suitability, but which now had to be considered. For example, in the early 1940s, stability became an important consideration in the design of military fighting aircraft under operational conditions. Aerodynamic stability was poorly defined, and there developed the recognition that control stick forces developed by the pilot required standardization. To solve this problem, much more precise weight and balance control was required. A great deal of effort was expended to simulate standardized control "feel" between one aircraft and another, using springs and bungees, until the growth in aircraft size required boost devices and artificial control systems with commonality of operating response.

Under these circumstances it became important to specify contractually, among other new characteristics, the maximum allowable weight, and even to couple this requirement with financial incentive-penalty clauses to assure compliance with the specification. In order to demonstrate compliance with the specified requirements, methods of measuring the specified variable had to be established. Demonstration of compliance with weight requirements posed no particular problem, since scales were already available.

The Introduction of Weight Control Procedures

Methods of weight control during the design and development process, however, had to be developed. In this case, satisfactory theoretical methods for estimating weight based on engineering drawings were already available; the volume of parts could be easily calculated, and the density of materials was known; given this data the weight is easily calculated. A data base consisting of the weights of existing components and systems also existed. The major problems involved the establishment of policies, the design of the application, monitoring and enforcement procedures, and the assignment of responsibilities within the design engineering organization.

Given effective specification and design control procedures and methods of measurement, the cost effectiveness of weight control still had to be determined. The application of these new procedures increased the development costs by a significant amount, and the expenditure of these additional funds had to be shown to be worthwhile. Requirements for weight control based on an analysis of past experience were therefore first negotiated into a few carefully selected development contracts where weight was a particularly important consideration. Estimates were prepared of the cost of compliance with the new requirement and balanced against the value of the resulting benefits. Since it was shown that the benefits exceeded the cost, the application of the technique was expanded to other contracts. With the general acceptance of the application of the new weight control procedures, manuals and handbooks were developed to standardize the procedures throughout the industry.

1.3.2 Engineering Handbooks and Manuals

Acceptable engineering models and procedures are normally recorded in engineering handbooks and manuals. Most engineering textbooks consist of the presentation of the derivation, description, and application of engineering models in a specific field. Many codes and standards (Section 3.2.2) describe the engineering models and demonstration procedures that may be mandatory or acceptable for contractual or regulatory purposes in specific applications. All military services publish a series of design handbooks; these are based on the peculiar environmental operating requirements and needs of each service. Products designed and tested in accordance with the models and procedures in these handbooks normally will be acceptable to the service involved.

In each basic engineering discipline there will be a number of generally accepted handbooks that provide encyclopedias of available models and demonstration procedures.

Individual companies develop their own handbooks in specific fields (structural analysis, heat transfer, etc.) which represent subsets of generally available models most suitable for a company's products (see Section 9.3). In addition, these company manuals may also contain proprietary models and procedures that the company may have developed at great cost and that give it a competitive advantage in the marketplace. Many company handbooks provide designers with data on performance versus quality versus cost, to enable the designer to achieve the desired performance and quality at the lowest cost.

New engineering models are usually generated by individuals or research groups (120) and made known to the engineering community in the form of technical papers presented at meetings of engineering societies. Most engineering societies have peer review procedures to assure the technical quality of the papers presented at their forums and published in their journals. It is common practice for individual engineers to collect significant technical papers directly applicable to their work and to assemble them into a manual for personal use. The general acceptance and use in a given organization of a particular paper will lead to the inclusion of the pertinent models in the company's manual. From this point the new model may find its way into industry codes and standards, into academic textbooks, and finally will be included in encyclopedic engineering handbooks.

1.3.3   Introducing a New Design Control Technique

The introduction of control of a new variable in the system development process will involve the following six steps:

1. Recognition of the variable to be controlled and its precise definition.
2. Development or selection of methods of specification and of measurement, including the creation of an appropriate data base.
3. Development of design-control procedures.
4. Selection and performance of test applications.
5. Analysis of the test applications and determination of cost effectiveness.
6. Development of manuals and handbooks in support of general application.

The introduction of every new variable in the system development process follows this procedure. For example, in addition to weight control, the analysis of the static strength of airframes has completed this cycle. Stress analysis today is such an accepted design control activity in airframe development that it is difficult to remember that its application was a very controversial subject as recently as 1930.

Many currently important variables have not yet completed the full cycle of the six steps noted above. In strength analysis, the full cycle can only be associated with static loads. Available models involving dynamic loads do not predict the performance of structures under repeated loadings with the same accuracy as do those for static loads. Grover (36) contains an overview of the available knowledge in this field, but provides little advice on procedures for providing quantitative assurance against fatigue failures. Smith (37) presents a summary of basic fatigue design principles, but this document does not have the status of a handbook. Although the fatigue problem has been studied for over 100 years, that is, ever since power machinery has been strong enough to impose significant repeated dynamic loads on metal structures, a generally validated and accepted data base and design manual still does not exist.

1.4 SOFTWARE MODELS

The procedures discussed in this book are based on the rigorous application of the scientific method as they are universally applied in the development of sophisticated hardware. The general procedure for the construction, testing, validation, and use of models is well understood and followed in the development of such hardware.

An advanced system development project, however, involves *hardware, software (procedures), and management structures (organizational relationships)*. All three are essential to operational success as well as to assure safety, quality, and reliability, and are of comparable economic significance. Yet only the hardware is subjected to rigorous design and quality control procedures; these usually are completely lacking in the other areas. Only recently, for example, has software become recognized as an essential component of complete operating systems; the term is not widely understood.

The most common use of the term "software" today is associated with computer programs, that is, with instructions to the computer to perform required operations. In this book the term "software" is used in its original general definition to encompass all instructions, procedures, manuals, data, reports, and other paperwork associated with systems development and operation. No reference to computer operation is implied or intended.

The development of models in the software area is just beginning. Related laboratory testing is still restricted to the demonstration of simple theories. The scope of the variables to be controlled has not been determined. There is practically no laboratory testing to demonstrate that software systems comply with design requirements. The concept of quality control of software is almost completely lacking.

This book attempts to extend the proven approach of hardware models to the field of systems-engineering management through the use of software models. With the increasing size of system-engineering projects, the quality of the technical project administration has become a critical factor, often determining success or failure of the project.

The author hopes to satisfy a new critical need with the definition of the software models presented in this book, and to stimulate the development of more intensive laboratory testing to validate and improve these software models.

1.5 MANAGEMENT STRUCTURES

In the development of management structures, only vague concepts are available; the concepts of models or of laboratory testing are completely unknown. Management structures are designed with reference to case studies by trial and error in the most casual manner, and decisions normally are made by means of the time-limited conference. The responsible individuals are assembled in a conference room and given a limited amount of time, say two hours, to arrive at a decision. If things do not turn out as expected, a subsequent conference is called, and the management structure is rearranged. Under such circumstances, a single strong personality can obtain decisions that may not be best for the organization or for the project.

Available management textbooks picture the manager as a reasonable man trying to solve problems in accordance with the principles expounded in the textbooks. His objectives are always clear and well-expressed. In actual fact, many organizational structures are built around the strength of dominant personalities. Such organizational structures serve as power bases fully as much to serve the personal needs of strong individuals as they do to serve the needs of the corporation or its customers. It is not always easy to identify the factors that make an organization tick or that make a great football team; or is it always possible to establish what forces motivate an individual manager in a particular power structure.

In their discussion of organizational structures, it would appear that standard textbooks generally omit consideration of the four following basic principles:

1. The Peter Principle: In a hierarchy every employee tends to rise to his level of incompetence (6).
2. Parkinson's Law: Work (including paper work) expands so as to use up the resources and to fill the time available for its completion (7).
3. Murphy's Law: If anything can go wrong, it will.
4. Coutinho's Conclusion: From the set of inappropriate but possible systems, we must strive to fashion the one which is least lousy.

These four principles define an environmental framework that automatically precludes an ideal or optimum system. Success is associated with a system that works. A bad system will work if the people involved want it to work and will devote sufficient time and effort to make it work. An acceptable system is one that people are willing to put up with. In striving to achieve the "least lousy system," we admit to our faith that progress is possible and worthwhile.

2

Systems Procurement

The word "system" has different meanings for different individuals. Every individual will have a viewpoint typical of his own scope of interest and activity.

Air Force Regulation 375-1 (38) defines a "system" as "a composite of equipment, skills, and techniques, capable of performing and/or supporting an operational goal." In analyzing this definition, Medlock (197) comments that the implementation of this broad concept requires skillful execution by a well-planned management "system," and he himself defines the word as "a formalized way of pulling together elements of the process of management into an integrated network, designed to achieve certain established results." Feigenbaum (198) carries this management system concept one step further. He identifies vertical and horizontal frameworks in an organization. Vertical frameworks are those associated with a particular discipline or function. Horizontal frameworks are concerned with end product characteristics and contain within one organizational jurisdiction all of the functions and disciplines necessary to produce a specific characteristic of the end product. Feigenbaum associates the word "system" with a "horizontal organizational framework" which integrates all of the various functions required to achieve a desired product characteristic.

There are many other definitions of the word "system." In one source (199), a system is defined as "the interacting equipment and humans directed by an information subsystem to achieve a specific objective." A transistor manufacturer will insist that a transistor is a system, and a complicated one at that. For the radar manufacturer, the transistor is an element or a subcomponent; the radar is the system.

In all of these definitions, the common denominator is the objective-oriented integration of lower-level interacting components, techniques, procedures, skills, information, and personnel into higher-level units for the purpose of achieving the preconceived objective. However, even this generality is not universally accepted. There are some specialists who define a system as a device or assembly of devices that require maintenance, while a "component" is not subject to maintenance.

Computer-Based Systems

Goode (8) is probably most representative of those engineers who classify themselves as "systems engineers." His definition requires a system to include a computer, that is, its capability must exceed that of a human in terms of speed in executing functions and/or manipulating volumes of data.

An early example of this concept was the American Airlines Reservation Center at LaGuardia Airport in the early 1950s. Up to that time, this airline had a very sophisticated and well-organized manual reservation system; nevertheless, during the mid-day hours when the telephone calls exceeded 2000 per hour, the manual system could no longer keep track of sales. The same seats were being sold simultaneously to different customers. The solution to this problem was found by putting the airline seat inventory on a computer. Each sales clerk was provided with a console with the capability to search the entire inventory and to record sales on an instantaneous real-time basis.

According to Goode, this computer capability is the heart of a "system," and without it an assembly of components is not a "system." Note that size is not an explicit factor in this definition.

Hardware Systems Concept: Size, Models, and Unique Components

The author finds the above computer-based definition wanting and prefers to define a system on an economic basis, whereby size is an explicit factor. The word "system" as used here is hardware oriented and leans on AFSCM 375-1 (38), as does Chestnut (4). It applies to an assembly of interdependent equipment, information, procedures (software), personnel, facilities, and a management structure, assembled and integrated for the purpose of accomplishing a set of interrelated missions or providing a range of services within a given scope of environmental conditions. A distinguishing characteristic of a system is that it requires an expressed or implied model to predict the interrelationship of all of the components; specifically a model relating the performance characteristics of individual components to system performance. The mathematical solution to a system problem is accomplished through the solution of simultaneous equations that express the impact of varying component performance on one another and on system performance. Individual components cannot be optimized individually; improvements in an individual component will normally impact on the performance of the other components, destroying the balance within the system and degrading the overall performance. The problem, therefore, can be solved properly only by the use of simultaneous equations. The difficulties in systems engineering are associated with providing the required degree of completeness in the model as well as with the lack of adequate techniques for identifying, defining, and solving the simultaneous equations.

One special feature of this systems definition is that at least some components are designed and produced for exclusive use in that system. The system, therefore, must be large enough to provide a market that will absorb all of these special components, while permitting the component manufacturer to make a profit.

Some Well-Known Examples

The telephone system with its capability to absorb specially designed telephone instruments is a good example of the systems concept discussed in this book.

Another system that fits this economic definition is the nation's railroad system. This system was designed well before the day of the computer, and it affected markedly where and how large numbers of people lived and worked in the United States for decades before other means of extensive travel and communications were developed. In this country the railroad system consists of a number of separate and competitive, more-or-less giant corporations. Nevertheless, satisfactory system operation depends greatly on the "commonality" concept which often permits indiscriminate mixed usage of rolling stock of the various corporations over the rails of competitors.

Still another well-known example is provided by our automobile transportation system. The system goal might be defined as providing safe, reliable, comfortable, and stylish point-to-point transportation between any two roadside points on a continent, at any desired time of day or night, in any kind of reasonable weather, at reasonable cost and time. This system consists of automobiles; roads and gasoline stations; car maintenance and repair facilities; road maintenance; driver training and certification; motels, restaurants, and rest facilities; off-road parking; traffic control; insurance and security (police and courts). The manner in which these components are designed, integrated, and balanced one against the others will determine the "quality" of automobile travel. Should any one specific component, say a particular police force, be optimized locally to achieve 100% parochial performance, apprehending all traffic violators in their jurisdiction, the performance of the overall system is affected in a complex manner. Although system safety may increase locally because of the elimination of accidents, the volume of traffic will decrease, and there will be a further slowdown as bottlenecks develop from motorists competing with each other in a cautious attempt to avoid violations. The traffic load in adjacent communities will increase as through traffic attempts to by-pass the slowdown in the affected community.

System safety as measured by the number of annual fatalities is a sensitive indicator of the balance between system components. The automobile transportation system in the United States is balanced at a level of about 50,000 fatalities per year. In contrast, the worldwide scheduled airline system operates at a level which costs about 800 lives annually. The safety of both systems could be improved, but effectiveness would decrease and costs would increase. Both systems are therefore balanced at a specific level of cost effectiveness versus safety which our society has been willing to tolerate.

2.1 SYSTEMS QUALITY

A general difficulty with the application of the systems approach is that missions often are conceived within too narrow a scope. Missions usually are defined in terms of what a supplier can conveniently provide, rather than of what the customer needs and wants. Hence, we find the airlines providing excellent transportation service between take-offs and landings while the customer, stuck in traffic between city and airport, might well wonder what the systems approach is all about.

This example also points out that the systems approach is not exactly a handy management tool for the individual stuck in traffic. It is a management tool for the design of large systems, and it begins with the formal analysis of the complete needs and wants of the user, including safety, reliability, maintainability, support and logistics requirements, life-cycle costs, and trained personnel. The execution of such research and analysis, and the development and preparation of correct specifications, requires considerable technical sophistication, normally available only to large organizations such as the military, the airlines, and the telephone and power companies.

2.1.1 Excellence in Systems Technology

Surprisingly, there is little pressure for professional excellence in many applications of systems technology. As mentioned in Section 1.5, any system will work as long as its users and operators want it to work, regardless of whether it is a good or a bad system. Because we are considering rather large systems, in many cases there is no effective competition, and users have no choice but to use the system available to them. Flying from New York to Los Angeles will save time over the train ride or any other available mode of transportation, even if the trip to the airport consumes an unreasonable amount of time. People will put up with it, and there are no procedures for them to exert effective pressure to stimulate any real improvements in point-to-point transportation from the total customer viewpoint.

In fact, a good system may not work as well as a bad one if users are not sold on it and refuse to use it. This happens quite frequently when the user is required to learn a new skill to take advantage of the capabilities of the new system, such as occurs in many new applications of computer technology. Sometimes a good system is rejected because "it wasn't invented here" or because "it's not our way of doing things." Men will fight to the death to preserve their "way of life," that is, their way of doing things or their "system."

The difference between a good system and a bad one is that the good system will provide a service in a more effective, economical, timely, or safer manner. The bad system may provide the equivalent service, but at a higher cost or with less comfort. It is, therefore, at this point that effective cost control should be applied initially. In many applications this is called cost effectiveness, and the systems analyst speaks of a cost effectiveness or a cost-benefit analysis.

2.1.2   Free Research Procedures

A good way to obtain insight into the systems approach is to consider what it is not. Actually, the systems approach is the exact opposite to the normal and time-respected *free research procedures* that are responsible for technological development. Systems engineering is objective oriented, and as such it is the antithesis of pure research, which has no other objective than the search for truth for its own sake.

At the time that the telephone was invented, there was no requirement, no established need for it. The same goes for the automobile, the electric razor, and the electric toothbrush. Today, many products from ladies' stockings to cigarettes are sold attractively wrapped in cellophane. At the time that this material was developed, there was no demand for it. The developer spent a sum equal to almost twice the original development costs of this new material to create a new market. This kind of sales effort is part of the history of many products without which we cannot get along today, products that truly enrich our lives.

It has been this undirected free research that has largely created our wealth. Today we have the technical capability to do almost anything we collectively decide to do. On the other hand, we do not have the resources to do everything that is desirable. Hence we must set our objectives and pursue them in an efficient manner. It is precisely in this area that the application of the systems approach is most effective.

2.2 CUSTOMER VERSUS CONTRACTOR RESPONSIBILITIES

The basic framework for the application of systems engineering design and development is provided by the system procurement procedures and contractual arrangements as practiced in both government and industry. These systems procurement practices and procedures are fundamentally different from normal commercial practices, because of the nature of the products and the high risks involved in their development.

New advanced systems are, by definition, associated with new technology; they involve *high risk, high costs, long lead times, and small production quantities.* Such systems cannot be procured through regular commercial channels. Competition often is restricted to the proposal and advanced development stage; the final engineering, integration, and initial production then takes place under contract on a single source basis. *The customer does not have the protection of the competitive marketplace.*

The procedure for the procurement of technically advanced systems from a single contractor is depicted briefly in Figure 2.1 and compared with the common distribution of commercial products.

2.2.1 Development of Commercial Products

Figure 2.1*A* indicates that a producer of commercial products assumes the lead in analyzing the market. He recognizes future consumer needs, and proceeds at his own risk to design and develop his products, to create a demand, and to offer his products in the public market under competitive conditions. To be successful in the competitive environment, his products and associated services must have some characteristics that are better (or that the customer believes to be better) than those of his competitors and that will attract and hold customers. The customer need not be aware of his own needs until the product has been offered for sale; and then he can select, from among the various offers, whatever available product appears to be most suitable for his own particular needs. In general, there need be no long waiting period between the customer's awareness of a need and the possible procurement of a suitable existing product with known characteristics.

2.2.2 Single Source Advanced Systems Procurement

In contrast to this procedure, Figure 2.1*B* illustrates the acquisition of systems from single sources, where there is a long waiting period between the announcement of a need by the customer and the acceptance of a system. The distribution of the risks between customer and contractor is quite different from that associated with the distribution of commercial

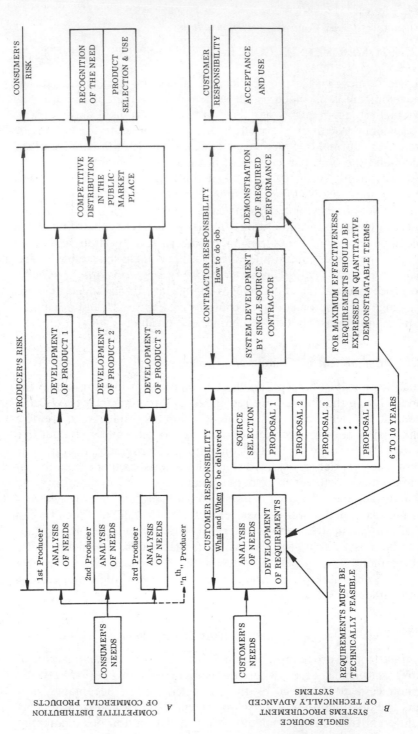

Figure 2.1. Comparison of the competitive distribution of commercial products with the procurement of technically advanced systems from single sources.

24

products. *The customer cannot rely on the normal protection inherent in the public marketplace, based on a choice between several different existing systems which are being offered in competition.* The customer must negotiate with a single contractor who will develop and construct the system, and will not have it ready for delivery until many years in the future. The success of the new technically advanced system will be highly dependent on the capability of the customer to define what is wanted and of the contractor's ability to extend the state of the art.

Therefore many contractual, management, and engineering procedures have evolved to protect the interest of the customer and to provide him during the long development period with reasonable assurance that his requirements are being met during design and construction. These procedures include both analyses and tests that demonstrate that the requirements are satisfied.

The systems approach begins with a recognized need, and a well-defined objective and schedule. All activities are planned, funded, and organized to achieve the objective on schedule as *required by the contract.*

2.2.3   The Systems Approach

The systems approach encompasses all engineering functions associated with the conception, definition, development, and deployment of a system. A system will include many types of materials, parts, components, and equipment. Some of these will be the equivalent of commercially available products, while other equipment may be unique, technically very advanced and sophisticated. The procedure used to procure a particular item, therefore, will be tailored to the nature of the item itself. The procurement practices applied in any one systems project will encompass a spectrum of practices. This spectrum will extend from the ideal commercial practices depicted in Figure 2.1*A*, to the ideal procedures for the procurement of a technically advanced piece of equipment as shown in Figure 2.1*B*. This latter procedure for procuring technically advanced equipment can be thought of as the extreme end of a spectrum of procurement practices, where the spectrum covers all shades of equipment complexity and reliability.

In an earlier work the author has attempted to describe the ideal procurement practices for advanced equipment at the extreme end of the spectrum (149). The discussion in this work is based on these software models and provides a base representing ideal conditions, which in any practical case must be modified to suit the particular existing circumstances.

2.2.4   Interlocking Responsibilities

Within the ideal business framework, customer and contractor each have very specific but interlocking responsibilities for the development of an advanced system. Figure 2.1 defines a possible division of responsibilities from a logical, technical viewpoint. The customer establishes system requirements; he is responsible for analyzing his own needs, for specifying what is to be delivered, and for the delivery schedule; that is, *what* is to be done and *when,* but not *how* to do it. The *how to do it* is the responsibility of the contractor; it is the element for competition and, in some cases, may have proprietary or trade-secret aspects. As soon as a customer specifies how to do a job, the contractor can no longer be held fully responsible for the results. Similarly, the customer retains the responsibility for the quality of the requirements; that is, it is his responsibility to specify those system characteristics that will assure that the product will be satisfactory for its intended use. This is a responsibility that cannot be delegated.

The contractor can only be made responsible for meeting the contract requirements as demonstrated by test under controlled conditions; not for how the system performs its missions when it is under the customer's control after delivery for deployment. *The contractor's responsibility is to produce equipment that will pass the demonstration tests specified in the contract; the customer's responsibility is to make sure that those tests are comprehensive and meaningful in terms of his intended use for the system.*

This is not the biased attempt of a contractor to get out from under responsibility. It is a fact of life; a principle. The meeting of demonstration requirements is a contractual obligation; if the product does not satisfy each and every demonstration requirement, it is not acceptable unless the customer grants a waiver.

An important point is that both customer and supplier, although they have different defined responsibilities, are part of the same development team. This is a significant difference between advanced system procurement and normal commercial practice.

There are other differences between commercial practices and single source systems acquisition procedures which extend beyond the time scale of the chart of Figure 2.1, but which are not included here for the sake of simplicity. For example, in the commercial case, the producer determines the extent of the service organization and the spare parts provided to the customer during the life cycle of the product. In single source systems procurement, the customer may define the extent of the support and the spare parts to be provided, as part of the basic contract.

2.2.5 Government Furnished Equipment

Unfortunately, in defining contractor responsibility in military contracts, it is not always possible to give the contractor full control over all of the elements that determine the performance of the system for which he is responsible. For example, every military advanced system contains a good deal of government furnished equipment (GFE); that is, equipment common to a number of different systems. Aircraft engines and many electronic subsystems fall in this category. They are developed under prime contracts to government agencies for use in a number of systems, proposed systems, or systems under simultaneous development, and their characteristics will in turn determine some of the performance characteristics of any system of which they are a part. The system contractor must accept GFE as supplied to him. Although he and other contractors scheduled to use the GFE may have had little influence and control over the development and characteristics of the GFE, the customer will still attempt to make the system contractor responsible for the performance of the entire system.

It is therefore important that a contractor monitor closely the development of GFE scheduled for use in his system, even though he has no contractual obligation to do so. It is also to his benefit to detect design deficiencies or incompatibilities early, either in the GFE or in his own design, and to arrange for corrective action while it is still relatively easy to do.

Some of the difficulties involved in the use of GFE are illustrated by the following example. Navy aircraft contract detail specifications (Navy terminology for specifications for the aircraft itself) often require the use of GFE engines that are under development. Hence, the detail specification will include only a preliminary estimate of the weight of the engine. The Navy structural design requirements obligate aircraft manufacturers to provide specified levels of structural integrity in the airframe system and to confirm the provision of those levels of strength and rigidity by analysis and tests. Manufacturers are required to obtain customer approval of structural test planning reports prior to performing the tests. In one case, the Navy's review of plans for dynamic tests of a new airframe design disclosed that the test plans provided for simulating the preliminarily estimated weight of the engine. However, it was known at that time that the engine would be considerably overweight. As a result of its review, the Navy required the test plan to be revised to include simulation of a more appropriate engine weight. This Navy review action accomplished the needed coordination of the manufacturer's design, analysis, and test groups that should have been accomplished by the manufacturer's project manager.

2.2.6   Impact of Economic and Business Considerations

In the actual procurement of advanced systems there is great variation from all baseline definitions made in the ideal customer/contractor relationship of Figure 2.1. One important reason for the variation is economic.

In a country with excess advanced technical capability, more advanced products can be developed and produced than can be funded. This is a desirable situation and can be explained by analyzing the balance between available funding and the ability to produce new systems. Industry normally will reinvest some of its net income to develop new capability in the hope of winning more and larger contracts. Sometimes, a small percentage of the total contract price is allocated for research. Productive capacity will grow faster and will lead the ability of customers to obtain funds for their new developing needs. Hence, customers of new advanced systems always will be short of funds compared to what they think that their needs are and to what the market can make available.

A dramatic example is provided by the Apollo Program, which was in fact generously funded. Nevertheless, the technical productive capability of the Apollo contractors always exceeded available NASA funds. Throughout the program, contractors were always ready and able to perform many desirable and apparently necessary tasks for which funds were not available. Hence, even Apollo Technical Administrators always felt that they were short of funds.

Leading comedians amuse us by asking military aviators how they feel about flying airplanes built by the lowest bidder. The suggestion that contractors will skimp to increase profits produces humor, but is a bit misleading. Skimping in commercial practice enjoys a certain freedom because no formal demonstration tests are performed on the product as they are in the procurement of advanced systems; see Figure 2.1. Moreover, the usual situation that the contractor can and normally wants to do more than the customer can afford, changes the complexion from "skimping" by the contractor more toward "economy measures" by the customer under very open and controlled conditions.

2.2.7   Funding Limitations

A customer of advanced systems has at least two good reasons for modifying the ideal division of responsibilities of Figure 2.1:

a. Lack of adequate funds to procure all of the systems for which he has requirements.

b. Lack of adequate funds to develop the in-house technical capability to specify and monitor his programs in accordance with the ideal procedure.

In this environment, customers, in order to stretch their limited funds, occasionally start more programs than they can support with available resources. In order to induce contractors to accept such underfunded contracts, specifications are deliberately vague and less exacting than required by good practice. *In particular, engineering requirements relating to proof-of-design, systems effectiveness, safety and quality assurance are abbreviated or omitted.*

Contractors, who in this environment underbid in order to win their contracts, face the problem of having to induce the customer to provide more funds to enable them to do an acceptable job. One procedure is to provide evidence that the original specifications were deficient, and to submit a new cost estimate for doing what the contractor believes to be an exacting job. In order to hold these proposed costs down, the customer steps in and tells the contractor *how* to do the job at less cost, thereby assuming part of the responsibility for the quality of the product. To obtain the funds necessary to continue those programs that the customer decides to support, he discontinues less promising projects and reallocates these funds.

In a country with an advanced technology, there is intense competition for the available funds on the part of industry; and an equally intense effort on the part of the customer to get the most for his money. Any given contract is formulated on the basis of a battle of wits between intelligent customers and industrial contractors. Each contract is separately negotiated, no two contracts are alike (although the contract format and pattern may be similar), and no contract completely conforms to ideal procedures or to the relationships of Figure 2.1.

This economically oriented procurement environment becomes even more complex when it is noted that the procedures are repeated on several customer-supplier levels, such as:

1. The customer dealing with the contractor.
2. The contractor dealing with subcontractors and suppliers.
3. Contractors and subcontractors dealing with lower level manufacturers or vendors of materials and standard parts.

2.2.8   Initial Funding Versus Downstream Payoffs

Another complication is introduced in government contracts by the procedures providing for appropriations on an annual basis. Government systems under development at any given time will not be operational for 6 to 10 years after authorization has been granted for the contractor to proceed with the design. The development of a reliable system requires more funds in initial design than a less reliable system; however at a later time, as a result of the higher reliability, there will be savings in operational costs that may more than compensate for the higher initial cost. However, these cost savings are far in the future. In some cases, as with airlines that compete on the basis of equipment life-cycle costs, future savings provide an incentive for increased funding in the initial design phases; but such is not usually the case with political organizations.

This situation is unfortunate, but it is common. Local governments, for example, faced with traffic congestion, buy land for expressways. The wider the right-of-way they buy at first, the easier it will be to expand road capacity later on. Yet, they are torn by the necessity of keeping current tax increases to a minimum and removing as little land as possible from the property tax rolls. Furthermore, funds available at the outset of the program may be quite inadequate, not only to buy the extra land, but also to build more lanes to satisfy foreseeable needs—even if more land was available. Without a good mechanism for making realistic assessments of future needs, and means for convincing fiscal people of those future needs, we are forced to let the future take care of itself.

One conclusion to be drawn from this discussion is that personnel responsible for assuring system effectiveness must understand the impact of economic and business decisions on technical procedures. In a highly competitive society, first priorities are assigned to equipment performance. Traditional economic realities assign to systems assurance procedures a relatively low priority, until the need becomes irrefutable.

2.3 DOMINANCE OF SCHEDULE CONSIDERATIONS

The development time of a large advanced system is 6 to 10 years or more. A system tends to be limited contractually to the state of the art as of the date of contract award as illustrated in Figure 2.2. It is impracticable with limited funds and limited time to include in a developing system all the technological advances being made in parallel with system development; the introduction of each new concept may require extensive redesign with concomitant effects on scheduling and funding. If a system is in a constant process of redesign it will never be completed and eventual costs could be relatively astronomical. Some advances can be included in the system development (dashed line), but not all. The difference, at time of deployment, between the available state of the art and the

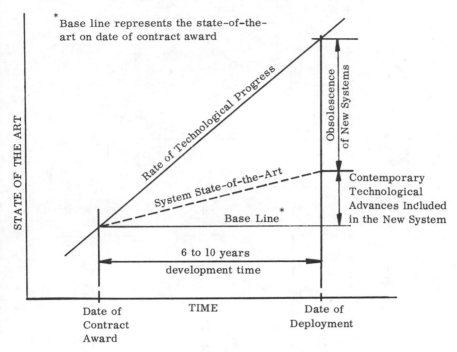

Figure 2.2 Obsolescence of new systems as a function of development time.

state of technology employed in the new system, is a measure of the obsolescence of the new system at the time of deployment.

Development schedules should, therefore, be established on the basis of estimates of the shortest time needed to perform adequately all required tasks. Every effort should be made during development to maintain these schedules so as to minimize obsolescence.

2.4 CORRELATION OF REQUIREMENTS AND TEST RESULTS

Within the established schedule time frame, demonstration testing is a dynamic feedback control loop to the design and development process (Figure 2.3). Testing must provide evidence to the customer that the system being offered for acceptance conforms with the contractual requirements. Therefore, there must be some direct quantitative relationship between test results and contractual requirements; preferably they

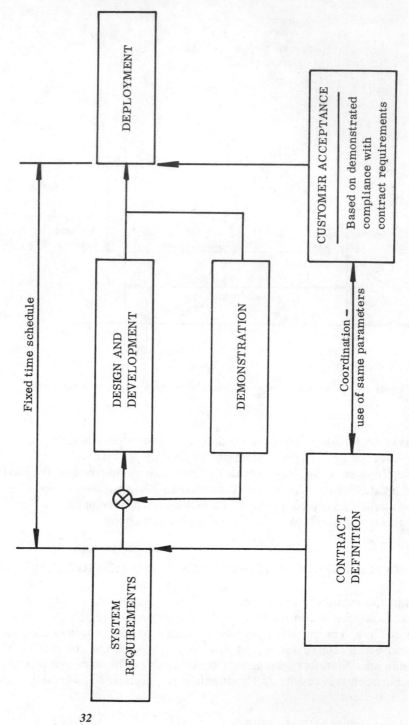

Figure 2.3. Integration of demonstration into design and development schedules.

should both be expressed in the same terms, to permit a straightforward comparison.

2.4.1   One-of-a-Kind Systems

One-of-a-kind systems should be reviewed, inspected, and tested at predetermined key milestones to determine that requirements are met at a time when effective corrective action can still be taken if any deficiencies are uncovered. These milestones should be scheduled at key points during design, fabrication, assembly, and operation, and should include operator training.

Common examples of one-of-a-kind systems are provided by production facilities for many mass-produced items from light bulbs to automobiles. Some of these plants constitute technically sophisticated systems. Not only is advanced technology required in the manufacturing process, but the product must be produced at a cost that permits distribution at a competitive price.

One of the best and most widely recognized examples of a commercial assurance program is the ASME Pressure Vessel and Boiler Code (196). This code is applicable to power plant equipment and is discussed further in Section 3.2.2.

2.4.2   Major Equipment

Other major equipment systems and their components are built in small quantities, and in some cases, no two configurations are alike. There may be no more than one test specimen available, and that specimen may not be quite like the final article. This necessitates that analyses properly take into account the effects of variations between the test specimen and the production configurations, and that compensations be made during tests and in the interpretations of test results for these variations.

2.4.3   Qualification Testing

In the defense industry it is the practice to subject one or more units to a qualification test to demonstrate the supplier's capability to comply with all of the requirements of his contract. These tests apply to the model type, not to the units being tested. A simple but important example is provided in aircraft development by the weighing of the first part made to demonstrate that the measured weight does not exceed the maximum weight requirement specified in the contract. The successful test specimen serves then as the standard for acceptability of follow-on units. The task then is to demonstrate that all critical characteristics of all follow-on units are equal to or better than the corresponding characteristics of the specimen that passed the qualification test. This is an easy task with a simple nondestructive test such as weighing; but the task becomes more complex with flying qualities where the aircraft is subjected to a variety of complex tests, some of them dealing with survivability and combat effectiveness.

2.4.4   Hard and Soft Requirements

Contractual requirements that can be demonstrated quantitatively before acceptance, such as weight, are sometimes referred to as "hard" requirements. Requirements that cannot be so demonstrated are then known as "soft" requirements. Some of the economic reasons for writing deliberately vague requirements are discussed in Section 2.2.7. However, sometimes a customer will call out soft requirements in a Request for Proposal (RFP) to determine if competitive bidders have an understanding of the technical problems and to give them a free hand in proposing meaningful hard programs. See Sections 5.1.1 and 5.1.2.

2.5   CHAPTER SUMMARY

In this chapter reference is made to advanced technology systems that do not exist at the time that the need for them is recognized. A contract must be awarded for their development, which may require up to a decade to complete. The procurement procedure for such systems is quite different from the procurement of existing commercial products in a competitive public market, where the availability of competitive products offers the customer a degree of protection.

To minimize the risks in this advanced systems development procedure, both customer and contractor must play active roles; each has specific responsibilities. If either side does not live up to its responsibilities, the likelihood will be lessened that the system will turn out to be satisfactory.

Some of the key factors influencing the customer/contractor relationship are discussed: economic and business considerations; schedules; and assurance demonstrations.

These concepts provide the foundation for the discussion of the systems development and demonstration cycle in the next chapter.

3

The Systems Development and Demonstration Cycle

A generalized flow diagram of the systems development and demonstration cycle is shown in Figure 3.1. This cycle applies to any new development, military or civilian, to a one-of-a-kind system as well as to systems scheduled for production.

The presentation is greatly simplified to show only the relationship of the main critical functions. However the diagram is complete to the extent that the functions shown are generally those necessary for assuring the probability of success of the project.

Every development organization groups these functions somewhat differently and uses its own characteristic terminology. This book uses general descriptions to describe the functions of Figure 3.1 in the hope that they will be more easily understood by a wider readership. Specialists in some functions may have difficulty in associating the general descriptions with their specialized terminology. However, specialized terminology is constantly changing, whereas it is hoped that the general descriptions provided here will maintain their validity over a longer period of time.

Most individuals involved in the systems development cycle are usually involved in only one phase of the cycle and often are not aware of the relationship of their role to the other phases. This recognition is difficult to acquire because of the organizational barriers between the various groups involved, and also because of the time factor, since the development and demonstration cycle shown often exceeds a decade.

This chapter includes a brief description of each critical function shown in Figure 3.1. Most of these functions are then discussed in greater detail in the following chapters. Figure 3.1 therefore represents a "roadmap" for the rest of this book.

The functions of Figure 3.1 are grouped under the following headings:

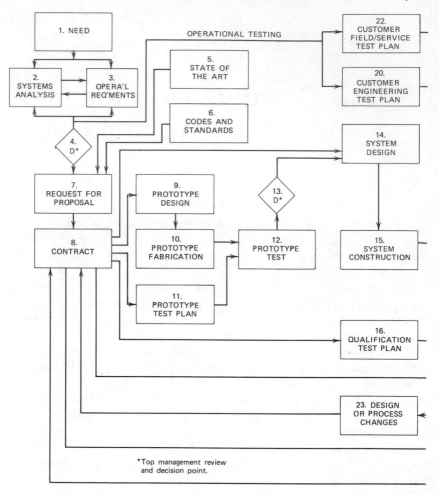

Figure 3.1. Overview of the systems development and test cycle. The letter "D" denotes a top management review and decision point.

Concept formulation

Contracting

Development and qualification

Customer operational tests

3.1 CONCEPT FORMULATION

The concept formulation phase encompasses all activities leading to the issuance of a Request for Proposal (RFP), namely boxes 1 through 4 of

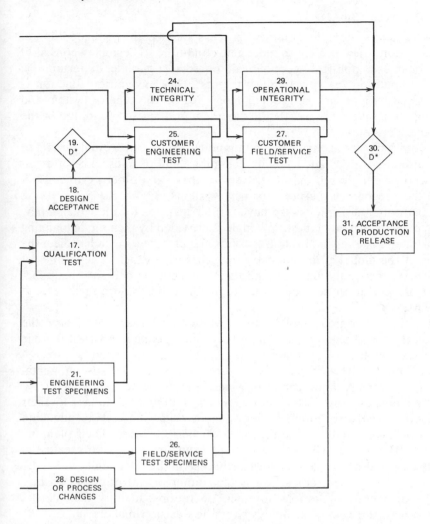

Figure 3.1. This process can be extremely complex, takes much time, and is associated with great volumes of documentation. It is not possible to treat this phase adequately within the scope of this book. Hence, except for this section, which briefly describes the basic principles as required for an understanding of the entire process, there is no further discussion of this material in the rest of the book. Some discussion of this phase is included in Chestnut (4).

3.1.1 Need (Box 1)

An advanced system development process starts with the recognition of the need for a new system to meet new conditions. These may consist of new business, political, geopolitical, or socioeconomic opportunities; new competition; or new military threats. This recognition function is performed by a variety of groups, depending on the type of system and organization involved, and should include a preliminary definition of the proposed system.

The military has agencies that continuously monitor external threats to the nation as well as advances in technology and obsolescence of available equipment. Based on an analyses of these and other pertinent factors, they recognize the need for new systems (39).

Industrial organizations also have formal agencies for recognizing new system needs. An outstanding example is provided by the unique planning departments of a few major U.S. aircraft manufacturers which project the needs of the global airline industry for years in advance. The airlines of the world, their governments, and their engine and accessory manufacturers all rely on these projections in planning their future facilities and systems.

Although other areas may not be as well organized as the military, the airline, the telephone, and the oil refinery and pipeline industries, there is always some formal mechanism for establishing the need for a new system, be it a new bridge, a new subway system, a transcontinental railway, a trans-Atlantic cable, or a new energy power plant.

Large corporations, which invest many millions of dollars in their production and distribution systems, must be assured of an adequate return on investment for the life cycle of their facilities. Their planning and advertizing departments recognize, promote, and even create the needs required to assure that their facilities will be adequately utilized.

Smaller companies, cities, towns, and other organizations often hire independent consultants to study their operations, identify their needs, and prepare the necessary analysis for the decision makers.

The documentation associated with the need phase should, as a minimum, include separate discussions of the following subjects:

Need for the system
System concept (performance)
Prospective relative effectiveness
Cost estimates (upper limits)
Schedules (time frame)
Technical risks
Environmental impact

3.1.2 Systems Analyses (Box 2)

The term "systems analysis" enters into various phases of the development cycle; at this point it encompasses the analysis of the following three basic areas:

The operational concept
The development concept
Economic feasibility

The analysis of each of these areas should result in the preparation of one or more written reports.

Operational Concept

The analysis of the operational concept examines how the new proposed system will be used, the skills and resources required (manpower, facilities, support, and logistics systems), the life cycle time frame, and may include an environmental impact statement. The impact of realistic reliability and maintainability assessments should be evaluated, and consideration should be given to alternative systems designs. This may become a very complex analysis; for example, the need for a new subway system should be evaluated against other forms of mass transportation including improved surface systems, within the given time frame.

Development Concept

The analysis of the development concept includes an examination of the required design, development, test, and support cycle. Critical issues should be identified: unknowns to be resolved, technical risks, schedules and milestones, required demonstrations, special support requirements.

Economic Feasibility

The economic feasibility analysis should establish the development and life cycle costs of the various proposed system configurations and determine the savings and other economic advantages that will result from the introduction of the various alternative proposals.

The impact of the new system on the customer's total resources (affordability) should be investigated, and trade-off studies between alternative solutions or systems should indicate the advantages and disadvantages of each proposed solution.

Scenarios

Modern techniques of systems analysis make extensive use of computerized scenarios of system operation in the anticipated operational environments as visualized for the system life cycle. Since an advanced system development program requires 6 to 10 years to produce operational systems, and the useful system's life normally exceeds 20 years, the scenarios should cover 10 to 30 years in the future.

The interaction of the proposed system with the other equipment that will be in the field at that time should be determined. The number of unknowns required to establish these scenarios and operating modes is very large; yet these analyses are of value from the following two viewpoints:

1. They assure that requirements are established on the basis of the best available current information.
2. They provide a record for the source of the requirements. When it later becomes apparent that the actual scenarios differ from the initially assumed ones, those responsible for updating requirements will have a better understanding of the initial source of the requirements and can more easily establish the justification for making the needed changes.

3.1.3   Operational Requirements (Box 3)

A set of operational requirements is developed in conjunction with the preparation of the systems analyses as soon as the results of the trade-off analyses become available and one or more basic configurations have been defined. The two sets of activities supplement one another.

The final operational requirements report summarizing all available technical data defining the system may become as technical and detailed as justified by the circumstances. All significant required quantitative performance parameters must be included, and they must be compatible with the assumptions made in the systems trade-offs and economic analyses. A test plan should be developed to demonstrate that the system meets the requirements.

The requirements document should also define the time frame for the development, and outline the operational, organizational, and logistic concepts involved. An assessment of the technical risks and an estimate of the development costs should be included.

If a positive decision is made to proceed with the systems development, this is the document that will be used to develop the request for proposal (RFP) and should, therefore, be sufficiently complete for this purpose.

3.1.4   Decision to Proceed (Box 4)

The systems analyses and operational requirements are reviewed in detail by the highest responsible organizational levels to arrive at a decision to proceed with the project. On a major system such as the Apollo project, this review may involve the President of the United States and the Congress. Decisions involving bridges and metropolitan subway systems may be studied and discussed sometimes for years by legislative bodies. Lesser systems are evaluated and decided upon at lower levels. Open lines of communication should be maintained between the decision makers and the analysis agencies.

Needed Completeness and Integrity of Analysis

The completeness of the systems analyses and operational requirements, and the care, professionalism, and technical integrity with which they have been prepared, have an impact on the decision makers (9).

The decision makers are generally not technical people; they may be statesmen, politicians, legislators, administrators, bankers, pension fund administrators, or businessmen. The concept formulation phase provides a basic interface between the engineering profession and the political and business decision makers. In this interchange, engineers and scientists must be professional and honest, especially in areas of technical risk assessment. Technical personnel who embrace political philosophies and who bias their testimony in favor of a preferred political position are a disgrace to the profession and perform a great disservice to the community (121).

The integrity and completeness of the concept formulation phase are of particular importance because the resulting operational requirements provide the basis for the RFP discussed in Section 3.2. If a requirement is not listed in the RFP, it is unreasonable for a customer to expect that the system will include the desired characteristics, unless the contract is changed at a later date with corresponding adjustments in cost and schedule goals.

Although a great deal of effort is expended on the concept formulation phase, normally it is the technically weakest aspect of the entire advance systems development process. Among the major reasons for this is the customer's inability to forecast the future operational environments and conditions used in the systems analysis scenarios which may represent conditions 10 to 30 years in the future. There will be a great deal of disagreement on these matters among the customer's own technical experts. As a result, there is a tendency among top management personnel to limit the scope of the scenarios or to bypass certain phases of the systems analysis process altogether.

Conflicting Requirements

A fundamental consideration is that every large advanced systems program is constrained by the following four basic interrelated variables (200):

Performance
Cost
Schedule
Risk

The equation that establishes the relationship between these variables on a given advanced systems development program is unknown. Hence, a customer may specify any two; the third should be negotiated with the contractor. Given three variables, the fourth is a complementary unknown that will seek its own level. Without specific knowledge of the relationship between the four variables, they will conflict with one another; there is no way that the fourth variable can be specified in a realistic manner. Nevertheless, all four variables will be specified in the contract.

Government agencies attempt to provide for program unknowns by including contingency funds in their requests to Congress or other legislative bodies. However, contractors are fully aware of the resources that have been made available, and in accordance with Parkinson's Law (7), they make their plans to consume all available resources. Hence, the customer is now boxed in. He can control three variables, but cannot prevent the fourth from floating.

In these circumstances there is no way that a program manager can assure that all of the requirements of the contract will be met, any more than he can reverse the laws of gravity. Under these operating conditions, an advanced systems development program becomes an almost continuous series of crises, requiring one trade-off after another. In making each trade-off, the program manager must retain a broad viewpoint, keeping in mind the user's priorities while generally attempting to achieve a system with the lowest life cycle cost in a future still unknown environment. The contract invariably becomes a dynamic instrument, undergoing almost constant change which, at any given point in time, records the status of the agreement between customer and contractor.

3.2 REQUEST FOR PROPOSAL (RFP) (BOX 7)

Given the operational requirements of Box 3, Figure 3.1, and the decision to proceed, Box 4, the next step is to prepare and issue a request for proposal (RFP), Box 7. This is a technical job which must be done by specialists in the field. The military services maintain special commands or bureaus to prepare such procurement documents in their specialized fields. Large corporations, both public and private, have similar groups; smaller organizations rely on consulting engineers.

The RFP reflects the knowledge of the customer's experts of what is technically feasible within a given time frame. It will therefore not always correspond to all of the operational requirements of Box 3 which may have been prepared by technically less sophisticated user-oriented personnel.

The RFP must describe the system desired in technical detail; all desired parameters must be listed together with the required tests to demonstrate that the requirements are met. The desired development schedule must be specified, indicating design reviews and other milestones. Quality control procedures and delivery conditions should be outlined. System management, manufacturing, reliability and maintainability, subcontractor and supplier control plans, integrated logistics and support plans, safety and training plans, and customer monitoring provisions should be specified. Data requirements must be spelled out.

The RFP must be complete. If a desired system characteristic or provision is not listed in the RFP, it is unreasonable to expect that a bidder will include it in his proposal. The format of the proposal must be carefully specified; specifically, all paragraphs of a proposal must be arranged and numbered in the same sequence and manner as those in the RFP and any referenced documents, so that competing proposals can be easily compared, paragraph by paragraph, with the RFP and with each other.

In rewriting the operational requirements into an RFP, technical personnel are restrained by the following two factors:

The state of the art
Codes and standards

3.2.1 State of the Art (Box 5)

The specialists preparing the RFP must be knowledgeable in the state of the art. Often new systems require new applications of advanced technology, or may even require developments which are currently nonexistent. The RFP reflects what the preparing agency believes is realistic and possible to achieve; this may be something less than specified in the operational requirements of Box 3.

3.2.2 Codes and Standards (Box 6)

All military services as well as all industries have codes and standards with which their systems must comply for safety, economic, or compatibility reasons, and which must be called out in the RFP. The quality of these codes and standards varies considerably between industries and between military commodities. Many of the safety and reliability problems associated with current systems can be traced to inadequacies in the RFP. In many industries these defects are the result of a lack of good and complete regulatory codes and standards, as well as of effective enforcement procedures.

One of the most successful commercial sets of codes and standards is the Boiler and Pressure Vessel Code of the American Society of Mechanical Engineers (ASME) (196). Under these procedures, a power company may decide to build a facility and establishes the desired output requirements. In most jurisdications of the industrialized world, building permits will not be issued unless it is demonstrated that the boilers and pressure vessels are designed, constructed, inspected, and tested in accordance with the code. Authorized inspectors provide step-by-step inspection as an item is being constructed. In general, insurance companies will not insure a plant within the scope of the code unless it conforms to the code requirements.

An agency preparing an RFP must be thoroughly acquainted with all the existing codes and standards applicable to the system they wish to acquire, and they must call them out in the RFP. If the available codes and standards structure is inadequate, the agency should really prepare new standards for the purpose. Many times, however, the preparation of new standards takes more time than that available for the preparation of the RFP. Like the state of the art, these codes and standards constitute a restraint on system design and may prevent the RFP from reflecting all of the operational requirements listed in Box 3.

3.2.3 Source Selection

Normally, as indicated in Figure 2.1, the RFP is sent to a number of qualified suppliers who are given a stated period of time in which to respond with a proposal. Often the availability of the RFP is advertized in the trade press, and other interested suppliers may request a copy of the RFP for the purpose of submitting a proposal. Normally, at least three proposals must be received in order to proceed with a competitive evaluation. The procedure leads to the award of a contract, Box 8. This subject is discussed more extensively in Chapter 4.

3.3 DEVELOPMENT AND QUALIFICATION

The control of development and qualification constitutes the keystone of this book. These functions are described briefly in this section and then discussed in greater detail in the following chapters.

3.3.1 Prototype Development

Large new advanced systems invariably include features whose technical feasibility has not been adequately demonstrated; proceeding with full-scale development at a given time may involve considerable technical risk. To minimize this risk a prototype phase involving one or more different prototypes may be included in the development plan. The purpose of a prototype is to reduce the technical risk by demonstrating the feasibility of the advanced features. Every effort is normally made to reduce the cost of a prototype by omitting as much as possible all those features known to be technically feasible. Hence, a prototype only models a portion of the final system; it often is a piece of operational hardware that demonstrates the desired advanced features as much as possible under actual environmental conditions. It is a test specimen constructed specifically to fill gaps in design data that must be available before full-scale system development can proceed.

In large and very advanced systems a separate contract may be awarded for a prototype, and the contract for the full-scale system development is deferred until the prototype is successfully tested. In smaller systems involving less risk, the requirements for the prototype may be included in the same contract as the full-scale system development as shown in Figure 3.1.

Program Phases

Large contracts involve large sums of money, and provisions must be included to assure the customer at all times that the funds are being properly utilized. The contract is broken down into a number of phases and subphases, and each subphase must be successfully completed before funds for the following subphase are released. An example of the major phases is shown in Figure 3.1 by grouping the activities between the major management decision points, Boxes 4, 13, 19, and 30. The entire program can be cancelled or reformulated at these points; but there should also be provisions for providing the customer for continuous visibility and control at all times between these major review points.

Design Control

In Figure 3.1 the contract includes both specific prototype design requirements, Box 9, and the corresponding demonstration requirements, Box 11. The contract should also provide for design control, which is defined as follows. For each requirement, such as power consumption, range, maximum weight, safety, reliability, maintainability, and other performance characteristics, the proposed design is analysed as soon as committed to paper to provide paper assurance that it will satisfy the specified requirements. Copies of the analyses are provided to the customer for approval. The customer must have the technical staff to evaluate these analyses properly and to exercise the customer's options in the scheduled design reviews. If the customer does not agree with the analytical predictions or the analytical procedure, the contractor must change the design or the procedure until the customer is satisfied. The contract should specify the analytical procedures, the report submittal, and design review schedule. As soon as a phase of the design is approved by the ·customer, the respective drawings and specifications may be released for fabrication, Box 10.

Test Plans

The contract should also require customer approval of contractor prepared test plans, Box 11. A major system development requires hundreds of formal demonstration tests, and a test plan must be submitted for each test. Customer approval of a test plan is normally considered a contractual commitment by the customer to accept the design if the specified test specimen passes the tests outlined in the plan.

Each test plan should refer to specific prototype requirements of the contract and to the respective analytical model on which the design is based and specify what parameters to measure under what environments, and by what means. Every test should be designed by reference to an analytical predictive model. If a predictive model is not available, statistical test design may be required. Since a prototype is normally designed to advance the state of the art and generate nonexistent design data, the test procedures will be breaking new ground and differences of opinion are bound to arise between a knowledgeable customer and an expert contractor. It is not unusual for a customer to reject several successive modifications of a test plan before approval finally is granted.

Test Reports

After approval of the test plan the customer releases the funds to proceed with the test, Box 12. The contract should provide for the presence of a customer witness who observes that the test proceeds in accordance with the test plan and who includes his observations in, and certifies, the test log.

In all cases of advanced equipment it is to be expected that unanticipated troubles will occur during the tests, and provisions should be made for corrective actions such as changes in design, processes, test equipment, or test procedures. All such occurrences should be fully recorded in the test log, and the affected drawings and specifications should be corrected accordingly. A test is not successfully completed until it has been demonstrated that the test specimen complies with all applicable requirements, unless a waiver is granted by the customer. A full report of the test results should be submitted to the customer for information. The original test log should be included as an appendix to the test report.

Decision to Proceed

Upon the satisfactory completion of all prototype testing, a management decision is made whether or not to proceed with the full-scale development, Box 13. One of the problems involved at this point is that, in order to meet schedules, waivers may have been granted during prototype demonstration, and all of the data required for system design may still not be available.

For the sake of simplicity, the above and following discussions consider the prototype and the system as undivided units. Actually, a prototype or system consists of a number of subsystems, assemblies, and components. The contractually stipulated prototype and system requirements must be broken down through various levels into requirements for each lowest level component, and these in turn are further broken down into part or material requirements. Much design feasibility or design verification testing of material or lower level parts and assemblies may have to be conducted to provide the data needed to support the design activities. All of this becomes a very complicated process, and it is necessary that at all times the customer maintains visibility and design control of this process. Hence all critical analyses and test results affecting design decisions must be reported promptly to the customer for approval.

3.3.2   System Development

Boxes 14, 15, 16, and 17 of Figure 3.1, applicable to the system, correspond directly to Boxes 9, 10, 11, and 12, as applicable to the prototype, and all the remarks made above for the latter also apply to the former, especially with respect to customer maintenance of design control and the necessity of providing for effective corrective actions as a result of the unanticipated troubles arising during the qualification tests. These tests, Box 17, are normally scheduled as early as possible in the development cycle, since it makes little sense to fabricate copies until the design is approved, subjected to a documented baseline, and configuration control established.

Acceptance Data Package

It is important to recognize what has been accomplished in the development this far and what the customer is actually buying at this point. The qualification test, Box 17, has demonstrated that the contractor has the capability of producing one or more test specimens that have complied with the requirements of the contract as interpreted in the approved test plan, as amended by whatever waivers may have been granted. In other words, what exists is a design that satisfies the amended contractual requirements. The evidence is incomplete that the amended contract adequately expresses the requirements for the needed operational system, Box 3, or that the contractor can reproduce the qualification test specimen on a production basis.

The design acceptance procedure, Box 18, is based on an examination of a contractor prepared acceptance data package (ADP) which consists of three independent sets of documentation as follows:

1. The approved qualification test plans and the corresponding test reports indicating that the specimens complied with all contractual requirements.
2. The drawings and specifications used in the fabrication of the qualification test specimens (as-designed records).
3. The inspection and test records describing the actual configuration and performance of the qualification test specimens (as-built records).

If there are no unacceptable discrepancies between the as-designed and the as-built records, the design is accepted. A management decision is now made on whether or not to continue the development, Box 19.

3.4 CUSTOMER OPERATIONAL TESTS

Two new customer testing groups now become involved in the development as follows:

1. A customer engineering test agency which conducts engineering tests, Box 22.
2. A customer field or service test agency which conducts field tests, Box 27.

These agencies provide a new test environment. They are no longer concerned with contractual requirements, they revert to the original documentation of Boxes 2 and 3, which may have been written 10 years ago, and they develop their own test plans based on their expert current knowledge of the developing and future needs of the customer, Boxes 20 and 25.

The initial test specimens provided to these agencies, Boxes 21 and 26, will be faithful copies of the updated qualification test specimen, Box 15. Engineering and field tests are not necessarily conducted in sequence; they may overlap or may be conducted in parallel.

The qualification test specimen of Box 15 was designed to pass the tests of Box 16; now the specimens will be subjected to the new environments of Boxes 20 and 25 which have little or no reference to the requirements of Box 16. To make matters even worse, to meet schedules and/or available funding, the customer may have waived some of the original requirements to enable a deficient specimen to pass a qualification test. Now we enter an entirely new test situation, and it cannot be expected that the initial test specimens will perform satisfactorily without some redesign and/or process changes. Hence, provisions must be made in the contract, Box 8, for timely and effective contractor support, Boxes 23 and 28, of the customer tests of Boxes 22 and 27. As soon as failures, deficiencies, or other troubles are uncovered, corrective action should be taken and incorporated in a timely manner in the equipment undergoing the customer tests. Strict configuration control must be maintained, and redesigned components may have to be requalified. To assure the necessary controls, all changes affecting interchangeability must be accomplished by formal procedures (engineering change proposals, ECPs). See further discussion in "Reliability Growth Cycle," Chapter 11.

3.4.1 Customer Engineering Tests

Customer engineering tests, Box 22, are conducted by a customer engineering test agency for the purpose of establishing technical integrity, Box 24, and to provide the technical data necessary to support the management decision on production release, Box 30. See Section 11.1 for further discussion.

3.4.2 Customer Field Tests

Customer field or service tests, Box 27, are conducted by customer field personnel to assure that the product can be properly operated and supported by the customer's operating personnel. Human engineering, maintainability, personnel training requirements, and logistic support considerations are reviewed in detail. These tests also generally require additional design and process changes, Box 28, to make the system fully acceptable to the customer, Box 29. See Section 11.2 for a more detailed description.

3.4.3 Customer Acceptance or Production Release

Upon completion of these two customer test cycles there is another management review, Box 30. If the system is one-of-a-kind, it is either accepted or modified until it is acceptable. If the system is scheduled for production, a decision is made whether or not to release the design for production, Box 31.

Production Redesign

Depending on the system, a production redesign cycle may be indicated to reduce cost and increase reliability and maintainability. Such redesign is, of course, not feasible in a one-of-a-kind system. But for systems to be produced in large quantities a thorough production redesign cycle is mandatory. For complex systems produced in small quantities over a long period of time, such as aircraft engines, a continuous redesign and retrofit effort may continue throughout the life cycle of the system. Many such changes are made to increase safety, reliability, and maintainability; or to decrease costs. Cost decreases, however, are more likely to be the result of better or more sophisticated tooling than of product changes. A redesign that appreciably increases performance is normally indicated by a new model designation (Model A, Model B, etc.).

The original designers of a new advanced system are primarily concerned with meeting difficult operational requirements. The first systems are built with simple tooling to meet schedules. Most machined parts, for example, are made from bar stock.

Once the operational requirements have been met, designers can concentrate on more economical configurations and manufacturing processes. Several machined parts, for example, may be combined into a single forging or casting. The system may be broken down into smaller assemblies, so that more people can work on each assembly simultaneously. The hands-on exposure of motivated shop personnel results in many suggestions for cost reductions involving redesign or improved manufacturing processes, and for increased maintainability.

More sophisticated tooling, jigs, and fixtures assure more uniform quality and further reduce unit costs.

Redesign for High Quantity Production

Production redesign-to-cost is most significant for systems mass produced in very large numbers and where interchangeability is required, as with ordnance items. Once such items are released for production, no changes can be made except at astronomical costs. Small savings per unit may result in substantial sums when large quantities are involved. Hence, all design deficiencies must be corrected before final production release, and rigorous quality control must be maintained. This production redesign cycle may be combined with the design of the manufacturing process and may require more time and funds than the original development cycle. See Chapter 11 for a more detailed discussion of some of these items.

3.5   CHAPTER SUMMARY

This chapter reviews the systems development cycle, from the time that a need for a system is recognized, to the time that the system is accepted or released for production. Most individuals engaged in systems development are usually associated with only some specific part of the development cycle, and it is difficult for them to have visibility over the entire cycle.

Each essential function in the cycle is described briefly to provide an overview of the systems development process. Those functions that affect systems assurance or systems safety are discussed in greater detail in the following chapters. This overview provides a "road map" indicating how the discussions in the following chapters fit into the overall cycle and relate to the other development functions.

4

Contracting

Figure 2.1 outlines the general procedure for the procurement of technically advanced systems. The discussion of Box 8, Figure 3.1, explains how the procedure is controlled by a contract. The contract describes in detail *what* the contractor will deliver, *when* he will deliver it, and specifies the customer's monitoring procedure.

Design to Performance

In moderate risk systems, where the customer assumes the full financial risk, development contracts are normally of the single source, cost plus fixed fee type. Competition may be limited to an evaluation of paper proposals. Such contracts are discussed below under the heading "Design to Performance."

Single source, cost plus fixed fee contracting is not considered the most desirable way of doing business; unfortunately in many cases involving the introduction of advanced technology there is no alternative. Many successful and superior systems have been developed under this type of contract, but the costs and schedule slippages associated with the superior performance of major systems have become so large that they are a source of major national concern.

Design-to-Cost

In an effort to control costs, a great deal of emphasis is being placed on design-to-cost contracts, which require that the same consideration be given to cost factors during design and development as are required for technical contractual parameters. Cost considerations become a factor in every design decision. In many such contracts the term cost refers to life cycle cost, which may include all development, production, operating, maintenance, and support costs over the system lifetime. Sometimes the term *cost-of-ownership* is used.

A limited form of design-to-cost contract is known as *Design to Unit Production Cost (DTUPC)*. This kind of contract is usually awarded for systems that will be produced in quantity, that require little or no maintenance, and where life cycle cost is not a major consideration.

Contractor Motivation

The problem still remains of motivating a contractor to achieve the established technical and cost goals on schedule as provided for in the contract. A commonly used technique is to provide for the payment of incentive or award fees at predetermined design review or demonstration milestones. Follow-on production contracts often include a *value engineering* clause (40) to provide further incentives for cost savings during production. Such cost savings are shared for a stated period of time, such as three to five years, by the customer and contractor. The procedure decreases the customer's cost while increasing the contractor's profits. The contractor's share of the cost savings is sometimes called a "royalty." Many cost reductions in production are achieved by improvements in procedures and in manufacturing methods and processes rather than in product redesign.

The effectiveness of financial incentives is questioned by many procurement experts, and a great deal of interest is being generated in competitive parallel developments (150). Since competing advanced development contractors want to win the single source follow-on production contract, they are motivated to achieve superior performance during the competitive design and development phase.

Enforceable Contractual Requirements

A contractual requirement is enforceable when the requirement is specified in the contract and associated funding is provided. If no funding is provided, the requirement is not enforceable. Especially in the field of reliability, customers sometimes specify contractual reliability tests in order to motivate the contractor to provide an effective reliability program. If the customer is satisfied with the contractor's results during operational testing, he normally has no intention of funding MIL-STD 781 (92) reliability tests.

Paperwork

These procedures create mountains of paperwork and contribute to the maintenance of a huge bureaucracy (see Section 9.9.5). Nevertheless, there is no easy way to create a new advanced system. The necessary paperwork and bureaucracy required by these procedures, especially the customer's in-house technical expertise, is much less costly than abandoning projects in midstream because it has become apparent that they will not satisfy the customer's needs.

In the following sections, the more important aspects associated with each of these procedures are discussed in greater detail.

4.1 DESIGN-TO-PERFORMANCE PROCEDURES

The critical elements of Figures 2.1, 2.3, and 3.1 are rearranged in Figure 4.1 to explain the design-to-performance contracting procedure. The development of the RFP has already been described in Chapter 3. The RFP, Box 1 of Figure 4.1, is a technical document that defines quantitative, demonstratable requirements, and indicates a delivery schedule. As noted in Figure 2.3, a properly written RFP and contract makes it possible for the customer to determine by test, when the system is built and offered for acceptance, and before formal acceptance and final payment, that all specified requirements are met.

Box 2 of Figure 4.1 indicates the various proposals submitted in response to the RFP. Competition is normally limited to an evaluation of paper proposals, although individual competitors may already have done considerable advanced development work including the design, construction, and testing of some key prototype or operational subsystems, and they will include these data in their proposals. Customer laboratories may also have performed advanced developments of government or customer furnished equipment (GFE) that they wish to be included in the competing proposals. The technical risk involved in the introduction of advanced technology should not exceed that which can be evaluated on paper or by examination of the data submitted (see "Date of Contract Award," Figure 2.2).

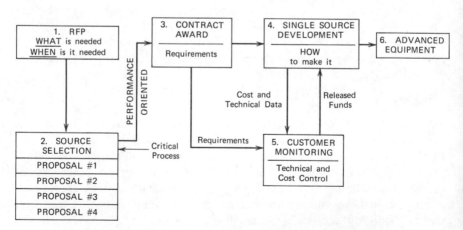

Figure 4.1. Design to performance.

Customer Priorities

Customer priorities tend to be in the following order:

a. Performance
b. Schedule
c. Cost

The customer selects the proposal that appears to best meet the performance requirements, giving due consideration to schedules and cost in the order of priorities listed above, and awards a contract to that offerer, Box 3. This contractor now becomes the single source for the development of the system, Box 4. At this point the customer lacks the protection provided by competition in the ordinary commercial marketplace. To furnish him with some assurance that the development will proceed in accordance with the contractual requirements, the customer actively participates in the process by monitoring the development, Box 5. The contractor is required to submit periodic cost and technical progress data at specified milestone dates for approval. The customer exercises technical and cost control by reviewing and approving the data submitted on a predetermined schedule, and then releasing funds for successive work elements. In addition to this technical review procedure, the contract may contain provisions for financial incentives or penalties related to stipulated requirements at specified milestones.

The procedure results in the acquisition by the customer of the advanced equipment, Box 6.

Success Orientation

Contracts tend to be success oriented, that is, they assume that every requirement will be met on schedule. Since this normally will not be the case, the contract is a dynamic instrument, subject to continuous renegotiation and change as the project progresses; sometimes these changes are substantial (see "Conflicting Requirements" in Section 3.1.4).

Advances in the State of the Art

The development of a new system incorporating some advances in the state-of-the-art by definition involves tasks that have never been done before; hence, there is insufficient specific experience for schedule and cost estimates. Although such estimates are necessary for budgeting purposes, they can do little more than to serve as guides in defining the planned scope of the effort.

A development program that requires any advances in the state of the art hardly ever proceeds as originally planned. The greater the required advances in the state of the art, the greater will be the deviation from the original plan (see ''Conflicting Requirements'' in Section 3.1.4). Program management becomes a joint customer-contractor effort. The mechanism is based on the function shown in Box 5.

Customer technical experts monitor the design and development progress by participation in design reviews and data submittal approval cycles. Decisions involving technical, schedule, and cost considerations are made jointly by customer and contractor personnel. The fact that these experts represent two sides and have different loyalties provides assurance that the best compromise is reached in a particular situation. Since the contractor does not have complete control over this process, he cannot be made solely responsible for the cost, and therefore some form of cost-plus contracting is indicated. A fixed-price contract is only applicable to efforts that can be completely defined, planned, and approved by the customer in advance, and where the contractor then has the sole and complete responsibility for the execution of the agreed-to program.

4.1.1 Source Selection

On major systems the source selection process (41) is very complex and may be broken down into several steps as shown in Figure 4.2. The first step is the technical evaluation of the submitted proposals by a board of technical experts. They will compare each requirement of the RFP with the corresponding paragraphs of each competing proposal and arrive at an evaluation indicating the relative merits of each proposal. Normally the offerer who meets most of the requirements at the lowest cost, and who has the required capabilities and facilities, will receive the highest rating. It is not necessary that the lowest bidder receive the highest rating, but selection of the lowest bidder is normally the easiest one to justify. This evaluation is usually submitted to higher authorities authorized to select a source. In cases of major military and space systems, the President of the United States and the Congress may become involved. These authorities may override the technical evaluation for political, economic, or business reasons. One of the main factors involved in a decision to override the recommendations of technical evaluators is the integrity of the contractor. Unfortunately, this factor is very difficult to include in the engineering evaluation.

The source selection process provides another interface between the engineering profession and the political and top business communities that requires the highest degree of integrity on the part of the technical

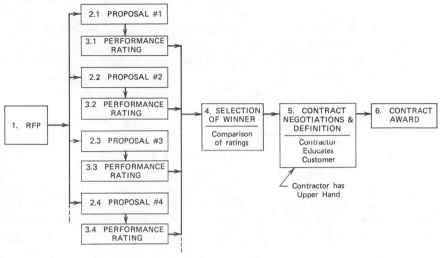

Figure 4.2.   Source selection—single source contracting.

evaluators. If the final decision makers arrive at decisions contrary to the engineering recommendations, they must be clearly aware of the technical risks and the penalties involved to achieve their other goals.

4.1.2   Contract Negotiation

Once a source is selected a contract must be negotiated. It is part of the game for the customer to request more in the RFP than he knows a contractor can deliver. The contractor, in the optimistic enthusiasm of a new project, sometimes promises more than he can deliver. Nevertheless, the proposal normally still does not comply with all of the items in the RFP.

Customer and offerer negotiating teams are organized and meet face to face to negotiate the contract. Each team consists of a management group authorized to sign the contract, and subgroups in each technical and management specialty involved in the contract. The number of customer and offerer representatives in each subgroup is usually equal.

In the negotiating sessions the offerer first explains his proposal and how he intends to comply with each paragraph of the RFP, at what cost, and on what schedule. The customer then points out the deficiencies in the offerer's proposal. Once the differences are identified, a negotiation takes place until agreement is reached on every item. Compromises must be made between the funds and time available, and what the contractor is prepared to offer. In some cases performance and schedules are more important to the customer than funds—and also create more difficulties for the contractor. Usually the customer's and offerer's subgroups meet separately each evening with their respective managements to discuss progress and plan strategy. If a particular subgroup cannot arrive at a resolution of a specific problem, the item then is discussed and resolved at the management level.

Contractor Has Upper Hand

In the negotiation process the contractor, who is the expert on *how* to do the job, educates the customer in what resources he needs to comply with the requirements of the RFP. Since there is no competition between contractors at this point, and the customer has nowhere to turn for counsel, this can be a painful experience for him. If the required compromises are too great and agreement cannot be reached, the customer may decide to select a new source or to postpone the development. In this case he will have to repeat the bidding process at a later date with a new RFP. This seldom happens because of the uncertainties involved in such drastic action. Both parties normally are anxious to get on with the job, and there will be give and take until agreement is reached and a contract awarded.

Wording the Contract

The difficulties involved in wording a contract cannot be overemphasized. The two teams, who in hard bargaining have negotiated the terms, know exactly what they mean by the words that they have written. However, a prime advanced systems development contract may run 6 to 10 years and involve many hundreds of subcontractors. The individuals who negotiate the terms are seldom those who execute most of the agreements. The contract must therefore be written so that other individuals, years later, will not misunderstand the original terms of the agreement.

4.1.3 One-of-a-Kind Systems

The unit price of any product normally decreases as production quantities increase. As a result of direct hands-on experience both systems design and manufacturing tools and processes tend to evolve with time; there is a "learning curve" that normally results in increased performance and decreases in costs.

On the other hand, the most difficult systems development projects to control for cost are the large single source one-of-a-kind systems, be they military, civilian, or commercial. There is no learning curve, possibly no repeat business. There are few, if any, built-in incentives to maintain cost estimates and schedules. Examples are bridges, subway systems, power plants. Even if little or no advanced technology is involved, they all tend to overrun with respect to cost and schedules.

A visible monument to cost overruns on a one-of-a-kind system, appreciated only by a few engineers who know the story, is the George Washington Bridge over the Hudson River in New York City. At the time that this bridge was designed it was the practice for the two suspension bridge towers to be encased in stonework following the Roebling practice, of which the Brooklyn Bridge is an outstanding example. Bridges tend to be among the longest lasting structures built by man, and they become key landmarks. Bridgebuilders through the ages have been aware of their esthetic responsibilities. The designer of the George Washington Bridge realized that there would be unforeseen cost overruns. He took pains to design the steelwork of the towers so that they could esthetically stand on their own, just in case the funds ran out before the stone encasement was completed. His foresight was correct, and the stone encasement was never built. However, the steel towers represent a satisfying design in their own right, and no one unaquainted with the original design proposals has realized that the bridge stands unfinished. Since the building of the George Washington Bridge, no other major suspension bridge in the world has ever been built with stone encased towers.

4.2 DESIGN-TO-COST PROCEDURES

In view of the high costs associated with cost plus fixed fee procurement, a great deal of thought is being given to design-to-cost techniques. Typical government documents describing the procedure are cited in refs. 43, 44, 45, and 46. Some actual applications are discussed in refs. 42, 151, and 152.

In addition to the acquisition costs, the costs of operations, maintenance, and support have become a matter of great concern. It is not unusual to hear of maintenance and support costs attaining values of 10 to 100 times the initial acquisition costs over the equipment lifetime. Terms such as "cost of ownership" and "life cycle cost" have been introduced to include all costs associated with development, production, and operation.

Priorities

In design-to-cost contracting, the customer rearranges his priorities from those cited in Section 4.1 to the following:

1. Operating cost
2. Production unit price
3. Performance

These considerations require that manufacturing (producibility) and logistics and support engineers become involved in the initial engineering development phase. Manufacturing facilities and processes and the logistic and support system must be developed simultaneously and in close cooperation with the design engineers. Hence there is a requirement for increased funding at the beginning of the design cycle.

Design-to-cost contracts are not compatible with large technical risks, since it is the problems associated with the introduction of advanced technology that make it impossible to prepare realistic cost estimates. However, if the technical risk is minimal, the design-to-cost contract may be quite effective.

The cost information required by the RFP normally requires the contractor to think through the entire development and production process and the support program before submitting his bid. Contractor bids are not only compared to the RFP and to one another, but also are evaluated against government baseline studies. In design reviews held during development examination of the contractor's management procedures and cost estimates are a priority item.

To assure that contractors meet their cost goals, design-to-cost contracts usually include incentive provisions (Section 4.3) or are associated with competitive developments (Section 4.4).

When life cycle costs are important, as with the airlines and the computer-rental industry, operating costs (reliability and maintainability) have a higher priority than unit production costs. Contractual reliability incentive warranties (RIW) also can be effective in holding down operating costs (153).

4.2.1   Design-to-Unit-Production-Cost (DTUPC)

The design-to-unit-production-cost (DTUPC) contract is a design-to-cost contract with a limited objective. It is used for the development of systems intended for quantity production, and which require little or no maintenance. Life cycle costs associated with maintainability are not a major consideration.

The objective is to design the product so that it will meet a specified cost goal at a given point in the production cycle. The price of any product normally drops as production quantities are increased. Normally the cost goal is set at a point after a small production lot has been produced and operational testing completed. Care must be taken to assure that a contractor, in his efforts to reduce cost, does not produce a design so marginal that it includes no growth potential.

DTUPC contracts usually include provisions for award fees to assure that the contractor meets his cost goals.

4.3   CONTRACTUAL INCENTIVES

To motivate contractors to meet their technical performance, schedule, and cost goals, many advanced system development and production contracts provide for the payment of fees in addition to the profits provided in the contract. The major fees are as follows:

1. *Incentives* are fees paid at predetermined milestones based on demonstrated performance. They may apply to any demonstrable technical, management, and cost performance.
2. *Awards* are fees paid at predetermined milestones for contractor technical, management, and cost performance that cannot be measured objectively. A customer "jury" committee of specialists evaluates contractor performance and, based on experience and judgment factors, determines the size of the award up to a contractually specified maximum. The monitoring committee has wide discretionary powers. Full award fees may be paid if the committee decides that the contractor has done an outstanding job, even though he may not have completely achieved all of his goals. On the other hand, the committee may withhold award fees if it determines that the contractor has not done as well as could be expected, even though he met all goals.
3. *Royalties* are fees paid for cost reductions or value engineering improvements achieved during production. Such cost savings result from contractor improvements of a customer owned design or process. The saving is shared by customer and contractor according to the provisions of the contract; the customer's cost is reduced and the contractor's profit is increased by the amount of the royalty for a specified period of time, say three to five years. After this time, the customer owns the improved design or process and no more royalties will be paid.

4.3.1   Effectiveness of Financial Incentives

Incentives, awards, and royalties must be large enough to motivate a contractor to respond to the customer's needs. Since government contractors must renegotiate profits over 15%, the fees should be fixed so that a contractor who does an outstanding job can come close to making the maximum allowable profit. However, depending on business conditions, additional profit may not be a prime motivating force. For example, in a depressed business situation, a contractor may be more interested in maintaining his staff and production base than in increasing his profits.

Although incentive and award fees can be effective in motivating a contractor to achieve his technical requirements and to reduce costs, they dilute the customer's control over the technical development process. If in satisfying the customer's wishes on any aspect of the work at a design review milestone the contractor loses his award fee, he can blame customer interference for his loss. Unless the contract is carefully formulated, the award committee evaluation may in effect replace the customer's traditional technical monitoring procedures.

Lack of Life Cycle Cost Data

In the administration of many design-to-cost contracts it is impossible to demonstrate that a contractor has met his life cycle cost goals, since it might take 20 or more years after a system becomes operational before the actual life cycle cost can be measured. Some life cycle cost data exists on components (154), but at this time there is no data available on large systems, since the procedure has not been in effect for enough time. This is why award fees are based on the best judgment of a "jury" committee.

DTUPC Contracts

The problem of determining award fees is easier when DTUPC contracts are involved. The final installment of the award fee is paid when the unit production cost is established after an initial production run is completed.

Multiple Incentives

The effectiveness of incentive and award fees and royalties is questioned by many authorities. In particular, the use of multiple *independent* incentives (technical, schedule, cost, management) is considered counterproductive, because the contractor can pick and choose among the incentives to suit his own purposes. Contractors normally understand how to manipulate incentives and they trade-off initial cost award fees versus value engineering royalties to maximize their profits; whereas customers are not normally as skilled in the use of award fees to minimize their costs. It is suggested that only one, or at most two, independent factors be selected as the basis for incentives or awards, and that the payment of any fees be made conditional on the contractor's meeting all other requirements of the contract.

Threat of Contract Cancellation

The threat of contract cancellation and the contractor's desire to win future contracts remain as the final effective means by which the customer can motivate an advanced systems development contractor to control costs and strive for superior performance. Much depends on the integrity of the contractor and the technical competence of the customer's monitoring staff. However, even in the very best organizations, there are excellent departments and weak ones; and top management is not always capable of compensating for the deficiencies.

4.4   COMPETITIVE DEVELOPMENTS

Parallel competitive developments were standard practice in the military aircraft business before World War II. However, at that time aircraft were relatively inexpensive. Since World War II there has been an explosive cost increase in advanced technology systems; the cost of parallel development for complete systems appeared prohibitive, and the procedure was no longer considered feasible. In the meantime, many experts have concluded that all other methods used to keep costs within reasonable bounds have not been very effective, and there is a trend toward introducing competitive development whenever feasible.

It appears that even if the initial cost of parallel developments is higher than single source developments, contractors can be motivated to reduce life cycle costs to an extent that will more than make up for the higher initial development costs. Usually, because of the cost factor, parallel developments are limited to key high risk subsystems to determine the best configuration of a subsystem. Such subsystems are then integrated into the complete system on a single source basis. However, parallel development contracts can also be quite effective when a new system is being developed to incorporate existing technology and when any competent contractor can be expected to produce a design with adequate performance. Such contracts may be combined with design-to-cost techniques, and the winning configuration will be the one that achieves the best performance at the lowest life cycle cost. The following discussion assumes that the competitive development is associated with a design-to-cost project.

The motivation provided by competition is extremely powerful. The ball-point pen, although a small item, provides an excellent example. The first such pens placed on the market sold for $15 to $20. Articles in the trade press at the time admitted that the price would decrease with time, but estimated that because of production difficulties associated with the precision ball and socket, the price could never be reduced below $5.00. In our competitive commercial environment, such pens are now available for 19 cents. The reduction in price in small hand-held electronic calculators offers a similar example of dramatic price reductions in a competitive market. Competition is therefore the strongest motivating force available to a customer for reducing cost while maintaining quality.

4.4.1 Characteristics of Competitive Contracts

The competitive development procedure differs from design-to-performance in three significant areas:

1. Source selection
2. Control of the parallel developments
3. Competitive evaluation

The competitive development negotiation procedure, shown by Figure 4.3 reverses the situation where the single source contactor, once selected, has the upper hand in the contract definition phase, as in Figure 4.2, Box 5.

In design-to-cost parallel developments the RFP, Box 1, Figure 4.3, specifies the maximum price per production system, and possibly also the operating and support cost, that the customer is willing to pay. It lists a range of performance requirements for selected system characteristics to

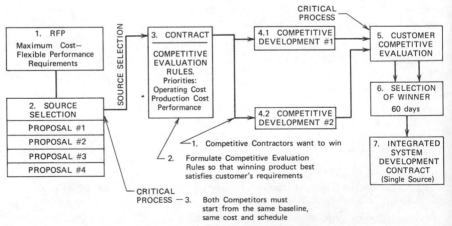

Figure 4.3. Competitive development.

allow bidders to propose the best trade-offs that can be achieved at the specified price. The successful contractor is then normally awarded a single source production contract (122).

The element of competition assures the customer that a contractor will do his best. Normally two contractors are selected in Box 2 to proceed with competitive developments (Boxes 4.1 and 4.2). The winner is selected on the basis of a competitive evaluation, Box 5. The evaluation process, Box 6, may require 60 days or more; this time period must be shown in the development schedule. The selection of two competitors in Box 2, instead of one as in Figure 4.2, requires basic changes in the source selection procedure which is discussed in Section 4.4.3.

Competitive Evaluation Rules

The key to success of the competitive development procedure lies with a contract that incorporates clear and complete competitive evaluation rules in Box 3 of Figure 4.3 (see Section 4.5). These rules must define all data to be submitted for evaluation as well as the parameters to be measured in the competitive tests, and must specify how these evaluations and measurements will be processed to select the winner, Box 6, in terms of the established priorities such as operating cost, production cost, and performance. The competitive evaluation rules must be formulated in such a manner that the winning configuration will best satisfy the customer's requirements. There must be provisions in the contract for relieving the customer of the responsibility of awarding a production contract, should neither design configuration meet the specified minimum requirements.

4.4.2 Competitive Negotiations

Two new major considerations enter the source selection process when parallel competitive developments (Boxes 4.1 and 4.2 of Figure 4.3) are anticipated:

1. Both competitors should start from the same baseline.
2. With the given schedule, the estimated development cost should be established on a competitive basis, and should be the same for both contractors.

Neither of these objectives can ever be satisfied in an absolute sense, but the negotiation procedures are designed to achieve them to the greatest extent possible.

For example, one offerer may propose a suitable and mature transmission design, whereas his competitor may have similar advantages with respect to his power plant. It becomes almost impossible to establish an identical "starting line" for both competitors.

4.4.3   Source Selection—Parallel Development

The procedure is shown in Figure 4.4. A number of proposals, Boxes 2.1, 2.2, . . . , are received in response to the RFP, Box 1. Each proposal is reviewed, Boxes 3.1, 3.2, . . . , and compared in detail to the corresponding requirements of the RFP. Any errors, omissions or deficiencies are noted, and the offerer is so notified with a request either for correction of his proposal or application for a waiver to the RFP requirement(s) involved. This procedure applies to all three areas of technical performance, cost, and management.

Under the pressure of competition offerers are usually willing to cooperate with the customer. To reconcile most discrepancies between the RFP and a proposal normally requires a series of face-to-face meetings between the offerer and customer's evaluators.

Customer Has Upper Hand

At this point the customer negotiator has the upper hand. Contractors in this situation have an entirely different attitude from when they have been already selected as a single source; they face customer evaluators who have access to all proposals, and who are negotiating simultaneously with several offerers. Any one offerer does not know what his competitors are proposing. Hence, it is the customer's negotiator who has the broad viewpoint and can set the terms in a competitive atmosphere. Customer negotiators, however, must be reasonable in their demands. A contractor can only do so much for a given number of dollars. Funding is required to demonstrate analytical conclusions; it takes a certain amount of testing to assure that a component is safe and reliable. If the customer cuts the funds allotted for certain tasks or tests below the minimum required, and forces the contractor to accept obligations beyond his financial capability, the result will be an overrun. The contractor in a competitive situation normally prefers an overrun to an incomplete job and an unsatisfactory product.

Offerer-Signed Proposals

After completion of the evaluation and negotiation, the fully negotiated proposal is signed by the offerer, Boxes 4.1, 4.2, ... There are no further negotiations after source selection, as in Figure 4.2.

Another view of the procedure is shown in Figure 4.5. The evaluation, Box 3, is shown as a closed loop control process that continues until the

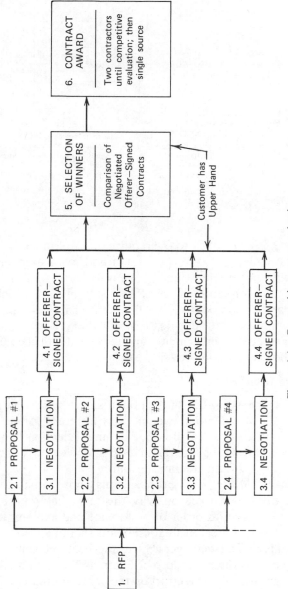

Figure 4.4. Competitive source selection.

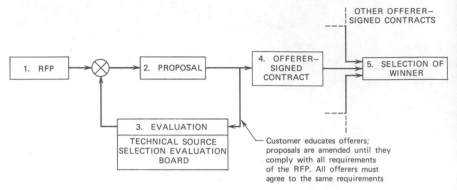

Figure 4.5. Competitive negotiating procedure.

proposal, Box 2, corresponds to the requirements of the RFP, Box 1. The purpose of the technical source selection and evaluation process is to produce a number of offerer-signed proposals, Box 4, all of which meet the minimum requirements of the RFP, any two of which could be selected for parallel competitive development. The individual proposals may still be based on different designs, but they all meet the minimum performance and cost requirements, and represent an acceptable "start" position for the competitive development.

Management Selection

The customer at this point has the option of selecting a single offerer as a single source, or of selecting two or more offerers for a competitive development as in Figure 4.4, Box 6. The final selection is made by top management who must be properly advised and who must have access to all the proceedings of the technical source selection group shown in Figure 4.5, Box 3. Although all proposals may now satisfy the minimum requirements, there is no reason why the technical source evaluation board may not list them in order of preference and explain their reasons for this order. The technical evaluation board should do everything it can to help the nontechnical managers make the best decision.

When compared to the procedure of Figure 4.2, design-to-cost negotiations require a great deal more effort on the part of both the customer and the contractors. However, it is believed that the procedure is effective in improving an offerer's proposal before source selection and that this will result in corresponding cost savings further downstream, particularly with respect to operating costs.

Limitations

This acquisition scheme also has its limitations. Each offerer promotes desirable features that appear only in his proposal. Customer evaluators must be very careful not to divulge to one offerer the contents of another offerer's proposal—neither the strengths nor the shortcomings. The winning proposal is then accepted as submitted by the offerer. At present, there is no way of taking the best features of the various offerers and including them all in one winning proposal.

In concentrating on a major critical subsystem, there may also be a tendency to neglect the definition of the interface with the rest of the system. In Figure 4.3, Box 7, the system integration is shown as a separate single source contract phase after selection of a winner in the competitive evaluation. Unless the total system is defined and the subsystem interface configurations frozen, the integration may involve major redesigns and result in a design configuration that may be inferior to the one which won the competition.

In military contracts a similar condition sometimes exists with respect to government furnished equipment (GFE). The contract should provide a firm list of GFE, and the acceptance procedures should verify the system/GFE interface.

4.4.4   Control of Competitive Developments

The procedure is based on the assumption that each competitive contractor in a parallel development wants to win. Given clear competitive evaluation rules that cannot be misunderstood, each competitor makes every effort to select and trade-off his design parameters to increase his chances of winning.

Under these circumstances customer monitoring differs from the loop shown in Figure 4.1, Box 5. This latter loop signifies active customer participation in performance, schedule, and cost trade-offs. In parallel competitive development, however, customer participation is counterproductive. If the customer exercises any influence whatsoever in the development procedure of one contractor and that contractor eventually loses, he will blame the customer for the loss. On the other hand, if he wins, the loser will claim that the winner had an unfair advantage.

Customer monitoring in a parallel competitive development must therefore restrict itself to impartial observation and must refrain from any active participation (such as approval cycles and design reviews).

Cost Monitoring

An item of primary interest to the customer during parallel developments is the contractor's ability to control costs. During the source selection procedure, every effort is made to assure that both competitors start the development on equal terms. Both are allotted the same resources for the development. Customer monitoring must assure that each contractor remains on schedule and within his budget allotments. This type of monitoring requires resident customer inspectors in the contractor's plants. Precise rules must be promulgated in advance for the treatment of schedule and cost overruns. It should be unacceptable for a contractor to use his own funds or resources to cover a schedule or cost overrun, since he is being evaluated on his ability to do a job within cost and schedule, and the use of unrecorded resources would give him an advantage over his competitor. The resident inspectors must provide assurance to each contractor that both competitors are being treated in the same manner. In unforeseen emergencies where one contractor may request special treatment, the request should be processed by a board consisting of representatives of the customer and of both contractors.

Customer Visibility Requirements

After a winner in the competitive development program is selected, Box 6 of Figure 4.3, he may be awarded a development contract for the integrated systems development phase shown in Box 7 of Figure 4.3. As a single source contract, this phase is administered under full customer participation as shown in Boxes 3, 4, and 5 of Figure 4.1. At this point customer monitors are required to have an intimate expert knowledge of the system. They can obtain this knowledge only if they maintain visibility over the design process during the competitive developments of Boxes 4.1 and 4.2 of Figure 4.3. Although there is no approval cycle, the customer experts who are to monitor the further development must have received during development periodic technical reports in sufficient depth and detail to give them a working knowledge of the design and of the justification for all significant design decisions and trade-offs. Although such technical reports are submitted for information only, the customer must retain the right to reject reports that are incomplete or do not provide sufficient visibility to permit the customer to administer a single source contract at a later date.

4.4.5 Customer Competitive Evaluation (CCE)

It has been emphasized that in competitive parallel developments the contract should contain the rules for the customer competitive evaluation

(CCE), since the contractor will make his design trade-offs so as to attain the highest possible score in the CCE. On the other hand, the CCE rules must be formulated so that the design receiving the highest score will also best satisfy user needs.

In the following material a quantitative scoring method is discussed, and examples are presented to illustrate how some parameters may be treated. These examples are included here only to illustrate the methodology; they cannot be applied mechanically to any given system. Each actual system has its own distinctive set of priorities, and the scoring system must be tailored to the reflect these priorities. The scoring procedure discussed here does not represent current practice. It is included in the hope that it will fill an existing need in current procedures.

Scope of CCE

The CCE usually applies not only to the design produced during the competitive development phase, but also to the contractor proposal for the initial production run at the unit price specified in the design-to-cost development contract.

Definition of Data Requirements

To prepare a valid scoring, it is necessary to have access to sufficient data. A scoring plan defines what data are necessary and sufficient and gives notice to those responsible, what data to gather systematically during development and test. Such data may otherwise be unavailable.

Objections to Scoring Techniques

The use of scoring techniques is a highly controversial matter. They are often suspect because it is believed that the parameters measured are not representative of the true value of a system resulting from complex trade-offs of a highly technical nature.

Many professionals object to scoring procedures because they feel that many critical factors are qualitative in nature and cannot be quantified with any degree of validity; it requires expert professional judgment to make an evaluation. Certain aircraft flying qualities involving complex trade-offs between conflicting characteristics such as maneuverability and stability provide a good example.

Customers often hope that under competitive development, contractors may develop new, imaginative, and desirable systems characteristics. They sometimes feel that a formal scoring plan may inhibit such imaginative thinking. This is not necessarily true, since the source selection authority may consider factors other than just those included in the scoring plan.

Considerations in Favor of Scoring

Although there is validity to these negative viewpoints, there exists in our American society a more overriding need to explain to other involved or interested groups the basis for a decision in terms they can understand and accept—particularly to the loser in the competition, to nontechnical higher business or civilian authorities, to the Congress, possibly to a court. It is believed that in this environment a scoring procedure provides the best method for justifying a decision. Many program management problems arise when nontechnical authorities do not understand or do not have visibility over all of the critical factors which influence a decision. Care must be exercised to assure that the procedure will not penalize a contractor with a truly superior technical potential and whose other shortcomings can be easily corrected.

The method requires that *all* known considerations affecting a decision be identified, arranged in order of priority, and quantified.

This is a very difficult problem. Within any group of experts responsible for developing a plan, there may be valid differences of opinion on every important issue. Nevertheless the value of the procedure rests in that all of the decision-affecting elements are identified, defined, and visible for all to see. Those who disagree with any phase of the plan can identify their precise disagreement and suggest revisions. The development of a scoring plan provides a discipline and a forum enabling those responsible for the scoring to integrate their ideas as they negotiate and modify the plan and its priorities until it is acceptable to the entire group.

An outside reviewer may also disagree with the assumptions, priorities, and scoring procedures; but he should be able to understand how the score was computed. He can then evaluate the quantitative effect of his disagreement and decide for himself whether or not to accept the decision.

Scoring Areas

A number of scoring areas have been selected as examples of the procedure for evaluating a high cost, low production advanced system. However, source selection will not be made on technical factors alone. Hence, in this example of the procedure, the scoring areas are not restricted to technical aspects, but are illustrative of the range of issues that might be considered.

The data required for the evaluation in these areas is more comprehensive than those data that can be generated during a CCE. Hence other sources of data must be tapped, both within the customer's organization and in the organizations of the competing contractors.

Although this plan is based on the use of valid quantitative differentiators, every effort has been made to define the criteria in as simple terms as possible. It is recognized that the more complex such a plan, the more controversy it generates. The plan loses its effectiveness if it is not easy to understand. The intent is to consider all significant evaluation areas and rank them in proportion to their importance to the customer.

In selecting the scoring areas and decision criteria, maximum consideration must be given to risk areas. The quantification of the decision criteria should be based on tests and analyses that demonstrate the extent to which the risks have been eliminated. The procedure should assure that the correct data will be acquired and that the minimum number of actually needed tests will be run.

The limitations of the scoring system are associated with risks not recognized and defined, and therefore not considered in the decision process. This is the reason why it is necessary to include and quantify all critical characteristics, no matter how difficult it may be to do so. It is also necessary to have the plan reviewed and agreed to by a group that contains experts in all applicable scoring areas to assure that no important considerations are omitted.

4.5   SCORING METHODOLOGY

In normal single source procurement, a product is inspected to assure that it meets the specified requirements. In a parallel competitive procurement, there is a different emphasis. Competing contractors strive not only to meet all minimum requirements, but also to exceed them within the limitations of available time and funding to provide a design that they believe best meets their customer's needs. Hence the purpose of a scoring plan is to provide, on the basis of data from the CCE plan and other supporting sources, a means for the following:

1. Determining that each competitive entry will satisfy the customer's minimum requirements.
2. Establishing which one of the competitive entries will best satisfy the customer's needs.

To distinguish between these two objectives, this illustrative plan differentiates between the following two classes of product characteristics:

a. Qualifying characteristics
b. Competitive factors

Qualifying characteristics are those that the system must possess to satisfy minimum customer requirements, that is, to satisfy objective 1, see Figure 4.6. The first part of this planned inspection, Sections 4.5.2 and 4.5.3, assures that all requirements for qualifying characteristics are complied with.

For convenience in the analysis, the term *critical characteristic* is introduced here. Critical characteristics are significant operational characteristics considered essential to mission performance; that is, without a critical characteristic the system cannot perform a specified mission.

Competitive factors are critical characteristics whose quantitative level or magnitude exceeds that of the qualifying level. The technical scoring is based on an analysis of competitive factors shown in Figure 4.6.

4.5.1 Competitive Areas

This illustrative example is not restricted to technical factors, but attempts to examine all areas that involve the customer's total needs. The

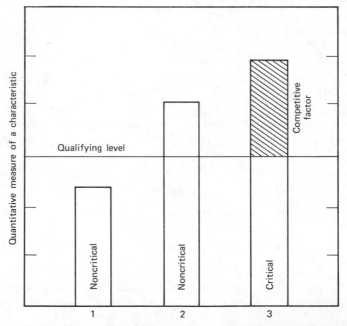

Figure 4.6. Classification of characteristics. (1) Noncritical characteristic does not meet qualifying level; project is not eligible for competitive evaluation. (2) Noncritical characteristic meets or exceeds qualifying level; project is eligible. (3) Critical characteristic must meet or exceed qualifying level. Competitive points are awarded for the amount by which the critical characteristics exceed the qualifying level.

second part of this plan, Section 4.5.4, is based on an analysis of the product with respect to six selected scoring areas considered in this example to be of major interest to the customer and to represent his total needs. In the scoring, a number of competitive points are awarded in each scoring area. The maximum number of points that may be awarded in each area, and which represent a typical order of priority, is assumed to be as follows:

Competitive Area	Points
Life cycle cost	P_L
Availability	P_a
Reliability	P_r
System performance	P_e
Program management	P_p
Performance growth potential	P_g
Total	100

Here $P_{subscript}$ represents the maximum number of points that can be awarded in each scoring area and should be representative of the value that the customer assigns to that area. The sum of $P_{subscript}$ should equal 100 for subsequent computational convenience. The above list of scoring areas will change depending on type of system and customer priorities. Other areas, such as human factors, environmental suitability, and survivability may be included. Allowance is made for other areas by including a line for "Other" in Table 4.1.

The contractor who receives the highest total number of points will normally be considered the one who best satisfies objective (2) in Section 4.5; that is, his system will best satisfy the customer's total needs; unless other unforeseen or unquantifiable factors compel a different decision. The contractor's competitive points are designated $p_{subscript}$.

This plan, therefore, breaks down into the following two parts: Section 4.5.2, "Qualifying Characteristics"; and Section 4.5.4, "Competitive Factors."

4.5.2 Qualifying Characteristics

Qualifying characteristics, unless waived, are those minimum critical and noncritical requirements that must be met before the prototype is allowed to enter the CCE process, or before consideration is given to competitive factors (Figure 4.6). The competitive procedure determines the superiority of one of the entries in best satisfying the customer's needs; the qualifying procedure assures that either entry would satisfy minimum customer requirements. The fulfillment of each qualifying requirement

Table 4.1. *Summary—Scoring of Competitive Factors*

Scoring Area	Maximum No. of Points Available	Points Awarded to		Reference
		Contractor 1	Contractor 2	
Life cycle cost	P_L	p_{L1}	p_{L2}	Fig. 4.8
Availability	P_a	p_{a1}	p_{a2}	Fig. 4.9
Reliability	P_r	p_{r1}	p_{r2}	Fig. 4.10
System performance	P_e	p_{e1}	p_{e2}	Fig. 4.11
Program management	P_p	p_{p1}	p_{p2}	Table 4.5
Performance growth potential	P_g	p_{g1}	p_{g2}	Fig. 4.16
Other	P_o			
Total	100	p_1	p_2	

can be judged on either a yes or no basis: on whether or not some minimum stipulated level has been met. Partial fulfillment is normally unacceptable. If a qualifying requirement is not met the entry is disqualified for evaluation and consideration of competitive data, unless a waiver is granted. Both competitors must agree to the waiver, since an advantage cannot be granted to one competitor without making a similar or equivalent concession to the other competitor.

A complete list of qualifying characteristics must be prepared based on a review of all contractual requirements and applicable specifications, and each item should be appropriately numbered, preferably by reference to the contract paragraph number. Each competing contractor should be required to submit a data package providing evidence of compliance with each item. The material in the contractor's data package should be arranged and numbered in the same sequence as the items in the list of qualifying characteristics.

In addition to the qualifying characteristics that apply to the system, there are a number of other organizational and support elements necessary to carry out a CCE. Hence, it is necessary to conduct a pre-CCE qualification examination to assure that all the necessary prerequisites for the conduct of a CCE are available.

4.5.3 Pre-CCE Qualification

A pre-CCE qualification review (Figure 4.7) should determine, before the start of the CCE, that all qualifying requirements are met, that is, that both competing entries satisfy all minimum customer requirements and that both contractors are ready for the competitive evaluation.

A customer review board, Box 2, should start meeting before the start of the CCE, say 30 days. The operation of the board is dependent on adequate contractor support. Furthermore, the board should approve a set of acceptance criteria to guide it in its decisions regarding the test specimens (go/no-go critical and noncritical qualifying characteristics) and the adequacy of the contractor support to be provided during the CCE.

Contractor's Data Package

The board arrives at its decisions after examination of a data package prepared and supplied by the contractor. One of the sections of this data package should consist of evidence that all the requirements for qualifying characteristics have been met. This evidence must be satisfactory to the board, otherwise the contractor must take corrective action until the board is satisfied. Reference 47, although prepared for a specialized field, provides a guide for evaluating a data package.

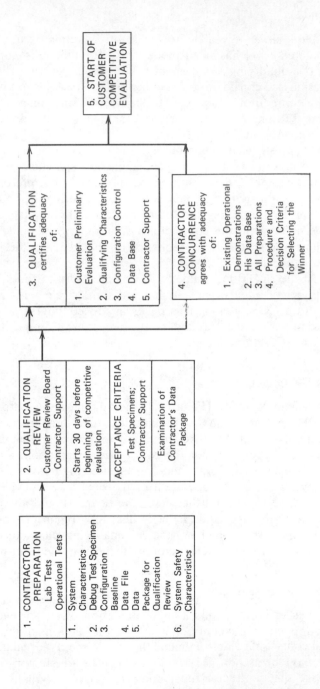

Figure 4.7. Pre-customer competitive evaluation (CCE) qualification review.

The contractor, Box 1, should be allowed sufficient time to prepare for the CCE; specifically, the development schedule should include an adequate number of months for the following:

Laboratory testing
Operational testing

During this time, the contractor should perform at least the following tasks:

1. Verify the operational characteristics of his system.
2. Debug his test specimens.
3. Establish his configuration baseline.
4. Establish a file of data required to support the CCE.
5. Perform the necessary inspections and tests and prepare the data package required for the qualification review board.
6. Identify any system peculiarities, especially those that might affect safety.

Board Determinations

The board, Box 3, should make the following determinations based on an examination of the contractor's data package:

1. The test specimens have passed all required customer inspections and tests and comply with all customer and other regulatory safety requirements.
2. All requirements for qualifying characteristics are met.
3. The contractor has established sufficient configuration controls to assure the following:
   a. No configuration changes will be made during the CCE tests without prior approval of the test board.
   b. The performance and quality of the production units will equal or exceed that of the CCE test specimens.
   c. The interface is adequately defined and simulated so that the subsequent integration with other system elements will not result in changes to the system that may render it inferior to the winning configuration.
   d. On government systems, that provisions for GFE are checked out and frozen.
4. The contractor's data base is adequate to support the analysis required to select the best design. Since the selection is not made only on the basis of performance, additional technical cost and historical data as defined in Sections 4.6 to 4.10 are required.

5. Planned contractor support is adequate to assure the successful completion of the CCE on the proposed schedule.

The customer review board should also formally determine and record that the contractor, Box 4, agrees to the following:

1. He has conducted sufficient laboratory and actual operational tests to:
    a. Verify the actual critical operational characteristics of his system (Box 1, Item 1).
    b. Debug his test specimens to the extent necessary for the competitive demonstration.
    c. Establish a configuration baseline, including the interfaces with GFE and other proposed system elements.
2. His data base is adequate to support the analysis of Sections 4.6 to 4.10 as required to select the design that best satisfies the customer's needs.
3. He has completed all the preparations necessary for the CCE and that he knows of no reason why the CCE should not proceed as scheduled.
4. The procedures and decision criteria for selecting the winner are adequate and, if he loses, he will not contest them. This item should be emphasized in the RFP.

The CCE, Box 5, can proceed as soon as all the conditions listed in Boxes 3 and 4 are met.

4.5.4   Competitive Factors

Competitive factors serve as the basis for a scoring system to select the system that best satisfies the customer's needs (Figure 4.6). In the following sections, illustrative procedures are outlined for assigning competitive points in the six evaluation areas selected in Section 4.5.1. The points awarded in each of these areas should be entered in Table 4.1 and added for each competing contractor. The contractor with the most points is considered to be the one whose system best satisfies the customer's needs.

Most of these scorings are based on data supplied by the contractors; hence it is necessary to require that all data used in these scorings receive prior approval by the cognizant customer's experts.

This method of scoring probably has little validity if the scores of both contractors are approximately equal. In such a case additional subjective factors must be considered to arrive at a decision. However, it is difficult to visualize a case where both scores would be approximately equal.

A more serious contingency would exist if the contractor with the best system could not supply the data that truly reflects his potential. A scoring plan of this type must therefore be considered as a guide to the final selection authority; the correct mechanical execution of such a plan does not automatically guarantee that a correct selection has been made.

4.6   LIFE CYCLE COST

In support of the life cycle cost scoring procedure, both contractors should be required to submit cost figures in a standard format, so that a direct comparison of each set of figures can be made. All estimates must be reviewed and approved by the customer. In addition, an independent customer estimate should be prepared in the same format to serve as a base for evaluating the contractor estimates.

The government procures many complex systems from a number of various contractors. To compare proposals on a meaningful basis, it is necessary that all contractors compute their costs in the same manner. The government has therefore established standard cost estimating procedures which are published in documents such as refs. 48 to 53. Because of their availability and wide usage, this life cycle scoring example is based on these documents. The major illustrative cost items considered are summarized in Table 4.2. To simplify the illustration, this example does not include development costs.

It is normally very difficult to obtain valid comparative cost data because of the differences in the accounting systems used by various contractors. Contractors cite many reasons for not complying completely with the work breakdown structure (WBS) of MIL-STD 881 (52). However, to make a valid cost comparison between contractors, it is essential that all involved use the same accounting system. Hence, deviations from the prescribed rules simply cannot be tolerated in a competitive evaluation.

4.6.1   Life Cycle Cost Breakdown

Life cycle costs should be broken down into the following categories:

Production costs
Operating costs and support costs

These costs should be established for the planned number of systems to be procured on the projected production schedule. The number of systems that will be allotted to each operating unit for various planned operational purposes, including personnel training, should be defined.

Table 4.2. *Life Cycle Cost Scoring Summary*

Costs Start at Beginning of Production System Life: X-Years	No Sunk or Fixed Costs Operational Plan	
Cost Item	Customer Estimate	Contractor's Estimate
Production costs Recurring Nonrecurring		
Operating and support costs Comp. operations Field and depot support Logistics support Personnel support Recurring investments		
Total X-year life cycle cost	C_g	C_c
$L_1 = C_c/C_g$		

Production costs should be broken down in accordance with the standard work breakdown structure, MIL-STD 881 (52).

Operations and support costs should be broken down by work unit codes, MIL-STD 780D (53).

Production Costs

Estimates of production costs for a given evaluation should be prepared in accordance with specified guides such as refs. 49 and 50. Operating and support costs should be prepared in accordance with guides such as ref. 51.

In this analysis it is recommended that both "sunk" and "fixed" costs be excluded. "Sunk" costs include research and development costs and all costs already expended. Fixed costs include corporate overhead, base/facility real property, road maintenance, construction and repair, and all other costs that are independent of the system and insensitive to the operation of the system.

Production costs are classified as recurring and nonrecurring. Non-recurring include tooling, engineering, trainers, and other one-time investments. Recurring costs include the system, spares, documentation, and other items for which recurring disbursements are required.

Operating and Support Costs

All costs should be calculated for a designated planned system life cycle period, say of 20 years. The costing period should start at the beginning of initial production.

Ref. 51 breaks down operating and support costs into the following categories:

1. Company operations
2. Base operations support
3. Logistics support
4. Personnel support
5. Recurring investments

Life cycle cost estimates are notorious for being affected by a number of uncertainties, and the longer the projected life, the greater the effect of the uncertainties. It is not unusual for projected life cycle costs to be underestimated by a factor of 6. Some of this can be circumvented by establishing high, low, and best values of the input cost parameters. The final cost figures should then be subjected to a risk analysis to obtain a degree of confidence. An example of a risk analysis program is provided by RISCA II which can be made available by the U.S. Army Logistical Management Center (54).

4.6.2   Evaluation of Life Cycle Costs

To qualify for the competition, all contractor estimates for each cost item must be equal to or less than the customer estimate. If a contractor's estimate for any cost item is higher than the customer estimate, either the customer estimate will be increased or the contractor estimate decreased through negotiation to equalize estimates. Competitive points are awarded to a competing contractor only for an estimate below the customer estimate, by means of the following procedure.

The L_1 is calculated from the data in Table 4.2 to indicate the ratio of the contractor's estimate to the customer's estimate.

The contractor should also be required to submit in advance of the start of the CCE a detailed estimate of the cost of the CCE. This estimate should be compared with the actual cost of the CCE. The ratio of the two values should be computed

$$L_2 = \frac{\text{cost of CCE}}{\text{contractor's estimate of CCE}}$$

The life cycle cost index L is defined

$$L = L_1 \times L_2$$

The number of competitive points awarded on the basis of the life cycle cost index is taken from Figure 4.8 for the indicated value of L.

In Figure 4.8 P_L competitive points are awarded if the contractor's estimate is one half of the customer's estimate ($L_1 = 0.5$), and if the contractor correctly predicted the cost of the CCE, $L_2 = 1.0$. This value of $L_1 = 0.5$ as a baseline is selected here purely for illustrative purposes. The contractor's estimate of the cost of the CCE is probably the only currently available test of the contractor's estimating ability. If the contractor underestimates the cost of the CCE, the number of points awarded are reduced accordingly. If the contractor's life cycle cost estimate is just equal to the government's estimate, that is, he does not exceed the qualification requirements, no points are awarded. If L is larger than 1.0 because of CCE overruns, no competitive points are awarded.

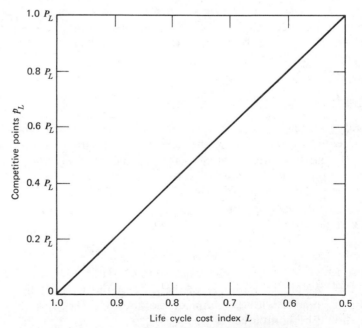

Figure 4.8. Award of competitive points—life cycle cost.

4.7 RELIABILITY AND AVAILABILITY

The following numbers are assumed for illustrative purposes:

Availability	A = 82%
Mission reliability	R = .986909—probability of completing a 1-hr mission
System MTBF	4 hr mean time between any system failure: failure rate $F = \frac{1}{4} = 0.25$
Maintainability	Maintenance Crew: n = 2 Mechanics

To meet the qualifying requirements, the system must be available for operation and complete its mission in $AR = 0.82 \times 0.987 = 81\%$ of all prescheduled operations when maintained by 2 mechanics. The system MTBF shall be at least 4 hours; that is, the failure rate $F = \frac{1}{4} = 0.25$. Competitive points will be awarded only for successful planned missions accomplished on schedule that are in excess of the $AR = 0.81$ and when the failure rate is less than 0.25. However, these reliability requirements may be modified for the CCE test specimen by the customer review board based on reliability growth considerations presented and agreed to by the competing contractors (see Section 11.3).

If an available system does not operate as scheduled within 10 minutes of starting time, the system should be considered unavailable for that operation, although the operation may start later and be considered a success from a operational test viewpoint. The same rule should hold for any test operation where more than two men are required to prepare the system for a scheduled operation; for this scoring such a system should be deemed unavailable.

The availability therefore should be reduced by the number of late operations and operations requiring excessive manpower, and the adjusted availability index should serve as the basis for the award of competitive points (Figure 4.9). Points P_a are awarded for an availability index of 100%, and no points are awarded for an availability index equal to or less than 81%.

Additional points are awarded in Figure 4.10 for failure rates below 0.25 (or system MTBF's above the contractual requirement of 4 hours). A maximum of P_r points are available for exceptional performance, and no points are given for a system MTBF of 4 hours or less.

The recording of reliability and availability data should be in an approved format. The product assurance representative for the program manager (PM) should use this data to establish mission availability and system MTBF.

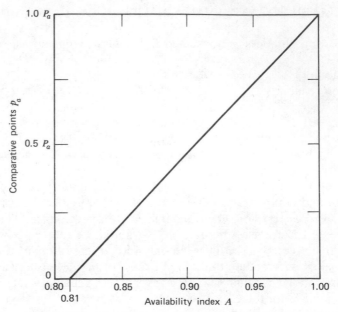

Figure 4.9. Award of competitive points—availability.

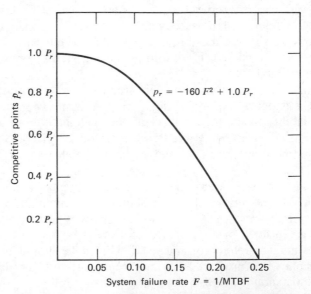

$$p_r = -160\,F^2 + 1.0\,P_r$$

Figure 4.10. Award of competitive points—reliability.

4.8 SYSTEM PERFORMANCE

The method for evaluating system performance is shown in Tables 4.3A and 4.3B. The method first determines the qualifying characteristics of the system as required to perform a number of specified missions. Each competitor must attain at least this level of system performance before he may be considered for the award of points. Competitive points are awarded for system performance characteristics above the qualifying level only if these characteristics contribute to those scoring areas important to the customer. For this example three such areas are selected as follows:

Reduction in life cycle cost
Availability
Reliability

Competitive points up to a maximum of P_e are awarded by means of Figure 4.11 to the extent that system performance characteristics have

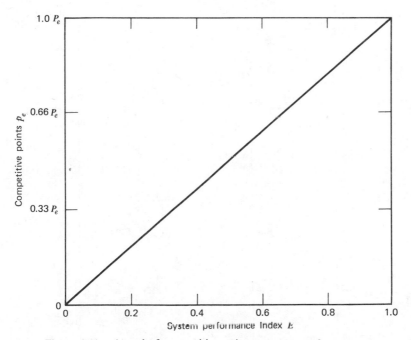

Figure 4.11. Award of competitive points—system performance.

Table 4.3A. System Performance Scoring—Worksheet

Mission	Mission Frequency (Fraction)	Critical Characteristic	Qualifying Value (1)	Measured Value (2)	Margin (2) – (1)	Marginal Contribution in Terms of Competitive Points Already Awarded					
						Reduction in Life Cycle Cost		Change in Availability		Change in Reliability	
						Fraction	Increment	Fraction	Increment	Fraction	Increment
From		1									
mission		2									
list		3									
		4									
		5									
		6									
		7									
		8									
		9									
		•									
		•									
		•									
Total mission increments							i_c		i_a		i_r

NOTE: One chart for each mission.

Table 4.3B.   *System Performance Scoring—Summary*

	Change in LCC	Change in Availability	Change in Reliability	Total
Total incremental points, *all missions*	I_c	I_a	I_r	
Competitive points awarded	p_L	p_a	p_r	p
Total increments × points awarded	$I_c\,p_L$	$I_a\,p_a$	$I_r\,p_r$	$\displaystyle\sum_{n=c}^{r} I_n p_n$
System performance index		$\displaystyle E = \sum_{n=c}^{r} I_n\,p_n/p$		

contributed to the award of the points already earned in the above noted evaluation areas.

4.8.1 List of Critical Missions

The various missions that are considered in this evaluation should be defined in the contract and a separate mission list should be prepared. For each mission, the list should indicate the mission frequency, the critical characteristics, and the associated qualifying values. For use in subsequent computations, the frequencies for all missions listed must add up to 1.0. A critical characteristic is one necessary to the accomplishment of the mission in the true operational environment. These characteristics should be included in the list of missions. The qualifying value is the lowest acceptable value of a critical characteristic required for mission accomplishment. All of these items are entered in the first four columns of Table 4.3*A*. A separate chart should be made out for each mission.

4.8.2 Impact of Competitive Factors on Scoring Areas

The measured value of a critical characteristic is recorded in the fifth column. These values may be obtained from any customer approved test including contractor tests or CCE. The "margin" or "competitive factor" in the next column indicates the difference between the measured value and the qualifying value, if any.

The next six columns indicate the contributions of the margin in the award of competitive points in the three selected scoring areas of concern, namely:

Reduction in life cycle cost
Availability
Reliability

For each instance where there is a margin, Figures 4.8, 4.9, and 4.10 must be reexamined and a determination made of how many competitive points p would be awarded in two cases:

p_q if the qualifying value is achieved
p_m if the measured value is achieved

If there is an increase in competitive points as a result of the margin, the fractional increase $(p_m - p_q)/p_q$ is entered in the "Fraction" column. The product of this increase and the mission frequency represents the incremental points assigned to this increase and is entered in the adjacent column entitled "Increment."

The increments are added by columns and summarized in Table 4.3*B*. On the line below these sums, enter the number of competitive points awarded in each of the three scoring categories and add horizontally to obtain the sum p.

On the last line multiply the incremental increases with the number of competitive points awarded in each scoring area and add horizontally. Divide this sum by p to obtain the System Performance Index E.

The number of competitive points awarded a given system performance index E is established in Figure 4.11.

4.9   PROGRAM MANAGEMENT

The ability of the contractor to supply and support his product, once it becomes operational, reflects on the product's cost and effectiveness. This contractor function is dependent on his success as a program manager. It is considered important, therefore, that a scoring of his managerial performance be made.

The program management scoring is based on a contractor's past performance on his three most recent programs. This scoring based on past performance is in contrast with the other scorings that are restricted to technical estimates and measurements.

This scoring is totally customer oriented, that is, it is based on *what* was achieved rather than on *how* it was achieved. Hence, no reference is made to organization charts, management methods, policies, tools, and so forth.

The four areas of customer interest selected here as an example of the basis for scoring are:

Schedule management (production)
Cost management (production)
Technical management (development)
Management responsiveness (production)

One of the areas, technical management, refers to the development phase; the other three refer to the production phase. If any program is still active, a cut-off date, say six months before start of the CCE, should be negotiated with the customer. Data older than six years also should not be used.

4.9.1 Schedule Management

In the scoring of schedule management, it is assumed that the customer is interested in how well the contractor maintains his schedules. The top line of Table 4.4 calls for the entry of the number of milestones in the master schedule of each program. In the second line are entered the number of milestones that were met as scheduled. The horizontal totals appear in the last column. The percentage of milestones met, B, is computed on the third line. Figure 4.12 then indicates the number of competitive points p_b that may be awarded for a given value of B. A perfect score is awarded when all milestones are completed on schedule, $B = 1.0$.

4.9.2 Cost Management

The cost management scoring is similar in concept and in processing. Cost is considered a matter of prime interest to the customer. The second set of entries in Table 4.4 requires, for the three programs under investigation, the number of contractual cost items and the number of these items delivered within the estimated cost. The ratio K for the three programs is calculated and entered into Figure 4.13, which indicates the number of competitive points awarded for a given value of K. A maximum of competitive points are awarded when all cost items are delivered within the stated cost.

If a contractor's record on these three selected past programs has been deficient, but he has made effective organizational and management changes that will assure better schedule and cost performance in the future, he may submit a description of these changes and whatever evidence he may have that the changes are effective. The customer review

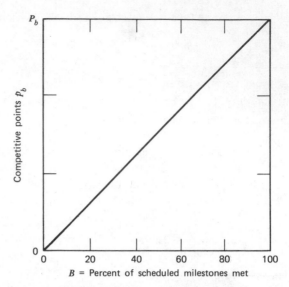

Figure 4.12. Schedule management.

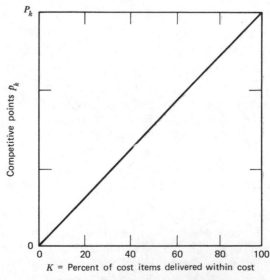

Figure 4.13. Cost management.

Table 4.4. *Program Management Scoring—Worksheet*

Item		Program 1	Program 2	Program 3	Total
Schedule Management (production)	Number of milestones—master schedule	B_{o1}	B_{o2}	B_{o3}	B_o
	Number of milestones met	B_{m1}	B_{m2}	B_{m3}	B_m
	Ratio of milestones met				$B = B_m/B_o$
Cost Management (production)	Number of contractual cost items				K_o
	Number of items delivered within cost				K_m
	Ratio of cost items delivered within cost				$K = K_m/K_o$
Technical Management (development)	Number of contractually required tests (qualification, structural demonstration, flight, etc.)				T_o
	Number of tests passed on first official trial with no waivers affecting initially required system performance				T_m
	Ratio of such tests passed				$T' = T_m/T_o$
	Number of tests included in T' passed on schedule				T_t
	Ratio of tests included in T' passed on schedule				$T = T_t/T'$
Management responsiveness (production)	Number of different failures or different unsatisfactory conditions reported by customer on production articles				H_o
	Number of different complaints corrected				H_m
	Ratio of complaints corrected				$H = H_m/H_o$
	Total number of complaint reports submitted				Q_o
	Average number of complaints per occurrence				$N = Q_o/H_o$

board shall have the power, at its discretion, to modify the mechanical award of points by means of Figures 4.12 and 4.13 as may be indicated by the contractor's presentation.

4.9.3 Technical Management

Technical management is scored on the basis of performance during the development phase of the three selected programs. The scoring is based on the concept that a perfect score. of P_t competitive points will be awarded if all contractually required tests are conducted successfully on first trial and completed on schedule, and no waivers affecting system performance have been granted. Table 4.4 calls for the entry of the number of contractually required tests on each program; these tests may be equipment qualification tests, structural demonstration tests, operational demonstration tests, and so on. For each program the number of tests that passed all requirements on the first trial and without waivers should be indicated as noted. The values for each program are added horizontally and T' calculated, indicating the ratio of tests passed successfully.

The last two lines in the technical management section call, by program, for the number of tests conducted on schedule included in T', and the ratio T of such tests.

Figure 4.14 indicates for any value of T' the number of competitive points P_t awarded. If all tests included in T' passed on schedule ($T' = T$), the upper solid line holds and the maximum number of points awarded is $p_t = P_t$ for $T' = 1.0$. However, if not all of the tests passed on schedule ($T_t < T_m$), then a new dotted line must be drawn for the appropriate value of T, lowering the number of competitive points awarded to a consideration of only those tests that were completed on schedule.

4.9.4 Management Responsiveness

Management responsiveness is measured by the number of effective corrective actions that a contractor undertook as a result of a customer complaint (deficiency reports of any type), as well as by the number of times a complaint was registered before effective corrective action was taken. The bottom set of data on Table 4.4 calls for the entry, by program, of the number of different failures or different unsatisfactory conditions reported by the customer on production articles. The term "different failures" is used here to differentiate the condition from one where a number of customer reports refer to the same failure. The next line of this

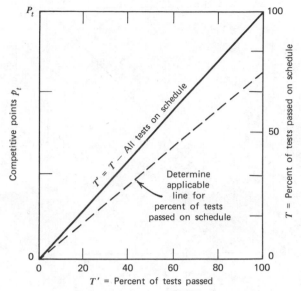

Figure 4.14.   Technical management.

set of data provides for the entry of the number of different items effec-tively corrected; finally H indicates the ratio of complaints that were corrected.

The following two lines indicate the procedure for calculating the aver-age number of times that an item was reported by the customer before effective corrective action was taken. Q_o in this table refers to the total number of customer reports submitted. Since H_o is the number of differ-ent failures that have occurred, $N = Q_o/H_o$ indicates the average number of complaints the customer submitted for each failure or unsatisfactory condition.

Figure 4.15 indicates the number of competitive points awarded for management responsiveness. The solid line indicates the points awarded for a given H, the ratio of customer complaints when $N = 1.0$. If N is larger than 1.0, as it normally will be, a new dotted line must be drawn for the value of $1/N$. This reduces the number of points awarded.

This method of measuring management responsiveness is based purely on past contractor performance—on what he has done. It makes no allowance for the difficulties involved and for how a contractor conducts his business in a competitive environment, since it is assumed that these factors are approximately equal for all competitive contractors.

Historical contractor data on cost, scheduling, and testing must be

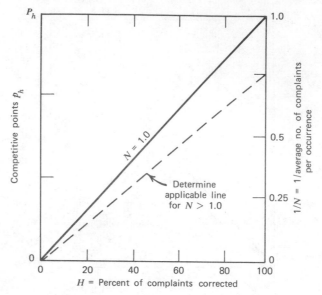

Figure 4.15.   Management responsiveness.

obtained directly from the individual contractors. This information should be made available to the customer review board for scoring of the items listed in Tables 4.4 and 4.5.

4.9.5   Summary—Program Management Scoring

The number of competitive points awarded in each program management evaluation area are entered in Table 4.5 and totaled to determine the number of points assigned to program management.

Table 4.5.   *Program Management Scoring—Summary*

Scoring Area	Reference	Maximum Points	Points Awarded
Schedule management	Fig. 4.12	P_b	p_b
Cost management	Fig. 4.13	P_k	p_k
Technical management	Fig. 4.14	P_t	p_t
Management responsiveness	Fig. 4.15	P_h	p_h
Total		P_p	p_p

4.10 PERFORMANCE GROWTH POTENTIAL

As soon as imaginative operators learn to use a new system, they invent new uses requiring performance beyond its initial capabilities. Every successful new system therefore grows to satisfy some new user expectations. Growth potential is therefore considered a significant product characteristic, and this scoring method may award up to P_g competitive points in this area.

The critical characteristics for the growth designs should be compared in Table 4.6A and B to the basic system (initial production unit) values and subjected to a LCC, availability, and reliability analysis similar to the one prescribed for the system performance evaluation, Table 4.3A and B. In a similar manner, reference is made in the first four columns to missions, frequency, critical characteristics, and the measured values as applicable to the initial production unit. The tables then call for the entry in the next column of new "Growth Potential" values, which reflect a proposed growth configuration. In the next column, "Margin" (competitive factor), is entered the increase in each critical characteristic, that is, the difference between the growth potential configuration and the measured values for the basic system.

In the next six columns, in a procedure similar to the evaluation of systems performance, the growth potential index G is computed for three selected scoring areas, namely:

Reduction in life cycle cost
Availability
Reliability

For each line of Table 4.6A where there is a positive margin, Figures 4.8, 4.9, and 4.10 must be reexamined and a determination made of how many competitive points would be awarded in each figure if the potential growth characteristics are achieved.

In the system performance evaluation, this determination has already been made for the measured values applicable to the basic system. Now, if there is an increase in competitive points as a result of the growth margin, the fractional increase is entered in the appropriate column of Table 4.6A. The product of this increase and the mission frequency is entered in the adjacent column entitled "Increment."

The increments are added by column and the sum entered in Table 4.6B. On the line below those sums, enter the number of competitive points awarded for the initial production system in each of the three evaluation categories and add horizontally to obtain the sum p.

Table 4.6A. Performance Growth Potential Scoring—Worksheet

Mission Frequency	Critical Characteristic	Measured Value Initial Production Units (1)	Growth Potential Value (2)	Margin (2) − (1)	Growth Factor Due to Growth Margin in Terms of Competitive Points Awarded					
					Reduction in Life Cycle Cost		Increase in Availability		Increase in Reliability	
					Fraction	Increment	Fraction	Increment	Fraction	Increment
From mission list	1									
	2									
	3									
	4									
	5									
	6									
	7									
	8									
	9									
	.									
	.									
	.									
Competitive points—basic aircraft					g_c		g_a		g_r	

NOTE: One chart for each mission.

Table 4.6B. *Performance Growth Potential Scoring—Summary*

	Reduction in Life Cycle Cost	Increase in Availability	Increase in Reliability	Total
Total incremental points, *all missions*	G_c	G_a	G_r	
Competitive points—initial production unit	p_L	p_a	p_r	p
Total increments × points	$G_c\, p_L$	$G_a\, p_a$	$G_r\, p_r$	$\sum_{n=c}^{r} G_n\, p_n$
Growth Potential Index		$G = \sum_{n=c}^{r} G_n\, p_n / p$		

On the last line, multiply the sums of the increments with the number of competitive points awarded in each of the three categories and add horizontally. Divide this sum by p to obtain the "Growth Potential Index G."

The number of competitive points awarded a given growth potential index G is established in Figure 4.16. Since $G > 1.0$, the value of $1/G$ is plotted on the abscissa. There is also a break in the curve at $(1/G) = 0.5$ in order to give more points for lower values of G.

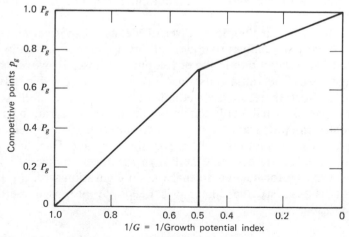

Figure 4.16. Award of competitive points—performance growth potential.

4.11 CHAPTER SUMMARY

This chapter discusses some of the factors that should be considered in the contracting function which, as indicated in Box 8 of Figure 3.1, is the controlling instrument of the development process. In advanced development projects the course of the development cannot be precisely established in advance; contracts, therefore, are dynamic instruments undergoing a constant process of change reflecting the current status of the agreement between customer and contractor.

The form of contract is dependent on the amount of advanced technology to be incorporated into the new system and the corresponding amount of risk. In the procurement of existing commercial equipment there is little or no risk involved, and a simple fixed price (lump sum) contract is the rule. If new technology is involved, the associated risk may be considerable, and the customer is required to assume the financial risk. He also participates in the development to assure that his risks are minimized and his funds are properly spent. Such contracts emphasize performance and are discussed under the heading ''Design-to-Performance.'' They tend to be single source contracts involving some type of cost-plus funding.

Cost-plus contracts are not considered a desirable way to do business. As soon as the risks are minimized, as with the start of production, contracts revert to the use of fixed or target prices. Design-to-cost, design-to-unit-production-cost (DTUPC), value engineering, and other techniques are used along with awards, incentives, royalties, and warranties to motivate the contractor to reduce production costs (as well as operating and support costs) while maintaining or improving performance and quality.

It is recognized that the most powerful means for motivating contractors is competition. Competitive parallel developments provide a means for introducing competition in the development phase. However the procedure increases the initial cost and involves considerable advanced thought and effort; therefore the method is used only on carefully selected major subsystems, usually only until a satisfactory configuration emerges. Since each competitor wants to win, the ground rules for the selection of the winner must be carefully outlined in the contract. Each competitor designs his configuration so that he has the greatest chance of winning. The rules must be formulated so that the configuration with the highest score will best satisfy the customer's needs. A proposed method of scoring is discussed.

5

Technical Requirements

Technical requirements are quantitative descriptions of product or system characteristics that a product or system must possess to perform its missions satisfactorily. After a contract is awarded, Box 8 of Figure 3.1, the term *requirement* takes on a different and more precise meaning from its use in Box 3 of Figure 3.1. It becomes an engineering term with a *specific technical and contractual meaning*. To avoid confusion, it is necessary to know who is using the word and to what program phase he is referring.

5.1 CONTRACTUAL REQUIREMENTS

Requirements may be expressed in contractually "hard" or "soft" terms, depending on many considerations such as complexity, amount of advanced technology involved, available means of measurement, and available funding.

In addition to performance requirements, this chapter contains a discussion of technical requirements involving producibility, maintainability, safety, and of the role of standards and commonality concepts. Quality assurance requirements are discussed in Chapter 10.

5.1.1 Hard Requirements

A "hard" requirement is one that is technically feasible and is specified in a contractual document in such terms that its compliance can be demonstrated before the product or system leaves the factory door; that is, before acceptance and payment. Clear accept/reject (go/no-go) criteria are established. If it cannot be demonstrated that the product meets the requirement within the specified tolerance, the product is rejected.

Rejection of an item must be based on measurements and on the clear accept/reject criteria. It should not involve any interpretive judgment. The acceptance criteria should be expressed in clear and definite terms

101

leaving no room for opinion or doubt regarding whether or not the requirements are met.

Required tolerances should make allowance for wear. If a system is designed to last a number of years, wear is an important consideration in establishing the initial acceptance criteria.

The formal demonstration tests should be capable of being completed within a reasonable time and at reasonable cost, since they must be completed before the product is accepted for deployment and final payment authorized. Each individual demonstration test should take place as early as possible during the program, to provide maximum time for any corrective action that might become necessary, should a failure occur during demonstration.

The preparation of a set of requirements that meets all of these conditions is a specialized technical activity requiring considerable experience and technical skill. The function must be performed with imagination and foresight, since any deficiency in the establishment of forward-looking and creative requirements cannot be compensated for in subsequent program phases, if at all, without great increases in cost.

5.1.2 Soft Requirements

If a requirement cannot be associated with such clear accept-reject criteria that nonconformance results in rejection, it is referred to as a "soft" requirement. Workmanship requirements are usually soft requirements, since it is difficult to reject a product solely because of marginal workmanship.

Although soft requirements are generally considered a less desirable way of doing business, there are circumstances that justify their use. The type of product and available time and funding sometimes determines the degree of hardness specified. Some contracts call for the development of a specific product on a given schedule, and all major requirements are hard; many experts believe that there is no place for soft requirements in this type of contract.

At the other end of the spectrum, a contract may be for an investigation, a study, or a "best effort" development, and the customer is willing to accept whatever the contractor can produce with available funds and time; such contracts contain few or no hard requirements.

5.1.3 Systems Model

In Chapter 2 it is pointed out that a distinguishing characteristic of a system is that it requires an expressed or implied model to predict the interrelationships of all the system components; specifically a model relating the performance characteristics of individual elements to overall system performance.

The basic system requirements are expressed in terms of required performance under specified environmental conditions. By means of the system model these requirements are translated into product characteristics and apportioned to the subsystems and various lower tier elements until the requirements for the lowest level piece parts or components are established. These relationships can be represented by a *parameter tree* or similar diagrams where the requirements for each component can be traced to systems performance requirements (74).

A parameter tree is a graphical representation of the system model. Actually, the term "trees" would be more accurate, since a single tree applies to only one system characteristic; otherwise the document would become too unwieldy.

The pertinent system characteristic is listed in a box at the tip of the tree. On the next level down, in boxes connected by lines to the box at the tip, are the next lower level subsystems involved in providing the system characteristic; the requirements for each subsystem are noted in the appropriate box. In like manner, on the next level under each subsystem are listed all of the next lower level components that contribute to subsystem performance, and the requirements for each component are noted in each box. The procedure is repeated through each subsequent lower level until the requirements for all of the lowest level parts are determined.

These parameter trees are of particular importance in establishing the required tests of, and allowable tolerances on, the characteristics of individual parts on any level. They are excellent tools for determining the minimum quality of lowest level parts required to achieve the specified performance. The graphical representation of the system model provides visibility into the relationship between the requirements among parts of the system, which cannot be attained in any other way.

5.1.4   Waivers

In the early development models of advanced complex equipment it is usually difficult to comply with all requirements. Where it is impractical to correct defective items immediately by redesign or reprocessing, the contract should contain provisions for the granting of waivers by the customer. Waivers are normally granted to keep the program on schedule; sometimes waivers are granted to save a piece of material in which a great deal of costly labor has already been invested. Waivers should be properly documented since they normally degrade performance and reliability.

In granting a waiver the customer should retain a contractual commitment that the defective condition will be corrected, and retested if necessary, before the production cycle begins. It also follows that a cost adjustment is justified when a significant waiver is granted. For customers and suppliers to do business in an amicable manner, it is necessary that both parties understand this division of responsibility clearly.

5.1.5 Pitfalls in Specifying Requirements

Categories of requirements often overlooked are: training operations and use of systems for the performance of missions originally not planned or foreseen. In the utilization of advanced military aerospace systems, it is not unusual for 85 to 95% of all missions to be training missions, which invariably include a higher percentage of extreme conditions than do service or combat missions. Constant training is required to maintain a flying unit in a condition of peak readiness, and new missions are conceived and developed during the 6 to 10 year period between contract award and system deployment.

Some desired product characteristics must be specified indirectly. One of the most important of these is system life. If a system is intended to last, say, 20 years, there is no point in including this number in a contract, since a 20-year life cycle cannot be demonstrated before acceptance, and the manufacturer cannot be expected to guarantee the system for this length of time. What must be done is to determine what critical measurable characteristics a product must have so that it will last 20 years (materials, finishes, drain holes, etc.). These characteristics may be determined by a critical comparison of similar systems that have lasted 20 years with systems that have not lasted this length of time. The failure history of the short-lived equipment should be analyzed, and it should be determined what product characteristics are required to avoid these failures. These characteristics should then be specified as demonstrable requirements in the new contract.

In one case a power company interested in life cycle costs traditionally procured its equipment with a one-year warranty. However, there is no assurance that equipment that survives its first year of operation will also perform satisfactorily during its expected service life. Studies disclosed that equipment that did not fail during the first five years of service normally operated trouble free for 40 years. Hence, the company began to specify a five-year warranty. Most suppliers refuse to provide a five-year warranty, because they cannot project economic factors that far in advance. Nevertheless, the power company has found a few suppliers who have been willing to provide a five-year warranty and is carefully monitoring the results of this long term experiment.

5.1.6   Design Trade-Offs

The basic difficulty involved in establishing requirements comes from the fact that many design requirements conflict with one another. To provide more of one product characteristic, it is frequently necessary to settle for less of another. A *trade-off* must be made between these conflicting requirements. In the design of complex advanced systems such trade-offs must be made among a large number of parameters simultaneously with a host of individuals and groups participating. An additional complexity is introduced by the required variety and range of missions and operating conditions. The many factors to be considered simultaneously make it a complex (if not actually impossible) procedure to establish a set of truly optimum design requirements for an advanced system. Among the most difficult trade-offs to make in this process are those involving safety, reliability, and quality versus cost, performance, and schedules. Reliability requirements must include consideration of such concepts as maintainability, shelf life, repairability, availability; permissible percentage of units down and awaiting parts; percentage of missions to be successfully accomplished; permissible accident and fatality rates; operations, maintenance, and support costs over equipment life time; availability of skilled maintenance personnel; and similar parameters.

Particular care must be taken to assure that the requirements are not so far beyond the state of the art that they cannot be achieved within available time and funding. Incompatibility of requirements and the design state of the art is one of the main, causes of equipment unreliability. It is not difficult to write requirements that cannot be met; in fact this practice is probably responsible for more premature contract cancellations than any other single reason.

It is a contractor's responsibility not to promise more than he thinks he can deliver, based on his estimate of foreseeable improvements in the state of the art. Just how far a contractor permits himself to be pushed in promising to extend the state of the art depends on foresight, experience, and economic considerations.

5.1.7   Trade-Offs Involving Extensions of the State of the Art

Section 2.3 states that system contractual requirements tend to reflect the state of the art existing on the date of contract award. Nevertheless, customers of advanced systems are aware that some new state of the art can be included in a development project (Figure 2.2), and they sometimes demand more performance in their RFPs than may be feasible at the currently available state of the art. Bidders who do not want to appear unresponsive in a competitive environment may agree to include some performance feature that may be unattainable at the time the contract is signed (see Section 5.4).

5.2 EXTENDING THE TECHNICAL STATE OF THE ART

Technology progresses in two basic ways: by extension of the state of the art, and by invention. The distinction between the two is significant. This discussion is limited to extensions of the state of the art; inventions are excluded because it is not practicable to invent on schedule and to predict the related effects on schedules and budgets. Inventions include for example, the introduction of radar, the transistor, and the jet engine.

Given a fixed schedule, a system development project cannot include more than a carefully determined extension of the state of the art—certainly not an extension comparable with an invention—and still be ready on the scheduled delivery date.

Attempts to predict, during the development period, whether a system will be ready on the scheduled delivery date, therefore, must include consideration of the feasibility of the extensions of the state of the art included in the technical requirements.

Extensions of the state of the art take place in relatively small steps, always based on extrapolation of organized past experience. On a large new advanced systems project, many small extensions of the state of the art will be made in a great many technical areas.

Even in the case of a dramatic new mission to be accomplished with a new vehicle as in the Apollo Program, the decision to proceed was made on the basis of expert opinion that essentially all of the needed basic technology was available and had been tested. The framework for this decision is provided by Figure 2.2 where the baseline represents the state of the art on the date of contract award. The amount of new technology developed during the Apollo Program, although significant and affecting a great many different technical areas, was relatively small compared to the large body of technology already existing at the time of the start of the design.

The amount of effort and resources required to achieve adequate reliability in a new system or product is to a great extent a function of the extension of the technical state of the art. The extension of the state of the art required to comply with the technical requirements of a development contract is, therefore, a significant parameter which has a paramount effect on the feasibility of the project. The relationship of these factors is illustrated in a general way in Figure 5.1, which is taken from ref. 155. The graph indicates the effect of the extension of the state of the art on two other parameters: reliability and required resources. Schedule factors are neglected in this simplified presentation; the impact on schedules is similar to that on resources and is of equal importance.

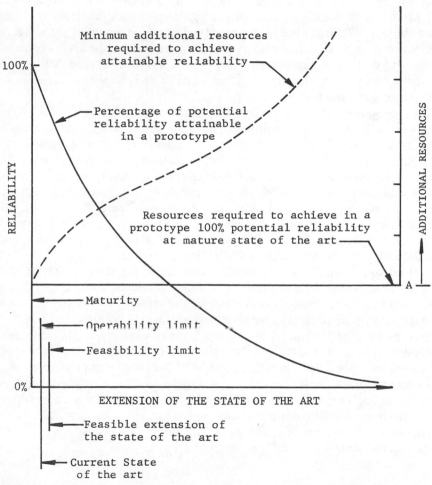

Figure 5.1. Effect of extending the state of the art on attainable reliability and required resources.

5.2.1 Maturity

A vertical line through the initial abscissa point of Figure 5.1 is designated as *maturity*. This line represents that level of the state of the art at which competent specialized personnel can attain 100% of the reliability inherent in the design while adhering to honest schedules and resources allocations. (Technical or management mistakes or incompetence that normally prevent achievement of 100% reliability in the application of mature technology are ignored in this theoretical definition.) As soon as the state of the art is extended beyond the maturity point, 100% reliability cannot be achieved. The greater the advances in the state of the art, the greater will be the drop in the percentage of potential reliability that can be attained in a prototype (solid concave line). It is assumed in the figure that the drop in reliability with extension in the state of the art is exponential, and that the attainable reliability rapidly approaches zero. This curve cannot be given in quantitative terms since there is no way to measure an extension of the state of the art.

By definition, at the maturity point it is known what resources (manpower, skills, funds, and facilities) will be needed to design and develop a product or system with 100% inherent reliability (point A in Figure 5.1; baseline for additional resources). As the state of the art is extended, the resources needed will increase for developing a prototype that exhibits the attainable percentage of potential reliability (dashed curve). At the point where the exponential solid curve approaches zero, this "additional resources" curve will approach infinity. This indicates that if the state of the art is extended too far, the product cannot be made to work (zero reliability), no matter what resources are expended.

Note that even for a very small extension in the state of the art, the reliability falls rapidly and the costs increase uncontrollably. The decision on how far to go in extending the state of the art for a new system or for a modified version of an existing system, therefore, is a critical problem that must be carefully balanced against available resources and planned completion date for the development effort. It is a decision that can spell success or failure for a project. In view of the lack of quantitative measurements of the parameters shown in Figure 5.1, this is a decision that is made under customer pressure by top management in a competitive environment on the basis of mature experience and judgment. The more limited the resources of a company, the more it will hesitate to try to extend the state of the art very far.

5.3 OPERABILITY AND THE STATE OF THE ART

All new advanced systems require some extension of the state of the art to increase their performance over that of existing systems; otherwise their development is difficult to justify. For the purpose of interpreting the graph of Figure 5.1, "state of the art" is defined as that level of technology which at a given point in time is commonly available from specialized personnel in a given field. Products supplied by such specialists do not always operate without failure; that is, they are not necessarily mature. The current state of the art, therefore, will be represented by a point on the abscissa of Figure 5.1, that lies to the right of the maturity or zero point. How far to the right this point can go, that is, the extension of the state of the art beyond maturity that can be tolerated in an operational environment, is a function of the required *operability* or *operational readiness*.

5.3.1 Trade-Offs Involving Operability

Operability does not require a complete absence of failures. If the failure rate of an element of construction is known, consideration can be given during development to redundancy and also to maintainability considerations, so that when a failure occurs in service, availability of spare parts or equipment and maintenance facilities are assured, and the design is configured so that repairs are easy to make. Operability requirements are established on the basis of complicated trade-offs between performance, maintainability, available resources, and similar parameters.

There is, however, a limit on how far these trade-offs can go. There are, for example, many devices that can be operated in the controlled environment of the laboratory by skilled personnel but that are not operable under field conditions by operating personnel. The state of the art of such devices appears to exceed the operability limit. On the other hand, it may be more cost effective to invest more in crew training and let them operate more advanced equipment in the field. This concept has paid off in the aviation industry, where pilots are highly trained and well paid.

Better trained and educated operators can do much to operate equipment more effectively and avoid failures and breakdowns in the field. There should be some relationship between the required educational levels and skills of operators, the value of the equipment they are operating, the operational difficulties and severity of the environment, and the consequence of mission failure or equipment breakdown. Human factors and reliability considerations in the initial design stages must be traded-off against logistic supply and field maintenance facilities, which are extremely costly.

Curiously, for advanced systems, there often seems to be a lower limit on permissible extensions of the state of the art. In the unlikely event that a new advanced system does not encounter the troubles normally associated with a system of its type in early deployment, the customer might assume that the development was too costly or too overdesigned, or that the contractor did not do all that he could to attain the maximum possible performance by extending the state of the art as far as he should have from the viewpoint of operability considerations.

5.3.2   Measures of Operability

The operability of a product or system becomes unacceptable if unexpected failures occur often and require time-consuming corrective action that prevents the system from the timely completion of required service missions with available support and resources. Unacceptable operability, therefore, is associated with unexpected failures that require corrective action of significant unforeseen magnitude. One measure of the magnitude of a corrective action is the time required to eliminate the deficiency. For the purposes of this discussion, operability requirements can be defined in a simplified manner as a variable that is dependent on the number of unexpected failures and the time required to eliminate them (down time).

The concept of operability depends on service conditions as well as on the characteristics of the hardware. Technically sophisticated facilities supported by competent specialized maintenance crews can cope with a greater number of unexpected failures than can less sophisticated facilities with less maintenance and support capability. However, for any given level of sophistication, the time required for corrective action to eliminate a failure is a valid measure of operability with respect to that environment.

5.4   FEASIBILITY

A project is feasible if it is possible to meet the requirements of the specification within schedule and resource allotments, even though an extension in the state of the art is involved. Feasibility includes considerations involving both technical and management skills. The feasibility limit, Figure 5.1, includes a greater extension of the state of the art than does the operability limit.

Feasibility is determined by how much support the operator is willing to supply. If the operator has at his disposal an excellent logistics support system and well-educated management and maintenance personnel, the feasibility limit may be extended beyond the point that could be accepted with less sophisticated resources.

5.5 THE REALITY OF HAZARD

Hazard, as a measure of the "unexpected failure rate," is a function of the state of the art. As long as the technical requirements for advanced systems require an extension of the state of the art beyond the maturity point as shown in Figure 5.1, the inherent reliability will include an expected constant generic failure rate which cannot be reduced below a constant level except by invention or by basic improvements in the state of the art (123, 155).

All elements of an advanced system that are beyond the maturity point are designed to some extent by trial and error or manufactured by processes that are not fully understood. In this development process all risks cannot be eliminated, and the technical requirements should make provisions for coping with the residual risks.

Unfortunately, this is a fact many procurement agencies fail to recognize, and many development contracts are defective in these areas.

One of the objectives of a development agency is to strive to attain a minimum constant level of residual generic failure rate. The logistic agencies then must provide for spares and support services to maintain the system in an operational condition at this established failure rate. Where safety is involved, adequate provisions must be maintained to cope with contingencies involving possible loss of life or property. These provisions are discussed further in Section 9.15 under "Risk Management."

5.5.1 Implications of Constant Failure Rate

The level character of the generic failure rate is an indication that a program is doing as well as can be expected from a technical viewpoint at the existing state of the art. An advanced systems development program is a very large operation involving hundreds of thousands of individuals and thousands of specialized suppliers spread geographically over a vast continent. This system is balanced at a given level of craftsmanship and technical competence. An isolated failure is a random disturbance in an otherwise carefully balanced system. Two basic elements of this balance

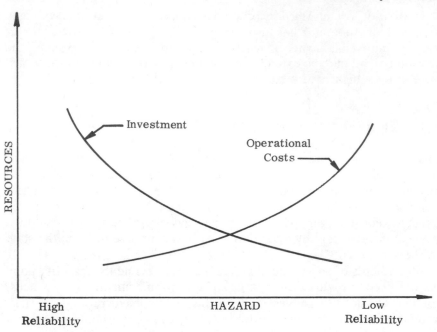

Figure 5.2.   Hazard as a function of investment versus operational cost.

are depicted in Figure 5.2, which shows the relationship of hazard to investment and operating cost.

The investment required to achieve zero hazard (high reliability) is shown to be unattainably high. At the other extreme, if the hazard is sufficiently great, a relatively small investment can bring about considerable improvement.

The second curve shows operational costs. These are lowest when the hazard is low, but they tend to increase at an intolerable rate as the hazard becomes large and reliability low.

These two curves intersect at a point below which additional investments to reduce hazard will not result in proportional decreases in operational costs. There would, therefore, appear to be little economic incentive to reduce the hazard below a certain point, unless safety is involved. It must, therefore, be assumed that all future advanced systems operations will always be associated with some hazard, and that, barring significant technical breakthroughs (inventions), the present hazard level will only be reduced slowly over a long time as the general state of the art improves.

5.5.2   Need for Reliability Engineering

The continued existence of hazard in advanced systems establishes some specific requirements for the inclusion of a reliability group in a design engineering department with a specialized capability for:

1. Monitoring and analysis of hazards.
2. Failure detection of unpredictable and unexpected failures.
3. Trouble shooting, isolation, and identification of failed parts.
4. Diagnosis of causes of failure.
5. Bringing about effective corrective action to prevent recurrence of the failure.

5.5.3   Comparison with Health Hazards

In ref. 123 the hazard for American males is computed from the American Experience Mortality Tables (20). Between ages 10 and 50 the hazard is quite constant there is no wearout. This condition has led to the establishment and maintenance by our society of medical services that have all of the capabilities (in terms of human failures) listed above as requirements for a reliability group.

Recognition of the condition represented by this comparison is important because technical administrators who have no reservations about supporting medical services to cope with health hazards, have difficulty in recognizing the need for a properly organized and motivated reliability group in their own technical organizations. Specifically, it is difficult for them to accept the fact that hazard exists. They feel that the use of qualified parts should guarantee that no failures occur. They look upon every failure as an exceptional case which should not have occurred, and cannot accept hazard as a natural aspect of life for which adequate provisions must be made. They feel that by paying for qualification, they are entitled to receive failure-free operation.

5.6   RELIABILITY REQUIREMENTS

The foregoing discussion has made it clear that all advanced systems incorporating some extension in the state of the art will contain a generic failure rate corresponding to the inherent reliability that can be achieved by the particular design.

The establishment of reliability requirements can also be approached from the systems analysis viewpoint. A scenario representative of the operational situation is constructed, and the reliability requirement for the system to accomplish its mission is determined. This reliability require-ment may have no relationship to the generic failure rate mentioned above; it may in fact, require a higher level of reliability than it appears possible to achieve. This is an indication that additional design effort is required, possibly a redesign cycle based on different concepts. It is permissible to allow for reliability growth between the prototype and the production units, see Section 11.3.2.

All major advanced systems contracts today contain quantitative reli-ability requirements. These reliability requirements are an important factor in determining the allowable extensions of the state of the art that can be incorporated into a new advanced systems design. The RFP usually asks that a plan be submitted detailing how the contractor proposes to control his design procedure to assure that the reliability requirements will be met, and how he proposes to demonstrate that his prototype meets the requirements, MIL-STD 785 (55). These techniques are discussed in greater detail in Section 7.6.9.

5.7 PRODUCIBILITY

This section is applicable to high-value large advanced systems produced in relatively small quantities over a long period of time. One-of-a-kind systems are excluded. The concept of increased producibility is as-sociated primarily with cost reductions.

The system RFP should require that proposals include a manufacturing plan and a description of the facilities and manpower skills available for the development and production of the system.

The development of an advanced system normally advances through three critical stages, advanced development, engineering development, and production.

5.7.1 Advanced Development

Every advanced system includes a number of new critical subsystems that constitute the key building blocks of the new system. For example, many current advanced systems include a dedicated minicomputer with special embedded software. These critical elements are normally developed by basic scientific laboratories, both in industry and government. Once such critical elements become available, it is possible to incorporate them in the design of a new advanced system with superior performance.

5.7.2 Engineering Development

The second phase is the development and testing of one or more working prototypes by an advanced engineering group. This group is project oriented; it may often control extensive dedicated production facilities and have its own subcontractor and supplier procurement organization. This group is mainly concerned with meeting the performance requirements on schedule. Costs are sometimes secondary considerations; the group cannot afford to take the risk of not meeting the performance requirements. Parts are hand made to the greatest possible extent; there is a minimum of special tooling, assembly jigs, and fixtures. Little concern is given to interchangeability of parts and other factors that become of prime importance in the production phase. Paperwork concentrates on analysis, test plans, and test reports, particularly those that contractually must be submitted to the customer for approval. In commercial organizations to protect the contractor's patent rights and to maintain records that might be required later in possible liability suits, company lawyers require that design and test engineers keep complete records of their work in bound, numbered notebooks with numbered pages, and that official minutes of all design review meetings be kept on file. These documents are also used for technical program management. On the other hand, prototype manufacturing sometimes may proceed on the basis of informal (but technically complete) layouts.

5.7.3 Production

Once the prototypes are accepted, the third phase starts with the award of an initial production contract. Price now becomes a matter of major concern. A new group of production and cost oriented engineers reviews the prototypes and all associated documentation; they may redesign all or large portions of the system. The emphasis is on formal production drawings, process specifications, and acceptance test procedures. All excessive quality is eliminated in an effort to achieve the lowest price. Actually, an advanced system may be developed and may evolve to maturity over a number of production contracts where the performance, reliability, and maintainability may be increasing while unit production and life cycle costs may be decreasing.

5.7.4 Manufacturing Schedules

The system is broken down into subassemblies, then into lower tier assemblies, and finally into lowest level components and parts. Manufacturing schedules are prepared, starting with the contractual delivery date and working backward through final testing, final assembly, through each subassembly, down to each individual part. The production or procurement processes and schedule for each individual part is determined.

These manufacturing schedules are large, complex documents, and they provide the key to the control of the manufacturing process. Sometimes they are computerized to the extent that each foreman receives a computer printout each morning advising him of his tasks for the next working period and assigning him the necessary resources. At the end of the working period the foreman marks up the printout and turns it in as his status report. This may be processed during the night and the next morning the foreman receives a new printout with an updated assignment.

The manufacturing schedules are keyed to the production drawings and process specifications. For each part either a purchase order is issued or, if the part is to be made in-house, a routing document is prepared by manufacturing planning. This document refers to the latest revision drawing and process specifications, and specifies when and where every operation, inspection, and test will take place. Purchase orders for special materials are issued, and special attention is given to long lead time items. Provisions are made for planned process changes; for example, machined parts may be replaced on specified downstream systems by long lead time forgings or castings, at considerable cost savings. Plans for maintaining min-max stocks are reviewed to assure that standard materials will be available when needed.

5.7.5 Tooling

For the initial production contract, a minimum of special tools, jigs, fixtures, and facilities are available, depending on how much time and funding is made available for the development. As subsequent follow-on production contracts are awarded, the special tools, jigs, fixtures, and facilities become more sophisticated, the amount of manual labor decreases, parts are redesigned for ease of production and assembly, and production costs decrease. It is not unusual for the final production system to cost one half to one third of the initial production system, while simultaneously having improved performance and reliability characteristics. On high production systems the cost decreases are proportionately greater.

On high quantity production systems much more effort is devoted to tooling on the initial production run, since the savings will be greater and changes will be more difficult to make on subsequent follow-on contracts.

However, this is just a matter of degree, since production improvements continue to be made on any advanced system as long as it is in production.

Profit-making organizations are very aware of production costs. Regular or periodic redesign and retooling to save costs is an important part of their operations.

5.8   SUBCONTRACTING

Most contractors of advanced systems are "system houses" with specialized design, integration, and testing capabilities. These features involve a high overhead; to reduce system costs they prefer to buy as many services and components as possible from low overhead suppliers. Furthermore, government contracts require that subcontracts be distributed on a wide geographical basis, and that a percentage of purchases be made from small businesses (see Section 10.7.2). Hence, the production capability of a contractor will depend to a critical extent on his "stable" of subcontractors and suppliers.

5.8.1   Contractual Relationships

In preparing his initial production proposal, the contractor will have arrived at a number of agreements with his subcontractors and suppliers. Subcontractors are organizations that design and produce specialized subsystems or components; they extend and complement the contractor's own in-house design capability. When the contract is finally awarded, its terms invariably are different from those listed in the proposal. The contractor must then review all of his proposed make-or-buy decisions and renegotiate all of his agreements with his subcontractors and suppliers. Since tight schedules and long lead times are involved on many items, it is necessary to come to an agreement quickly with the subcontractors and suppliers so that they can start their work as soon as possible. Subcontractor and supplier negotiations, therefore, are carried on under intense urgency and pressure.

Since the system is not yet completely defined, the understandings with subcontractors and suppliers are often incomplete. Furthermore, in cases of sophisticated equipment such as some hydraulic control systems, it is not possible to completely define all of the required quality characteristics by means of engineering drawings and specifications. Considerable specialized knowledge of the subsystem and how it will function in the final system is required to interpret the specifications. Hence, the relationship of a contractor with his subcontractors and suppliers is a rather special one based on mutual understanding and confidence developed over many years of working together.

5.9   CONTRACTOR MOTIVATION

The government has used a variety of methods for motivating contractors to reduce or hold down costs. The most effective method for reducing costs is competition. However, areas where competition can be effectively used in the development and procurement of advanced systems are limited, and the government has a number of schemes applicable to single source procurement. Some of these follow.

5.9.1   Engineering Development Programs

Incentive Payments.   These are based on points earned for achieving test, production, and performance goals at specified milestones during development.

Design-to-Unit-Production-Cost (DTUPC).   The engineering development contract includes a negotiated target unit production price. Award fees are paid at scheduled design reviews based on the judgment of a jury committee which determines how well the contractor is doing.

Reliability/Maintainability Warranties.   Advance payments during the design phase in exchange for a contractor commitment to repair or replace all items later returned to him which have not met the reliability or life requirements.

5.9.2   Production Programs

Value Engineering Royalties.   Cost savings, achieved by product redesign or improved processes during production, are shared with the contractor for a given period, usually three to five years. After that time ownership of the new design or process is vested in the government.

Target Price Contracts.   A new lower target price is negotiated for each successive production contract. Within specified limits, the contractor's profit is increased if the cost is less than the target; his profit is decreased if his cost exceeds the target.

5.9.3 Operational Phase

Reliability/Maintainability Warranties. When items in service exceed the reliability and maintainability requirements and cost savings are achieved, these savings are shared with the contractor.

5.9.4 Advanced Planning

The use of a particular method depends on the immediate objective; however it is not always clear in advance which method will yield the best long term results.

Continuity of talent is an important factor in achieving cost savings in long-run production. To the greatest possible extent the manufacturing plan required in Section 3.2 should be prepared by manufacturing personnel who will remain with the project. During the development of the prototype they can provide support and guidance to facilitate the transition to the production phase. They can introduce production concepts in the prototypes and plan the subassembly breakdowns so that the initial simple assembly jigs and fixtures serve as the baseline for the production tooling. They can provide guidance during detail design to assure that available processes, tools, and skills are used to the best advantage.

5.10 MAINTAINABILITY

Most consumer products are sold competitively on the basis of initial cost. In the case of large high cost advanced systems, technically sophisticated customers are sensitive to life cycle costs. In some cases, operating and support costs over the life of the system run as high as 10 to 100 times the initial cost. In these circumstances the customer is willing to pay a higher initial cost if it lowers the operating cost. Airlines, for example, are very sensitive to life cycle costs.

An important technical requirement affecting operating and support costs is *maintainability*. Maintainability is a product characteristic established during design. It should not be confused with *maintenance engineering*, which is concerned with proper maintenance procedures and is discussed in Section 5.11. There is no one single quantitative measure of maintainability; requirements should be based on a scenario that best reflects the customer's operational environment. The customer must describe the anticipated operating conditions and expected equipment utilization. Given operating and maintenance personnel with known skill levels, a common requirement is expressed in terms of the *maximum number of maintenance man-hours per operating hour* required to keep the system in an operational condition. The maximum number of operators and maintenance men per system is sometimes also specified (see Section 4.7).

Other requirements that sometimes appear in specifications are: time between overhaul, minimum number of operating hours per day, turnaround time, maximum allowable down time per specified time period, and the tools and lists of equipment required to perform maintenance.

5.10.1 Maintainability Design Data and Models

When maintainability was first discussed, it was not clear how a contractor could discharge his responsibility for maintenance and support concepts. Then, quite independently, the reliability effort started to generate handbooks of failure rates of elements of construction. With these data it became possible to predict how often and in what manner equipment would fail in operational situations. This led to the development of maintainability models. In support of these new models additional handbooks were developed to indicate skill levels and time needed to perform various standard tasks.

One important applicable model consists of a work-flow diagram showing all the sequential tasks that must be performed between the end of one operational cycle and the start of the next one, as well as the time relationship of all tasks to one another. The system, the facility, and the support equipment must be shown and their interfaces defined, and all inputs and outputs to each operation must be indicated (158).

A design monitoring procedure similar to that outlined in Sections 6.4 and 6.5 was developed. The system-maintainability requirement is broken down by subsystem and component to individual replacement, repair, or maintenance tasks. In allocating repair time, materials, and other resources to individual tasks, the anticipated frequency of occurrence must be given consideration. During the development cycle the monitoring process assures that the allocations to individual tasks all add up to a number that complies with the overall system maintainability contract requirement. Consideration should be given to troubleshooting procedures, so that failed parts can be quickly isolated and replaced.

A great deal of work is being done on automatic test equipment capable of detecting incipient failures and isolating parts about to fail (see Section 7.6.4 and 7.6.11).

5.10.2 Maintenance Categories

Maintenance may be classified as scheduled and unscheduled. Scheduled maintenance is based on time between overhaul, replacement of expendibles, and safety checks. Unscheduled maintenance takes place when an unexpected failure occurs and corrective action must be taken to restore the system to an operational condition.

5.10.3 Field Data Requirements

To extend the time between overhaul for important components, it is necessary to have field monitoring and data analysis systems. If the development contract contains provisions for reliability and maintainability warranties as mentioned in Section 5.9, it is essential that contractor and customer agree contractually on the details of the reporting and data analysis system (Section 7.6.10).

5.10.4 Maintainability Specifications and Handbooks

The military has developed a comprehensive set of maintainability specifications and handbooks (56-61). These specifications call for the submitting of a maintainability program plan to be implemented during design (Section 3.2) and specify methods of predicting and demonstrating maintainability. These specifications are supplemented by a set of DD# 1664 Data Item Description Forms (62–66), which are contractual documents describing the reports to be submitted for approval. These references cover the Maintainability Program Plan, Maintainability Reports, Maintainability Mathematical Models, the Maintenance Support Plan, and Maintenance Engineering Analysis (MEA) Data. The MEAs are prepared during design to assure that the requirements are met. Reference 66 is a 35-page document describing the procedure in detail and including examples of the worksheets that must be filled out. A few of the available textbooks are listed as refs. 10–12.

5.11 LOGISTICS AND SUPPORT

One of the major causes of the need for new design controls in the decades after World War II was the increasing size and complexity of aerospace systems, brought about to a marked degree by breakthroughs in propulsion and in electronics.

Originally aircraft manufacturers were responsible only for the airframe and its flying qualities. Other airborne equipments were relatively simple and were generally procured separately and installed by the operator after delivery of the aircraft.

This situation changed with the advent of new sophisticated avionics, which had to be integrated into the airframe, giving air systems entirely new performance dimensions. The airplane manufacturer became a system contractor with the responsibility for demonstrating that his system could perform its mission, such as delivery of warheads to specified types of targets under stated environmental conditions with the required accuracy. Subsequently he was also made responsible for the design of the integrated logistic support system. All of this called for new aerospace industry organization and operations; in particular, a new relationship with a multitude of electronic subcontractors and suppliers had to be established.

To comply with these responsibilities, contractors have developed a number of concepts and procedures which are described in the following sections. These concepts are more associated with *maintenance engineering* rather than *maintainability*. For a more comprehensive textbook see Blanchard (13).

5.11.1 Maintenance Zones

The system breakdown for maintenance and support purposes is different from other breakdowns in that it is based on maintenance zones. A maintenance zone is defined as a part, component, subsystem, system, or region that is made accessible by partial disassembly, such as the opening of an access door, or the removal of equipment items or an engine. Progressive maintenance operations can be organized on the basis of maintenance zones. For example, instead of removing a unit of equipment from service for a prolonged time for a complete overhaul, certain partial types of scheduled overhaul can be accomplished by servicing progressively one maintenance zone at a time whenever some slack time becomes available between missions.

Another advantage of the maintenance zone is provided when a failure occurs calling for unscheduled maintenance requiring accessibility to normally closed areas. For example, the planned corrective action for a failure requiring an engine replacement assures accessibility to an enlarged, normally closed, maintenance zone. This provides an opportunity for the completion of regular maintenance tasks planned for a future time block.

5.11.2   Maintenance Cards

A great deal of progress has been made in organizing the accomplishment of maintenance tasks. For each individual task all individual operations are described on a *maintenance card* with sketches, photographs, appropriate instructions, and spaces for inspection records. Upon assignment of a task, the workman withdraws a package from the stockroom containing the maintenance card and all of the required tools, materials, replacement parts, and instruments necessary to perform the complete task. After completion of the work the mechanic returns the package to the stockroom before being given a new assignment. The returned package is carefully checked. The card now contains all the necessary inspection records and is filed as the record indicating performance of the work. The replaced parts, unused materials, and returned tools are inspected for indications of possible human error in any phase of accomplishing the task.

5.11.3   Maintenance Considerations in Initial Design

To establish a maintenance plan, the troubleshooting and repair procedures must be developed concurrently with the initial design. Components of limited life or parts that are more likely to fail or to require scheduled maintenance or replacement must be located in more readily accessible places than other parts. Such concepts can be demonstrated during design on appropriate mock-ups.

The designer must decide whether replacement parts should be repairable or throw-away, interchangeable, or whether they must be reworked to fit in the field. The economics of logistics are very complex; it is extremely costly to supply and maintain front-line stock rooms. The number of units in the supply pipeline can be reduced by standardization, but too much standardization can reduce systems performance, since individual parts will no longer be optimum for the application. The number of units in the supply system can be further reduced by increasing the size of the replaceable module.

5.11.4   Maintenance Levels

The operational situation normally is provided with several levels of maintenance: organizational, intermediate, depot, and factory. Organizational maintenance includes the shops associated with the front echelon operational units, similar to a neighborhood automobile service station. Depot maintenance is performed in rear echelon, factory, overhaul-and-repair facilities having the facilities for all types of major overhaul, such as overhaul of complete systems or engines. The intermediate level includes all of the shops between the organizational and depot levels. The factory level provides for the repair of highly sensitive and specialized units, such as gyros, which cannot be overhauled at other levels. Organizational levels are capable, for example of replacing a complete electronic assembly; intermediate levels may be able to replace integrated-circuit cards in the assembly. If the circuit cards are repairable, it may be necessary to send them to the depot for overhaul. In analyzing a potential failure, the level most suitable to perform the repair must be determined, and this analysis determines the stocking points for classes of spare and repair parts.

5.11.5 Automatic Checkout

In the design of trouble detection and isolation (failed part identification) procedures, a great deal of work is being done on built-in, automatic checkout features. These devices monitor continuously, or test periodically and automatically, the operation of a component and notify the operator as soon as a failure occurs. The failed unit generally will be identified on a higher replaceable level. The associated design principles are discussed in Section 7.6.4, which also lists a number of pertinent references. These procedures are gaining particular significance in aerospace systems involving safety of flight such as power operated flight-control systems, and also where the requirements for turn-around time are more stringent. In case of a failure in an aircraft in flight, the pilot can report the failure to his base and the repair crews, and replacement units can be made ready to effect the repair promptly after the aircraft touches down (159).

5.11.6 Ground Support Equipment

In addition to airborne equipment, an aircraft system includes considerable ground support equipment such as handling, test, and repair equipment. Some such equipment will be already available or planned for at the time of contract award and will be common to more than one model of aircraft, preferably to all. In such cases, it is necessary to design the aircraft so that it can be supported by the available or planned support equipment.

However, in some aircraft systems, particularly those advancing the state of the art, available common support equipment may impose constraints on the flight equipment that compromises performance to an unacceptable degree. It then becomes necessary to design special support equipment (SSE) to support the more efficient airborne configurations. Since the justification for the use of such SSE is increased system performance, there must be the closest coordination between designers of flight equipment and the SSE.

5.11.7 Product Support

The developments in maintenance engineering and logistics have resulted in the rise of specialized product support groups in the contractors' organizations. In brief, the requirements call for the contractor to submit an integrated logistic support program plan (Section 3.2), which should describe at least the following: program objectives and requirements; contractor-management controls; procedures and status reporting; maintenance plans and analytical techniques to be used during design; support program as needed during initial deployment; provisioning of spare and repair parts; maintenance personnel, required skills, and training programs; technical manuals, handbooks, and other documentation; facilities studies and site surveys; support equipment; and test plans to demonstrate compliance with contract requirements. The integrated logistic support program plan is subject to customer approval and must be compatible with the customer's field organization.

Contractor developed maintenance and support concepts form the basis for the initial determination of required maintenance resources; establish the depth and frequency of required maintenance actions; and identify the maintenance functions and tasks assigned to each level of maintenance. Typical considerations involve safety; reliability, repairability, and special handling requirements; adjustment and checking requirements; mean time to return to service; special support equipment requirements; new facility requirements; special training requirements; and costs of maintenance actions and other economic factors.

5.11.8 Provisioning Requirements

The provisioning requirements for spare and repair parts are normally established at a planning conference that includes representatives of the contractor, the procuring agency, and operating and support organizations, and which takes place as soon as possible after the design freeze. All design drawings are reviewed in detail, and all items to be delivered as spares are identified and the required quantities determined. In this process reference is made to analytical reliability reports, available failure rate data, and past experience with similar equipment. Special consideration is given to long lead time items, high-value assemblies, vendor repairable items (factory level), identical part coverage (identical units in different components or subsystems), and similar factors.

5.11.9 Training Program

The training program also must be developed in close coordination with the design to assure the earliest possible identification of requirements for personnel, training, and equipment. The contractor trains the instructors who in turn train the field maintenance and support personnel.

5.11.10 Trouble Reporting and Analysis

The field trouble data collection system, Section 7.6.10, plays an important role in verifying the effectiveness of, and improving an established maintenance and support program. Information that may be computed from the data provided by such a system include: types of maintenance actions; frequency of maintenance actions, time required for each type of action, and identification of actions performed at the various maintenance levels; total maintenance effort, material and part usage, and part-failure rates. In analyzing data provided by elaborate data collection systems, it is necessary for the analyst to be constantly aware of the natural and universal tendency for such systems to omit or edit out significant factors at lower reporting levels. For example, such systems seldom report properly a human error such as a workman dropping a sensitive instrument. On the other hand, human errors often are significant and can be minimized by design changes. In this connection it is essential to design components so that they can be reassembled only one right way, and to provide for adequate visibility and physical access.

Such field data are useful not only for reliability measurement of deployed systems but also for the provision of failure rates for use on new designs. They provide a means for correlating laboratory tests with service experience. This information is of critical importance to a responsible advanced system contractor who wants to know how good his analytical and manufacturing techniques are. There are several sophisticated field data collection systems in existence, however they are generally designed for special purposes, and their objectives are quite narrow. It would appear desirable to correlate and broaden these existing systems to provide greater management visibility into the quality of operating advanced systems, and in particular provide data for:

1. Monitoring the quality of technical requirements
2. New design requirements
3. Product improvements
4. Integrated logistics support

The procedure for designing the maintenance and support program, including many of the necessary forms, is given in some detail in ref. 67.

5.12 SAFETY REQUIREMENTS

Given a residual generic system failure rate or hazard as discussed in Section 5.5, it must be assumed that some of these failures will involve safety. Safety refers to failures that result in personnel injury, loss of life, hazards to health, or an economic loss exceeding a specified minimum amount.

As the definition of quality (see Chapter 10) changes with user wants and needs in a competitive market, so does the definition of safety change with public expectations and social conditions over time. Engineers designing systems with a life cycle of several decades are being required to give increasing attention to these changing social requirements (160). Safety is not concerned with systems effectiveness or mission accomplishment. A system may break down and fail to accomplish its operational mission, yet be perfectly safe. Although closely related to reliability in the design stage, safety has a different orientation, and it is essential that the RFP include specific safety requirements and call for a safety plan. A great deal can be gained sometimes by examining the procedures in other industries (161).

Since safety requirements involve the public interest, they are often established by a public regulatory agency responsible for monitoring compliance with its requirements. Building codes are a common example. In this case, execution of a contract involves a three-way relationship: customer, contractor, and regulator. The procedure is discussed in Section 9.15, "Risk Management."

Four kinds of safety requirements are discussed here: hazard identification and evaluation, controlled conditions, accident prevention, and motivation. In addition to safety analysts, it is desirable that human factors engineers be included in the design review process.

5.12.1   Hazard Identification and Evaluation

Safety criteria and a safety plan applicable to the advanced system should be specified by customer reference to specific applicable standards and by criteria developed in a hazard analysis conducted during the preliminary development process. The government has an extensive list of available safety specifications and design handbooks; refs. 68–72 are listed as examples, and refs. 14 and 15 represent two of a number of available textbooks.

These safety requirements provide design guidance. They also place certain constraints on the designer. When safety goals are established and hazards are identified, the designer is required to try to design specific hazards out of the system, or to modify the design characteristics to reduce the probability of occurrence to an acceptable level within the operational requirements of the system.

Hazard analysis is directed at predicting the effect of potential subsystem or component malfunctions and at identifying the interactions between these subsystems that may cause accidents or unsafe conditions. Hazard evaluation is directed at assessing the impact of these predictions.

The existence of desired system safety characteristics is confirmed through design reviews and tests. Specific scheduled safety reviews should be required as an integral part of the design, development, manufacturing, and test processes, and of the operational phase. The hazard analysis should be updated as more detailed information becomes available and should also apply to test and operational procedures, using data on the capabilities and limitations of the system as well as of the personnel who interface with the system. As changes are made to the system, a hazard analysis should be required to confirm that the inherent safety is not degraded.

Knowledge of specific performance hazards may also be used to modify the operational envelope if this will eliminate or reduce the hazard. Specific display requirements for parameters of safety significance should be established, and correctional measures should be planned and incorporated into the operational procedures.

Warning Signals

Most major accidents do not occur without some prior warning signals. If an identified hazard cannot be eliminated by design or operational procedures, the warning signals associated with the malfunction should be determined and a special list of these warning signals should be prepared. Operators should be trained to watch for these signals and to recognize them. Warning signs should be posted at all critical locations. Operators should be drilled on exactly what to do when they detect a warning signal: who to call, what valves and switches to turn off, where to go. These procedures should be as simple as possible. A man who has recognized a danger signal should not be required to perform any further analysis or do any thinking; he must be trained to act automatically, decisively, and quickly.

Section 7.6.11 discusses trend analysis as a technique where certain critical parameters are measured and recorded continuously or periodically to assure that the measured values fall within specified limits. A specific incipient failure can often be anticipated when a monitored parameter starts to exceed its prescribed limits.

The same procedure can be used for accident prevention in hazardous operations. Critical parameters should be identified and monitored systematically. Any trend for a critical parameter to exceed its specified limits is a warning signal requiring immediate corrective action to prevent system breakdown or accident.

Unfortunately, most currently available analysis methods do not consistently and completely cover the human/system interface, or are they capable of adequately evaluating the changes in the hazard levels resulting from activation, test, and operation of the system.

5.12.2 Controlled Conditions

MIL-Q-9858 A (73) specifies that all functions necessary for the production of an advanced system be performed under "controlled conditions," that is, in accordance with written work instructions on adequate equipment in a favorable environment. For safety reasons this requirement should extend to applicable operational phases. There must be some means for assuring that the written work procedures are followed.

Individuals working even under the most rigorously controlled conditions cannot be watched at all times. Considerable reliance must be placed on the integrity and sense of responsibility of individuals. However, even a normally reliable workman or operator, who on a particular day may be indisposed for any number of personal or family reasons, may drop a sensitive device and unknowingly cause an internal failure. The written procedure may require him to report all such unscheduled incidents, but in addition to his indisposition, he may sense that the climate in the organization is such that it is not in his own best interests to submit the required report. If no one else saw him drop the device, the failure will go undetected until it is too late.

The existence of this hazard has long been recognized by engineers who are fond of quoting Murphy's Law, namely, "if something can go wrong, it is certain that somewhere, sometime it will go wrong."

As a result of the respect accorded to Murphy's Law, a great deal of effort is expended in the advanced systems industry to design equipment so that the possibilities of wrong assembly, operation, and maintenance actions are minimized. Although such "mechanic-proofing" helps, it does not solve the whole problem. The human aspects are so extensive that they must be attacked from every conceivable angle. The objective is to contain the problem; there is probably no real solution; that is, there is probably no solution in the same sense that pure technical problems are solved.

Training and Discipline

Training and licensing of personnel, and the aggressive encouragement of craftsmanship, responsibility, and pride of accomplishment are all useful techniques. There are many schemes for incentive and bonus payments to align the objectives of the employees with that of the employer, and some organizations have had good experience with such extra payment schemes.

In spite of all of these techniques, there still remains the key problem of establishing personnel performance acceptance criteria for compliance with a set of written procedures; in particular, how to set the threshold for acceptable performance in marginal cases. The normal tendency is to open tolerances as much as possible, in line with conventional thinking that associates wide tolerances with lower costs. Thus, minor infractions and narrow escapes normally are excused on the basis that no substantial damage has been caused and no harm was intended, or that the infraction is no worse than the infractions normally committed by everyone else. Intent is an important factor in judging the seriousness of an offense.

It is assumed that people want to do a good job and that they are performing about as well as they can. Under normal circumstances no one gets reprimanded for a narrow escape. Punitive action is taken only in cases of major infractions, where gross negligence or intent to do harm is indicated; that is, cases involving personal injury, loss of life, substantial economic loss, or criminal activity.

In the absence of such dramatic pressures and evidence of absolute necessity, it is very difficult to motivate management to increase the discipline requirements involving personnel performance, just as it is impossible to improve public safety codes without first having a major accident.

5.12.3 Accident Prevention

The effect of tolerating discipline infractions can be illustrated by consideration of the basic principles of accident prevention developed over the years by safety engineers. In particular, Heinrich (15) discovered through the analysis of extensive accident records, that in common industrial situations there is a very specific ratio between major and minor accidents; namely, for every major accident involving a major injury, there are 29 accidents involving a minor injury, and 300 accidents involving no injuries (Figure 5.3A). This ratio appears to be universally valid throughout industry. An accident is defined here as an unplanned and uncontrolled event that results in personal injury or the probability of personal injury. An accident is the result of an unsafe act and/or exposure of personnel to unsafe conditions. The majority of accidents is caused by unsafe practices, that is, man-failures rather than material-failures.

A typical no-injury accident (narrow escape) occurs when an employee loses his balance on a slippery floor, nearly falls, but is saved by grasping a nearby support.

Domino Theory

To understand the significance of the ratio of major to minor accidents, it is necessary to analyze the anatomy of the causes of accidents. A major accident is never the result of a single cause. Nobody wants an accident to happen, and no one will deliberately take the simple action necessary to effect a major accident.

A detailed discussion of this subject appears in Heinrich (15). Briefly, causal factors can be classified in categories such as: the social environment, personnel deficiencies, unsafe acts, and mechanical and physical hazards. The contributing causal factors must occur in a fixed and logical order. One is dependent on the other and one follows because of another, constituting a sequence that may be compared to a row of dominoes placed on end and so aligned that the fall of the first domino precipitates the fall of the entire row (domino theory). An accident is the result of the sequence. If this series is interrupted by the elimination of even one of its several factors, the consequences can be avoided. On the other hand, given the situation where all of the individually unlikely contributing factors are present in the required relationship and the first event of the chain occurs, there is no way of preventing the accident. Furthermore, if there are a sufficient number of accident causal factors in a given situation, it is only a matter of time until they all appear in the required combination for an accident. This is a paraphrase of Murphy's Law, Section 1.5.

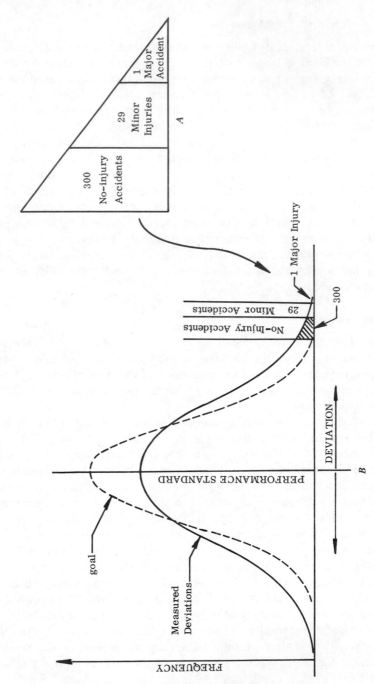

Figure 5.3. *A.* The foundation of a major injury. *B.* Frequency of deviations from performance standards.

Enforcement of Safety Practices

Generally each of the 300 no-injury accidents in Figure 5.3A. results from one or two causal factors, primarily man-failures, sometimes in an unsafe environment. This type of accident can be prevented, and it is the prime responsibility of supervisory personnel to do so. A basic function of supervision is to enforce production or operational rules, and this function is usually well done. However, if a supervisor has the power to control production or operational procedures, he can apply the same techniques to enforce safe practices. Supervisory action is the key to eliminating man-failures, such as providing and enforcing the use of mechanical guards to prevent accidents.

The basic theory of accident prevention rests on the premise that if effective action is taken to prevent the causal factors for minor accidents, then causal factors will never accumulate in the combinations required to produce a major accident. If people are trained and disciplined to follow procedures that prevent minor accidents, they will never perform the chain of man-failures required to support a major accident. It is estimated that 98% of all major accidents are preventable by this method.

To gain guidance for the development of an action plan for accident prevention, it is instructive to perform a statistical analysis by plotting the frequency of the deviations from performance standards, illustrated by Figure 5.3B, solid frequency distribution curve. The accidents shown in the triangle of Figure 5.3A now make up the two tails of the frequency distributions in Figure 5.3B. The goal is to eliminate the minor accidents by decreasing the dispersion as shown by the dotted curve. The problem becomes one of how to motivate personnel to adhere to standard behavior patterns in a more consistent manner.

5.12.4   Motivation

The RFP, Section 3.2, should require that the safety plan include provisions for motivating personnel to comply with safety regulations. Reference 162 contains a description of such a program, which has been successfully in effect since 1954 at the Johns Manville Corporation in their building materials division. The program involves careful individual personnel indoctrination, educational posters, motion pictures depicting right and wrong procedures, group discussions, posted process control charts, and charts showing the effectiveness of crews and individuals. Craftsmanship and individual responsibility are emphasized.

Awards Program

An important aspect of a motivational program is an awards system, both for suggestions for improving production or quality and for personnel performance. Different classes of awards may be given for different levels of accomplishment, such as: awards for a one-time exceptional performance; awards for sustained exceptional performance over an extended period of time; and awards for a one-time exceptional craftsmanship performance.

These awards are effective only if the criteria for granting them are generally known, understood, and accepted by all those competing for a particular award. This system provides a method of encouraging craftsmanship, instilling pride in workmanship, and fostering team spirit.

The criteria for granting the various classes of awards should vary from department to department within an organization. For example, the criteria within the creative design engineering department will be different from those in the accounting department, machine shop, maintenance, or operational organizations. However, within each of these departments or groups, each man must understand the criteria for winning an award. The criteria must be so selected that winning an award is within the capability of everyone in that department, provided he makes the special effort required. No one should be excluded from an opportunity to win an award.

Management and Union Support

It is also considered important for employees to be convinced that their own management as well as their labor unions support the motivational program; furthermore, it is also desirable to get the employees' families interested and involved.

Management and union support can be advertised in talks and educational posters which should be changed frequently to maintain continued interest. The president of the company may write periodic letters delivered by mail to the homes of the employees.

On the Apollo Program, the major awards to nonsupervisory personnel were presented by an astronaut who made the presentation an occasion to tour the plant of the award-winner as a morale-building device. Another major Apollo award for nonsupervisory personnel included a trip to Kennedy Space Center for the winner and his wife—when possible, to witness an actual launching.

It is quite apparent that the control of personnel performance is a difficult task and must be approached with tools other than those used for the control of technical quality-assurance factors. However, there are methods that are effective if approached with intelligence. Complete and clearly written directives and maintenance of discipline are basic factors. Personnel must be motivated to accept this discipline; they must understand the reasons for it; they must accept the objectives of the organization and identify themselves with it; they must be encouraged to develop craftsmanship and a sense of responsibility for their part of the output.

5.13   STANDARDS AND COMMONALITY

A convenient way of calling out requirements in a contract is by reference to publicly available codes, standards, specifications, handbooks, and manuals.

The term "standards" is used both in a generic and a specific sense. In its generic use it covers all of the above mentioned types of documents. In its specific meaning it is a document describing one or more procedures that are acceptable under specified conditions.

The concept of "commonality" refers to the use of special equipment on more than one systems development project. For example, government furnished equipment (GFE) such as aircraft engines or radar sets may be developed for use on several projects. Much of the following discussion of standards applies to some extent to commonality.

Standards (in the generic sense) are depositories of past experiences; they make available to the user a coordinated set of proven practices which an individual agency could hardly collect by itself. Correct usage of these documents usually results not only in better design but also in considerable cost savings. However, the benefits do not follow automatically. Standardization limits the options available to a designer and may provide constraints that restrict performance or increase cost. Hence, the extent to which standards and commonality are contractually required should be subject to a trade-off study balancing the benefits versus the penalties.

Standards may be mandatory or voluntary, and applicable to an industry, a company, or even a large systems project. Some voluntary company standards may be proprietary.

5.13.1 . Mandatory Standards

Mandatory standards are established by the government to provide an operating envelope for commercial transactions, to protect the public or the environment. Such mandatory standards include the use of the English language for communications and dollars as the medium of exchange, as provided for by the Congress. Our system of weights and measures is established by the National Bureau of Standards. The Environmental Protection Agency (EPA) and the Occupational Safety and Health Administration (OSHA) issue mandatory standards to protect individuals or the environment. Whenever monopolistic operations are in the public interest, as with utilities, business practices will be regulated by public bodies.

Such regulation creates a three-way relationship between customer, contractor, and regulator, as already mentioned in Section 5.12. In such cases the procedures discussed in Section 9.15, "Risk Management," are usually applicable.

5.13.2 Voluntary Standards

The United States is the only country in the world where the free enterprise industrial system relies on voluntary standards. In Russia, where all industry is state owned, all standards are issued by the government and are mandatory. In other industrialized countries there is some combination of government and voluntary industrial standardization. However, the tendency is toward government domination and manipulation of standardization for national commercial advantage.

As a result of the free interaction of creative interested forces and the lack of bureaucratic interference in the U.S. voluntary system, American standards include a high proportion of the best standards in the world, which, because of their quality, are accepted in many industrialized nations. For example, all international airports in the world use U.S. standard procedures, and communications with the control towers takes place in standard English terminology. On the other hand, because of lack of interest of those involved, U.S. standards in certain other areas are quite deficient.

There are over 100 voluntary standardization bodies in the United States, almost all of them associated with the American National Standards Institute (ANSI), which assures that all interested parties are involved in reaching a consensus. The list of these voluntary organizations includes, for example, the American Society of Mechanical Engineers (ASME), the American Society for Testing and Materials (ASTM), and the Society of Automotive Engineers (SAE). Examples of important standards are: screw threads, properties of structural steel, and properties of lubricating fluids.

5.13.3 Types of Specifications

Specifications are documents that spell out the properties, performance characteristics, quality control, test, and delivery procedures applicable to a product or a process. Specifications may apply to materials, parts, components, equipment, or system characteristics; to a design, test, or manufacturing process; to safety, reliability, maintainability, quality, operating, and support requirements; to data and documentation requirements, and other similar product or process characteristics. Requirements preferably should be expressed in quantitative, demonstratable terms; only when this is not feasible are qualitative expressions used.

5.13.4   Handbooks and Manuals

Handbooks and manuals are lists of acceptable principles, procedures, and models. Reference to such documents in a contract normally implies that if the specified analyses are prepared in accordance with the applicable handbooks they will be acceptable to the customer (see Sections 1.3.2 and 9.3).

5.13.5   Company Standards

For economic reasons, a manufacturing company normally limits itself to a number of standards from among all of those available. For example, few organizations make use of the full spectrum of available standard bolts; they select an assortment that best satisfies their specific needs and restrict their usage and inventory to this assortment.

Manufacturing companies also prepare a set of their own unique company standards whenever industry standards are not available to satisfy their particular needs. This situation occurs often when proprietary processes are involved and also in areas of rapidly advancing technology, since it makes little sense to issue standards that will become obsolete in a short time. However, as a technology matures and stabilizes, economic pressures encourage the merging of the various company standards into industry standards.

5.13.6   Military Standards, Specifications, and Handbooks

The U.S. Armed Forces have an extensive list of military standards, military specifications, and design handbooks which are routinely called out in all contracts for military equipment. Some of these standards are generated with the cooperation of the industries involved through such voluntary bodies as the Standardization Committees of the Aeronautical Industries Association (AIA) and the Society of Automotive Engineers (SAE).

5.14   CHAPTER SUMMARY

This chapter discusses the major factors that should be considered to provide for systems assurance when preparing a set of technical requirements for an advanced system.

To justify the development of a new advanced system, its proposed performance should exceed that of existing systems; therefore the new system invariably represents some extension of the state of the art. The extension of the state of the art will be small, but it will occur in many technical disciplines and affect many components. The extension of the state of the art is a major factor in determining the feasibility of the new system. A reliability program is required to provide assurance during the design phase that the proposed design ultimately will meet its operational requirements.

A number of other plans should be required in the RFP and included in the contract to assure system success. A manufacturing plan should provide assurance that production engineers are involved in the early design phases to a sufficient degree to assure that the system can be produced within the specified cost, and that adequate facilities for the construction of the system will be available as needed.

A maintainability program and an integrated logistics and support plan should be specified to assure that the system is designed to be operated and supported as planned in the proposed operational environment with the stipulated support, and that life cycle cost requirements are met.

A safety plan should assure that safety considerations are included in the production and operation of the system. Requirements for hazard analysis, controlled conditions, accident prevention, and motivation (training) are discussed.

6

Systems Engineering

In Section 1.2 it is noted that the word "system" has become popular and taken on a new meaning since World War II. This is a reaction to certain basic changes in technology which took place at that time in the development of advanced systems.

The Pre-World War II Project Engineer

Before World War II, in the environment described in Section 1.2, a project engineer had the capability to grasp all of the technical aspects of his project and to integrate them into an acceptable operating entity, no matter how large the project. He also clearly understood what technical and business requirements he was expected to achieve with his design. With the magnitude and technical complexity of advanced systems since World War II, the technical factors involved in developing and integrating the elements of a project to meet specified requirements became simply too great for a single project engineer to grasp and to control.

Evolution of the Project Manager

After World War II the project engineer started to acquire a staff, typically including an assistant project engineer for electronics and another for planning and scheduling, who had access to a computer. These were soon followed by an expert in operations research.

This was the time when airplanes carrying 20 passengers were being replaced by aircraft capable of carrying 120, then 400 passengers. Equipment costs increased by factors of 100 and more, and equipment utilization became a matter of major importance. Customers of advanced systems became technically more knowledgeable, and new concepts involving life cycle cost began to appear in contracts. Life cycle costs became as important to the economic survival of a customer such as an airline, as the initial manufacturing cost is to the supplier.

New groups of specialists appeared on the scene: reliability, maintainability, producibility, systems safety, and so on and on. Gradually these new additions to the traditional design engineering department were consolidated into a systems engineering staff. They replaced the old-time project engineer in relation to the detail designer, the man who designs a pump, an actuator, a transistor, a radar set, or a railway car.

The project engineer, whose responsibilities grew to include life cycle cost considerations and integrated logistic support, became a project or program manager. He coordinated not only the product engineering department, but also all the other resources required for an advanced development: contracting, funding, testing, planning, purchasing, manufacturing, quality control, public relations, and whatever else was needed. In the engineering hierarchy, systems engineers represent a new layer of management and technical resources control, between the program manager and the detail designer.

Design Control

In a manufacturing organization, engineering drawings and specifications are the basic documents of reference. In the introduction to Chapter 1 it is noted that, in the past, design engineers have been autonomous within their respective domains. Their authority was derived from a personal relationship of mutual trust and confidence with the project engineer. Communications between project engineer and designer were generally confidential, important matters were treated verbally, and personal code signals served as the normal means of communications. The design engineer proceeded on his personal interpretation of what had to be done, based on years of experience as a member of the design team. The procedure was very similar to that which we observe in a well-trained professional football team, with the project engineer acting as the coach from the sidelines.

Systems engineering provides a new controlled environment for the activities of the formerly autonomous design engineer. The new impersonal relationship requires a new form of communications based on extensive documentation. The systems engineer must record his analysis and in all important matters obtain prior customer approval. Gone is the single project engineer who was responsible for the success or failure of the project and who personally knew the history and reasoning behind every critical design decision and program requirement.

Growth of Systems Engineering

As a result of these developments the relative growth of our engineering departments has been in the systems area. The percentage of detail designers to total employees has not grown substantially. The engineering departments in advanced systems development houses have grown from about 10% of all employees to something over 30%. This growth has occurred primarily in the systems engineering disciplines.

6.1 SCOPE

The difficulty in defining systems engineering arises out of the fact that the discipline has taken over so much of the old project engineer's total responsibility for management of technical resources, for integration, and for assurance. This is an open-ended responsibility. There is only one sharply defined boundary to systems engineering: A systems engineer is not a detail designer. He thinks bigger.

Systems engineering, therefore, is involved in practically all of the boxes shown in Figure 3.1. However, except for some remarks to keep the discussion in perspective, the main thrust of this chapter is concerned with post contractual activities up to Box 18, design acceptance. The major systems engineering and analysis activities include the following:

1. The quantitative analysis and justification of operational needs.
2. The identification and establishment of operational (mission) requirements and environments.
3. The analysis of these requirements to apportion the performance, design, and test requirements to and through lower system levels down to individual components and elements.
4. The techniques for controlling the design, development, or selection of components to assure that they satisfy requirements (design assurance).
5. The techniques for integrating lower level components into all higher levels of assembly all the way to top system levels.

To be meaningful, performance and design requirements must both reflect operational needs and be correlated with test and demonstration requirements (see Figure 6.1). For technically advanced systems the design and proof-of-design requirements must encompass not only all of the idealized operational situations that probably will be encountered in planned usage with all systems and subsystems operating satisfactorily, but also must encompass those nonideal operating situations of reasonable probability resulting from: unplanned, inadvertent, or improper actions of personnel; less than ideal upkeep and maintenance; adverse environmental conditions; and critical malfunctions or failures of systems and subsystems. Furthermore, operational needs must be established on

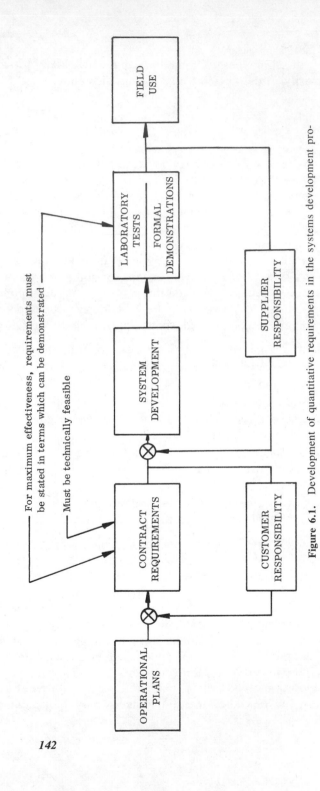

Figure 6.1. Development of quantitative requirements in the systems development process.

the basis of a realistic analysis of all of the significant factors involved, including trade-off studies between the benefits to be gained by specified reliability, maintainability, producibility requirements, and so on, versus cost, performance, and schedules.

Systems engineering is a never-ending "objective oriented," closed cycle that includes operations research studies involving the development of alternate strategies for most effective system utilization, planning of future systems, development of future operating plans, system definition, system development, subsystem-and-component development and integration, and system deployment. At this point the cycle can start over again (see Figure 6.2).

This procedure takes place in cycles which are repeated at various levels of system development. Although the principles remain the same at the various levels, application techniques vary with the nature of the problems.

It should be noted that the discussion of Figure 6.2 and other following figures is ultrasimplified and touches on main points only. The purpose of this discussion is to provide an understanding of the principles involved; it is not an exposition of detailed procedures.

6.1.1   Planning

In the *planning phase,* systems engineering is concerned with the selection of *requirements for future systems.* The work involves trade-off studies to justify the selection of strategic objectives, concept formulation, and feasibility studies.

The planning phase is concerned with the definition (assumption) of the operational plans required for the achievement of the strategic objectives and with the assumed future operating environments. Studies encompass the entire range of field operations, resources planning and management, integrated logistics and support, facilities, and personnel and skill requirements.

6.1.2   Systems Development

In the *systems development phase,* interest centers on the definition of the system required: system design, support facilities, operational doctrines, and operational environments. The work includes detailed mission analyses, systems synthesis, and effectiveness studies. For example, a number of feasible hardware configurations might be proposed, the advantages and disadvantages of each configuration identified with respect to the requirements of the operations plan, and trade-offs tried in a

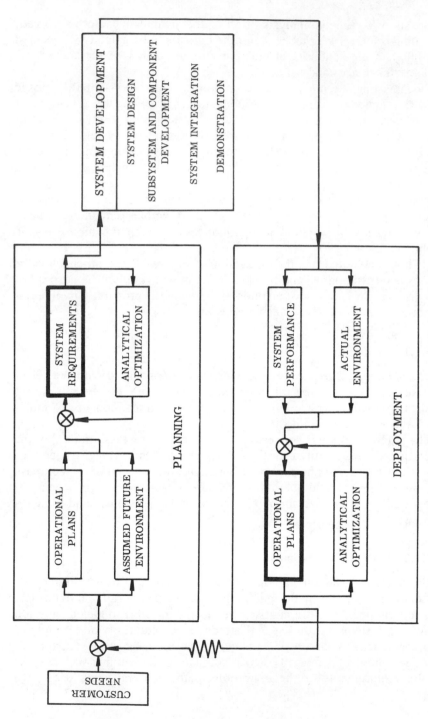

Figure 6.2. Scope of system engineering.

systematic way to determine the best configuration consistent with funding and other restraints. Generally these studies are supported by extensive computer simulation programs. The results of these studies is a set of design specifications for the various field systems required by the operations plan.

This general procedure is repeated on the subsystems, components, and lower levels, but more in terms of smaller hardware units. The requirements for each level are analyzed and broken down for the units on the next lower level. Furthermore, on each level there are procedures to assure that the developing hardware will meet the requirements. A technique for recording the breakdown of system requirements into subsystem, component, and lowest-level parts and elements is the "parameter tree" mentioned in Section 5.1.3 (74). This graphical technique summarizes and provides visibility over the analytical process, assuring "completeness" of the analysis.

System integration and demonstration complete this phase. These subjects are discussed in detail in the following chapters.

6.1.3 Deployment

Subsequent to hardware development, systems engineering continues to participate through operations research techniques in the *deployment phase*. This activity is similar in nature to the initial planning studies except that the roles of the variables have switched: in the planning stage, the objective is to optimize the performance of the planned system; in the deployment stage, it is to optimize the *operational plans* (Figure 6.2).

6.1.4 Feedback

Figure 6.2 shows analysis of deployment as the feedback control for planning activities, with a resistance in the loop. This resistance is a function of the length of the loop with respect to calendar time and number of organizational boundaries to be crossed, and is discussed in detail in subsequent chapters.

6.1.5 Hardware Orientation

Since this book is concerned primarily with hardware development, the remainder of the discussion centers on the subsystem and component level. Those interested in other aspects of system engineering are referred to the other books in this System Engineering and Analysis Series, especially Chestnut (4). Other viewpoints are given in refs. 8 and 16. These last two references are oriented toward the computer definition of a system as discussed in the introduction to Chapter 2 and, therefore, emphasize disciplines other than those represented in this book.

6.2 HISTORICAL ORIGINS OF SYSTEMS ENGINEERING

To provide a better insight into current systems engineering practices, a brief historical review of the development of these practices is included here. In line with the author's personal experience, these comments are based primarily on U.S. Naval practices developed for controlling the structural design and proof-of-design of Navy airplanes prior to World War II by the Structures Branch of the Naval Air Systems Command (NAVAIR).

The procedures are documented in MIL-A-8860 to 8870 (75) and reflect the end results of a gradual transition from empirical/intuitive goals to requirements more realistically related to pertinent operational situations and to human factors.

6.2.1 Quality Assurance Provisions

Fixed-price contracts prior to World War II for both prototype and production aircraft contained two very important, but not necessarily completely enforceable quality assurance provisions:

1. A "guarantee" by the contractor that the airplanes delivered to the Navy would be satisfactory in "all" respects.
2. A "correction of defects" provision whereby the contractor would have to remedy at his own expense, by redesign or replacement, all design deficiencies called to his attention within six months after delivery of the airplanes to the Navy.

6.2.2 Analysis and Testing

Under this design control procedure, and because of limited funds for aircraft procurement, structural proof-of-design leaned heavily on analysis to convert anticipated take-off, flight, and landing design conditions into design values of aerodynamic and inertial load distributions; and relied on stress analysis to assure adequate yield and ultimate strengths in all parts and elements of the airframe. Structural testing in the laboratory was limited primarily to tests of piecemeal critical structural elements or components, carefully selected to supplement design assumptions and to verify results of stress analyses. Manufacturers' flight testing of prototype and production models consisted of a few dives and pull-outs to prescribed limits, prior to acceptance for Navy trials, to provide reasonable assurance of safety of flight for Navy operations during the trials, and in particular, freedom from uncontrollable spins.

6.2.3 World War II Production Requirements

The limited scope of laboratory and flight structural testing was reasonable and economical during pre-World War II peacetime procurement and operations, characterized by the small number of flight articles of each model. Operational and environmental conditions were well known; the pilots were uniformly well trained. However, the imminence of United States involvement in World War II suddenly changed this basic environment. The procurement of various models of airplanes was stepped up from a dozen or so of each model to several hundreds or thousands of each for use by England, France, and other countries and for the forthcoming combat employment by our own government.

6.2.4 New Operational Requirements

All at once an entirely new situation was at hand, calling for a new approach to design control. Much more intensive and structurally critical operations could be anticipated by pilots with widely varying degrees of training and flight proficiency. It was foreseeable that Navy airplanes could be used in conditions for which they were not designed; fatal accidents would be inevitable; and traditional methods for "correction of defects" were no longer available.

To understand further developments, it is necessary to recognize that reliance had been placed on analytical determination of load distributions and on paper stress analyses to assure provisions of adequate strength in all parts and elements of the airframe, as discussed in Section 6.2.2. These procedures, based on hardware models, dominated the thinking of airframe structural engineers. Whatever limited testing there was, always was closely coordinated with the analyses; it was those items that in the analyses had been determined to be most critical that were selected for test. Each test was designed to confirm as much of the analyses as possible.

In the attempt to extract a maximum amount of information for each dollar from each test set-up, tests were generally extended to failure; that is, beyond the required levels to determine the actual margins of safety or the performance capability of the critical members of the structure.

On the basis of these concepts, it is possible to predict the behavior of a structure in a condition for which it is not designed, if the following are known:

1. Original design conditions.
2. Margins of safety and behavior of all structural members with respect to each original design condition.

6.2.5 Systems Testing

As soon as Navy engineers became aware that large numbers of their aircraft were likely to be subjected to flight conditions beyond the original design conditions, they immediately proceeded to devise means to obtain extensive and reliable data to permit them to calculate new load distributions, identify critical members under new loading conditions, and determine the new margins of safety. This led to the formulation of the rigorous concepts of "requirements," "specifications," and "demonstration" used in this book. In particular, they developed the need for more extensive testing to confirm more, or all, of the details of the analyses, that is, the detailed measurements of the structural characteristics of all possibly critical elements. Special significance was placed on verifying that the actual stress distributions in the airframe structure conformed to the analytical predictions. Such tests required the determination of the static and dynamic interactions between all structural elements and, therefore, involved the testing of complete airframes instead of selected isolated components. For this type of testing a major and very expensive expansion of existing laboratory facilities was essential (777, 163).

6.2.6 Funding for Structural Laboratories

The logical justification advanced by structural engineers at that time in support of such comprehensive testing is extremely interesting. They stated that if precise knowledge of all structural detail characteristics were not available and an accident were to occur in service, there would be no way of determining if the accident was caused by airframe deficiencies or by pilot error. In fatal crashes, lacking such detailed information, the tendency would be to assume that the cause of the crash had been pilot error. If the actual cause of the crash was an airframe deficiency, there would be no way to develop a corrective action plan, thus exposing additional pilots to an unnecessary continuing hazard. Naval authorities, when confronted with this logic, soon found the funds to support the proposed expanded structural testing programs.

6.2.7   Motivating Contractors

Most aircraft manufacturers, however, were less than enthusiastic about the new comprehensive structural proof-of-design requirements. They were preoccupied with expanding their production facilities in programs of unprecedented magnitude which already overtaxed all their technical and management resources. They already found the additional bookkeeping requirements and in-house government monitoring associated with cost-plus-contracting particularly onerous. They therefore felt that additional and comprehensive structural test programs would merely compromise their ability to produce aircraft of the quality and in the numbers needed by the government. However, when the Navy initiated preparations to perform the necessary extensive structural testing in its own laboratories, the manufacturers quickly found ways to comply with the new structural testing requirements without compromising their production.

6.2.8   Customer Monitoring

Whereas ref. 75 spells out the technical procedures in detail, ref. 76 specifies the documentation to be supplied to provide evidence that the technical procedures of ref. 75 and of a multitude of similar Navy documents have been properly followed.

The Department of Defense has expanded this approach to cover the entire scope of systems engineering. Reference 78 lists all of the data items that may be required contractually in a DOD systems acquisition program. Each item is described on a separate DD1664 Data Item Description form and assigned a DI-series number. All DOD contracts today specify the required documentation by listing the DI-numbers considered applicable to the particular acquisition program. Each Data Item Description form:

1. Refers to a number of MIL-Specifications and MIL-Standards that describe the required technical tasks.
2. Specifies the data and the documentation to be submitted.

In this approach, as pointed out in Figure 2.1, the customer describes what is to be delivered, and delineates the delivery schedule; that is, *what* is to be done, and *when,* but not *how* to do it. This procedure lends itself to formal progress monitoring in terms of milestones established in advance. Progress is measured by determining what documentation and hardware has been approved or accepted at given points in time. It is not necessary for the customer to review or monitor in detail *how* the job is done.

This procedure assumes that maximum cost effectiveness is achieved when minimum penetration takes place by the customer into the contractor's activities, and when the interface is maintained on a formal basis.

6.3   MAINSTREAM FUNCTIONS IN SYSTEMS DEVELOPMENT

Figure 6.3 shows the following functions as the mainstream of systems development:

- Specification of requirements
- Product engineering
- Progress monitoring by customer
- Manufacturing
- Quality assurance
- Service (support)

6.3.1   Product Engineering

Contract requirements have already been discussed. Given these requirements, the function of product engineering is to define, design, and qualify the product to be built or supplied (hardware and software) and to document the justification for all design decisions (data requirements), as shown in Boxes 9, 11, 12, 14, 16, and 17 of Figure 3.1. The output of the engineering department is mainly drawings, specifications, and technical reports, as well as technical support of other functions such as fabrication and construction as shown in Boxes 10 and 15 of Figure 3.1. (Boxes 13, 18, and 19 deal with design reviews and acceptance, and subsequent test and operational phases.)

6.3.2   Customer Monitoring

Product engineering activities should be monitored by the customer: all *significant** drawings, specifications, and reports should normally require customer approval before implementing action is taken. Customer approval of a set of drawings and specifications at a predetermined mile-

---

* It is extremely difficult to define "significant" in this context. Each case is different, depending on people and circumstances.

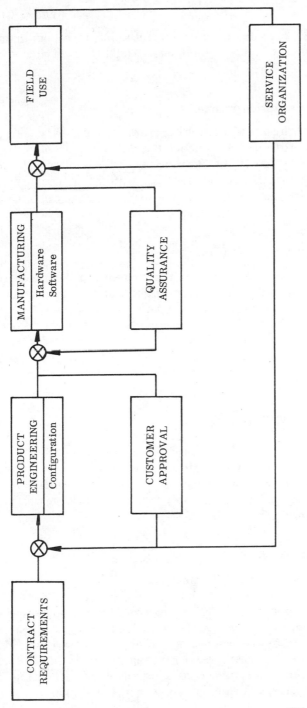

Figure 6.3. Ultrasimplified system development—mainstream functions.

stone is usually considered a commitment to buy a product that complies with the demonstration requirements established in the contractual documents.

6.3.3 Manufacturing Functions

The function of the manufacturing department (or corresponding software producing department) is to produce the product in accordance with the instructions provided by engineering. Quality assurance certifies in a formal manner that the product complies with the applicable engineering drawings and specifications and maintains the necessary records to justify these certifications.

6.3.4 Service Organization

The service organization supports the customer in his field use of the product with technical advice, support and training equipment, operator and maintenance personnel training, spares, and any other service needed to keep the product in an operating condition. It feeds back operational data for information and to facilitate corrective actions on quality and design deficiencies to the manufacturing and engineering departments, respectively.

6.4 SUBSYSTEM DEFINITION

As shown in Figure 6.4, the purpose of the subsystem definition phase is to produce a set of requirements for each component comprising the subsystem. When integrated these component requirements must equal the contract requirements. The proposed component requirements are subject to customer approval.

6.4.1 Subsystem Design

In designing a subsystem a designer arranges the available or proposed components in a manner that he hopes will best satisfy the requirements. A creative designer is normally able to propose many potential arrangements, but from experience and by intuition he and his consultants are able to discard many of them because of obvious inferiorities. However, after eliminating as many potential arrangements as possible by these means and by other qualitative considerations, there always remain a few competing arrangements from among which the "*optimum*" one cannot

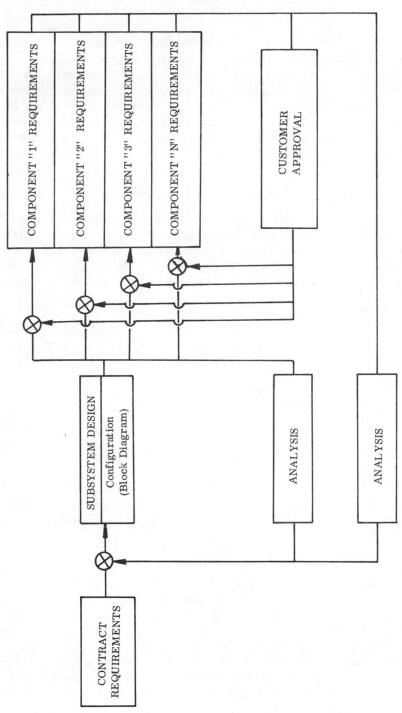

Figure 6.4. Simplified subsystem configuration development.

be selected by qualitative considerations alone. Each of these surviving competing proposals has different advantages and disadvantages under the various proposed combinations of operating conditions and missions. It is not obvious which arrangement will be the most suitable subsystem from an overall viewpoint. A quantitative trade-off technique, configuration analysis, is briefly described in Section 7.6.2; although this technique provides some guidance in making a selection, a *truly optimum* arrangement is difficult, if not impossible, to identify and define.

6.4.2 Apportionment of Requirements to Lower Levels

A number of other analytical activities take place simultaneously and in parallel with the design activity. Given the critical subsystem requirements, each of the proposed subsystem arrangements is analyzed to determine the design requirements for each lower-level subsystem, component, and element, and the results of these analyses may be recorded in a set of parameter trees.

The requirements for a particular component must be based on the most severe conditions imposed on the component. They should be selected from a safety viewpoint so that the component will not fail in service; they should include some margin for errors in the analysis, minor variations in manufacturing, and possible occasional operational overload conditions. The A-spectrum of ref. 75 represents recognition of the probable nature and frequency with which pilots are likely to "inadvertently" exceed authorized operating limits in routine service operations, particularly in training.

A subsystem requirement that is critical for one component will not necessarily be critical for another component. In actual operation there is usually no single condition that ever imposes a critical condition on all components simultaneously. Normally the operating requirements for any complex advanced subsystem encompass a broad range of conditions.

6.4.3 Technical Reports

All analytical and trade-off studies should be properly recorded and filed in report format. These permanent records should provide traceability of, and accountability for, requirements from basic critical system operational conditions specified in the contractual documents down to the component level, and these data constitute a record of the justifications for all design decisions. Should a failure or other unsatisfactory condition later occur during development tests or in service, the component design conditions can be checked and traced to the subsystem or system contractual requirements. It can then be determined whether or not an error was made in the design or in the analysis, or whether the system requirements were incorrectly specified. The feasibility of potential corrective actions can be evaluated. In either case, the experience then can be exploited to improve the design procedure and prevent such troubles in future systems design.

6.5   COMPONENT DEVELOPMENT PROCEDURES

Given a contractual system specification, the procedure for developing the components of the system is illustrated in a simplified diagram in Figure 6.5. The word "component" is used here to indicate a lower level unit resulting from any logical breakdown of the system into subsystems, major assemblies, components, elements, and so on. Depending on the intended purpose, there are many such breakdowns possible, each with a different emphasis and number of levels. A component is visualized here as a replaceable assembly of elements such as a unit supplied by a vendor.

The exact differentiation between subsystem, subassembly, and component is of little significance in this discussion, since the purpose is to describe the basic development procedure which can be modified to apply to any lower level of assembly.

Figure 6.5 shows the interlocking customer-contractor responsibilities, in particular the customer approval cycle at each significant milestone. The procedure is based on a rational relationship between design-and-construction requirements, design, analysis, and developmental and demonstration tests.

6.5.1   Component Requirements

Starting on the upper left side of Figure 6.5, the system configuration is developed by the contractor to comply with the customer's system requirements. Component requirements are then derived from the system configuration (Section 6.4.2). This is a complex and critical technical milestone and is, therefore, subject to customer approval. For his own protection, the customer should have the technical in-house capability to perform a critical technical review of all items submitted for approval. Reference 79 provides a standard for preparing component procurement and performance specifications.

Complex advanced system equipment is operable normally in a variety of modes with varying degrees of effectiveness over a range of missions. As mentioned above, design requirements for various conditions some-

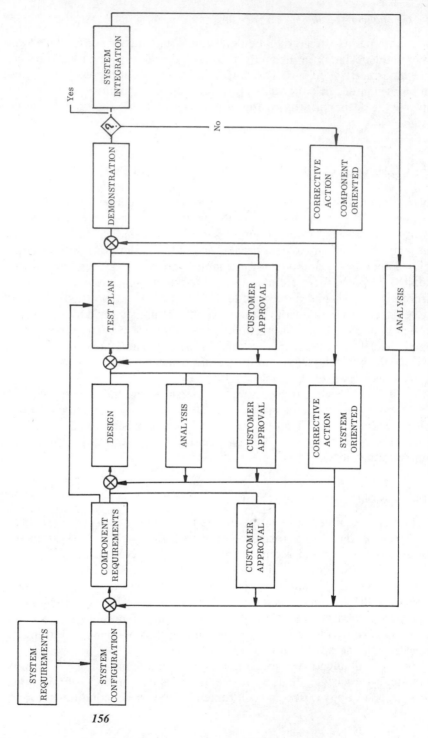

Figure 6.5. Simplified component development.

times conflict with one another and often are inadequately defined. In this complex matrix of modes, missions, and conflicting operating conditions, it is difficult to identify all of the critical design conditions.

Written specifications and contractual agreements always represent the state of knowledge at a particular point in time. As the design progresses in the hands of competent and creative designers, analysts, and test engineers, new knowledge comes to light. As this new knowledge develops, the original written specifications and contractual agreements become obsolete. On large projects the amount of paper work is so great and the changes occur at such a rapid rate, that it is physically impossible to keep the documentation continuously up to date.

There is never enough time or are there ever enough funds to investigate all potentially critical conditions, nor to demonstrate by tests all significant performance parameters in those critical conditions most likely to encompass (or which are representative of) a carefully selected number of the most severe operating conditions. The challenge is to make the best use of whatever time and funds are available.

6.5.2 Design Control

In the next step the designer lays out and investigates a number of feasible component designs and selects the best configuration with the assistance and cooperation of a number of analysts in various fields. It is not unusual to consult responsible customer personnel during this selection process. Each analyst reviews the various design proposals and assures so far as possible that the design selected will do the following:

1. Meet all component requirements.
2. Pass the required demonstration tests.

A critical function of the analysis is to predict during the layout stage, on the basis of engineering models incorporating all pertinent past experience, the expected performance of the proposed design during the demonstration tests. The importance of this function may be evaluated from the following consideration.

As mentioned above, the establishment of critical design conditions involves many compromises requiring the exercise of engineering judgment in making best use of available time and funds. Another critical factor is the correlation of performance during laboratory tests and under actual operating conditions. Laboratory test conditions always involve simplifications because of test equipment, time, and cost restraints. All of these factors add an element of uncertainty to the predictions of the analyst.

Among the many predictions that the analyst is required to prepare, usually only one can be physically checked before actual operation; namely the specific performance of a specimen in a demonstration test. If the specimen does not perform as expected in a formal demonstration test, all other associated predictions must be placed in doubt.

6.5.3   Customer Design Approval

Both the proposed design and the supporting analytical reports must be approved by the customer before the test plan is formulated. This procedure provides the customer with positive control over contractor output. If the customer wishes the contractor to perform a stress analysis of an airframe structure, the contract will require the delivery of stress reports by specified dates. The requirement outlines the report format and lists the parameters to be included in the reports; for example, a summary report of margins of safety must include identification of all critical structural elements and their margins of safety under specified design conditions. If the customer does not agree with the analytical procedures used, or the conclusions in the stress report, the report will be rejected and the report or the design must be modified until it attains customer approval. Great care must be taken to keep paperwork to a minimum; technical reports must be designed to perform the functions described here as efficiently as possible, but not to do more than is necessary.

6.5.4   Design Reviews

In practice, approval of analytical and design data sometimes takes place at a design review meeting (Section 7.6.12). Even when unlimited time is available, it is not always possible to prepare a set of drawings and specifications that are complete in every respect for complex equipment. Some agreements involving very complex technical trade-offs, which are too involved to be reduced to writing in the time available, may be developed between knowledgeable professionals on an oral basis. Under the pressure of deadlines, the supplier's documentation often includes treatment of only those aspects of the analysis and design that he considers significant or critical. The knowledgeable customer seldom agrees completely with this documentation. Sometimes there are also differences of opinion both in the contractor's and the customer's organizations. The design review meeting is an important device for obtaining resolution of all outstanding technical questions on a timely basis, and reducing all oral agreements to writing.

A given design is always configured on the basis of trade-offs and compromise between different or conflicting requirements. Members of the design review team should include representatives of all groups interested in the various requirements. Sometimes outside experts are also invited. All pertinent drawings, analyses, specifications, test plans, test reports, and other pertinent documentation are made available to all members of the design review team in ample time before the meeting for them to become acquainted with all issues; usually a minimum of 10 working days is specified.

Mock-Up Boards

Sometimes, in addition to drawings, a mock-up or physical scale model of the equipment may be available for inspection. In complex equipment mock-ups provide visibility with respect to dimensional compatibility and other interface features that may be difficult to visualize on the basis of technical drawings alone. If the mock-up is an important one, the review procedure may be known as a "mock-up board" meeting. A mock-up board usually approves a full-scale mock-up of either the entire system or at least its simulated critical features before proceeding with the detail design phase.

At the design review or mock-up board meeting, all differences of opinion are brought to light and agreements negotiated on all outstanding differences. All deficiencies in requirements or in the proposed design are recorded as "action items" or "chits" and these become contractual commitments as soon as approved by the board or design review meeting and confirmed by the contracting authorities. Action items involving funding or schedule changes must be subsequently reviewed by top contractor and customer managements for final contractual adjustments.

6.5.5   Qualification Test Plan

Given a customer approved design and the component requirements, a formal demonstration or "qualification test" plan is prepared and submitted to the customer for approval. Customer approval indicates agreement that the proposed component test conditions reflect the system operational conditions and that a component that survives the tests is acceptable for use in the system. Customer approval of test plans is a critical customer-supplier interface because of the many complexities involved in deriving component requirements from mission objectives, and the many compromises made along the way. The process of obtaining customer approval may require several cycles of submitting plans and customer review until approval is granted. (In one case, the author was involved in 14 such cycles until customer approval was granted.)

6.5.6   Qualification Test

If the qualification tests demonstrate that the specimen complies with all the requirements of the test plan, the component is released for system integration. Normally, actual formal demonstrations do not go according to proposed plans and schedules. Performance is sometimes less than required, and unforeseen breakdowns or failures occur. When such events happen, immediate corrective action must be taken to correct the deficiency. Quality defects in the test specimen and deficiencies in the test equipment or test procedure, are subject to direct corrective action; but corrective actions that affect either the design or the component requirements must first be reviewed from a system viewpoint effect on other components of the system.

6.5.7   Feedback Cycle

A final analytical cycle in Figure 6.5 indicates a review of the performance of the component in system integration and a feedback of this information into component requirements. This representation is quite simplified. The basic task is to monitor component-test results and field performance, to collate all pertinent information with existing experience, and to make pertinent information available in usable form for product improvement as well as for new and future design. This may be done by preparing new, or by revising existing specifications, design handbooks, or technical publications. Because of the length of time involved, this feedback is the least effective and most neglected aspect of the design procedure.

6.5.8   Adversary Customer/Contractor Relationship

Each one of the customer approval loops shown in Figure 6.5 involves the development of a set of decisions, often based more on professional evaluation than on complete, factual, supporting data. These decisions are made by rigorous adherence to formal procedures. It is interesting to note that the method parallels the decision making procedure in other important areas of human endeavor where error cannot be tolerated. For example, our judicial system provides a similar rigorous and formal procedure for arriving at binding decisions, often in the face of incomplete supporting evidence. Prosecuting and defending attorneys monitor each other to assure that the established procedure is followed. If one of them makes a misstep, the other moves for a mistrial, and the system provides for a new start, regardless of costs or time delay.

In the procedures of Figure 6.5 customer and contractor monitor one another, from their individual independent viewpoints, to assure that proper procedures are followed. If the contractor makes a mistake, the customer calls for corrective action. If the customer is dissatisfied with progress, and adequate corrective action is not forthcoming from the contractor, the customer must be ready to terminate the program.

6.5.9   Section Summary

The key analytical tasks in the procedure of Figure 6.5 are the following:

1. The establishment of requirements.
2. The prediction of the performance of the selected design in the qualification test.
3. The measurement of the performance of the test specimen to determine compliance with the requirements.

6.6   ILLUSTRATION—WEIGHT AND COST CONTROL PROCEDURES

A straightforward example of the procedure of Figure 6.5 is provided by the typical weight control program as practiced in the aerospace industry. All aerospace contracts contain requirements establishing the maximum acceptable vehicle weight. These requirements are always the results of difficult negotiations reflecting the difference in customer versus contractor views of past history and their estimates of attainable improvements in the state of the art during the development cycle of the proposed design. The requirement is a trade-off against payload, vehicle earning capability, performance, and other factors.

The system weight requirement is broken down and each proposed component receives an allocation, subject to customer approval, as indicated in Figure 6.5.

6.6.1   Weight Design Review

As the design progresses, a weight analyst continuously reviews a designer's proposed configuration and maintains an up-to-date record of his estimate of the predicted weight of the design. As long as the designer's proposals remain within his weight allowance, all is well. It seldom is. As soon as the estimated weight of the proposed design exceeds the allowance, the weight analyst notifies the designer and requests a redesign. If the designer cannot comply the management is immediately notified, and a decision is made either to continue to struggle with the redesign or to accept the overweight condition with its performance and other penalties. This is a trade-off involving the weight requirement and cost/schedule considerations. Periodic reports are submitted to the customer who is thereby also in a position to take timely corrective action at his discretion.

6.6.2 Timeliness

The analytical activity is shown in Figure 6.5 as a feedback control loop to the design tasks. This loop must have sufficient dynamic response to permit corrective action. The analyst must make his input to the design at the proper time; this is essential. If the analytical input is not synchronized with design progress, it will be ignored and will be quite useless. The necessity for maintaining the schedule already has been mentioned in Section 2.3. An aerospace system may have between several thousand and several million components, depending on the definition of a component. The development of these individual components must be coordinated so that each one will be available as needed for system integration on schedule. This is a hard fact which dominates the entire development process.

6.6.3 Weight Prediction and Demonstration

The weight of a piece of hardware that has been completely designed is easy to predict and to demonstrate; nevertheless, all aspects of the formal procedure are included in this simple example. Given the material, its density, and the geometric configuration of the part, the weight of the proposed part can be predicted analytically with considerable accuracy. In fact, the weight estimates of the weight control engineer during the hardware development stage are always accepted at face value. As soon as a part is built, it is subject to a qualification test that consists of a weighing, thus formally verifying the analytical prediction.

6.6.4 Parametric Procedures and Their Limitations

Although the model is simple, the actual administration of a weight control program becomes very complex because of the many conflicting requirements involved in the design of a complex system. For whatever the reasons, the weight of any actual airplane "as built" usually is substantially (and in some cases is ridiculously) greater than the weight "sold" to the customer during the contracting phase. According to reports in the technical press, all current experimental aircraft projects are having trouble meeting their weight requirements.

From the strict engineering point of view, the simpler the model, the easier it is to meet an associated requirement. However, even with the simple weight control model, it is very difficult to meet the requirements in a complex system. It is clear that as a model becomes more complicated, the more difficult it becomes to meet the underlying requirement. However, even with the simple weight model, it also becomes more difficult to meet requirements as the system becomes more complex. On the surface, the reasons for this contradiction are not apparent. Regardless of the complexity or size of a system, it should be no more difficult, technically, in a complex system than in a simple one to compute the weight of the bits and pieces and then determine the total system weight by adding these individual weights. However, weight estimates for proposed designs are not prepared in this manner; but by statistical and so-called "parametric" procedures. These procedures are quite sensitive to business conditions, to "brochuremanship," to the time available to prepare the proposal, and to similar influences. It would appear that the more complex the system and the tougher the competition, the more optimism is introduced into the statistical estimates in the planning stages of development. This is more a function of business policy than of application of sound engineering principles.

6.6.5   Effectiveness of Weight Control

The power of the weight control procedure lies in its capability to demonstrate compliance with contractual requirements by a simple, timely, nondestructive test, consisting of a weighing, as soon as a part has been manufactured. This should be kept in mind when specifying reliability or other complex requirements. The simplicity of the weight control procedure should serve as a guide when developing demonstration procedures for other technical requirements.

6.6.6   Cost Control

Another parameter that is easy to measure is cost. Dollars are easy to count, and there are sound procedures available for maintaining accounts, balancing books, and conducting tests to assure that the books represent the facts. Yet cost is another parameter that is difficult to control; there have been few large advanced systems engineering projects without overruns. The simplicity of the method of measuring cost merely makes it easier for all concerned to be informed of progress on a current basis. In view of the many considerations involved in a complex trade-off, knowledge of the fact that a cost requirement is not being or will not be met, by itself, is not sufficient to bring about appropriate and effective corrective actions.

6.6.7 Limitations on the Use of Models

A simple model is a necessary but not a sufficient tool for good control. Systems assurance and reliability models, for example, are generally quite complex and difficult for management and technical decision making personnel to understand. Reliability testing is costly and often cannot be conducted in a timely manner. As a general rule, it can be assumed that when, on a given project, design considerations are so complex and constraining that they make it difficult to meet weight and cost requirements, then it will be that much more difficult to meet the reliability and other similarly complex requirements.

6.7 DESIGN DATA REQUIREMENTS

It was stated in Section 1.3 that an analysis predicts performance on the basis of pertinent previous experience (engineering models). Many advanced components incorporate features that are beyond the state of the art and the available design data are not always adequate to support the required prediction, or the available time is not sufficient to permit an acceptable analysis to be performed. In this case it becomes necessary to supplement the design and analysis effort with development testing to generate the data necessary to complete the performance predictions. Some of this testing is performed in simulators as discussed in Section 6.9.

6.7.1 Categories of Development Tests

It should be noted that there are no standard classifications for development testing, and the terminology varies with different disciplines. The test classifications and the terminology used here are those that the author has used in his own development testing programs and are summarized in Figure 6.6.

Such a development testing program often consists of three main phases: feasibility tests, element tests, and design verification tests.

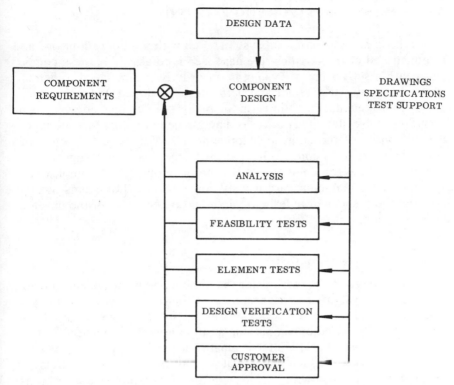

Figure 6.6. Design data requirements.

Feasibility tests provide some degree of evidence that the proposed concepts and configurations will meet the desired requirements.

Sometimes, design data are available only in general terms, and the specifics are inadequate to prepare a complete analysis. *Samples of the elements* in question are constructed and tested to determine their specific characteristics, and these values are utilized to complete the analysis.

Most analyses during engineering design, as differentiated from pure research, are based on simplifying assumptions. In complex advanced components the implications of these assumptions may not be fully understood. In many such cases it is desirable to support the analytical predictions with *design verification tests*.

Most development testing is conducted in an informal manner, but whenever the test results are used to support the conclusions in an analysis submitted to the customer for approval, it is wise to coordinate the test plans with the customer in advance.

6.8   SUBSYSTEMS INTEGRATION TESTING

Given the components of a subsystem, each with a set of established and demonstrated characteristics, the next step is to assemble these components into a subsystem or system in an orderly manner, and to demonstrate that they will interact and perform as predicted under the anticipated environmental conditions. In the majority of complex development programs, subsystem integration as diagrammed in Figure 6.7 is the most difficult phase of the entire development. Most of the unexpected and unpredicted troubles come to light at this stage, because of higher order interactions between components. In the development of complex subsystems, more man-hours are usually consumed in subsystems integration and the associated redesign than in any other phase of the development program.

6.8.1   Deficiencies in Models

There are several basic reasons for the difficulties in subsystems integration. The subsystems engineering models that apportion the system requirements and define the conditions at the interfaces between components are again based on simplifying assumptions, and the effects of the simplifications may be difficult to evaluate and predict. Furthermore, components will have characteristics that are not listed in the requirements and that may be unknown to the systems analysts, and these characteristics may conflict with one another when the components are brought together for the first time in an interacting relationship. There exist, also, physical processes that are almost impossible to model and predict quantitatively, such as the flow of temperature shocks and the vibration power levels through a complex structure. Furthermore, the interaction of a new system with a new environment, such as an aircraft flying for the first time at record speeds or at other new critical operating conditions, always contains new factors that cannot be predicted and that often bring surprises. It is in subsystems integration that for the first time in a development program the true extent of the difficulties associated with the program become unavoidably obvious. Unlike paper requirements and specifications, hardware systems-integration difficulties are nonnegotiable; the problems are fully visible as they arise and must be solved in terms of operating hardware.

6.8.2   Funding for Systems Integration

The difficulties involved in subsystem integration are always underestimated in advance and are reminiscent of the training a juggler must go

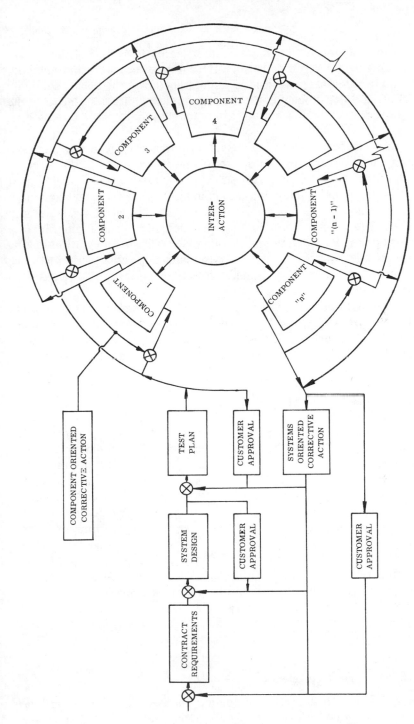

Figure 6.7. Subsystems integration testing.

through in attaining proficiency. He first learns to twirl a single baton, then he may even learn to twirl six different types of batons individually. But when he attempts to twirl all six simultaneously, the task is suddenly of a different order of magnitude. This example is quite descriptive of the normal order of difficulty certain to be encountered in subsystems integration.

Rigorous and thorough early subsystems integration testing is a critical requirement in developing a product so that it will be as trouble-free as possible in service. Yet subsystems integration testing is among the most difficult tasks to justify in the initial planning stage. As a result, adequate funding for subsystems integration testing is usually made available only after the existence of serious troubles has been uncovered in other phases of the program. By the time an adequate subsystems integration testing program is then launched, it is relatively late, corrective action is more time consuming and expensive, and corrective actions are more likely to introduce additional troubles of their own.

The presentations in Figures 6.5, 6.7, and 6.8 are simplified in that there is only one component level between the total system and the element levels, whereas every actual system has many more intermediate levels. The one level shown is, however, adequate to illustrate the main principles.

6.9   SIMULATION TESTING

In integrating a subsystem consisting of several interacting components, it is sometimes impossible or inconvenient to assemble a complete subsystem for test purposes. One or more components may be missing. If the performance of the missing components can be described with sufficient accuracy by a mathematical model, a simulator can be used for integration.

The term "simulator" has different meanings among different professional groups. Among programmers and computer oriented personnel, a simulator is a program of a mathematical model. When quantitative values representative of real variables are input to the program, the outputs correspond to the results that would have been attained in the real situation that the model describes. No hardware is involved other than the computer. Whenever hardware becomes a necessary part of the setup, it is known as an "emulator."

6.9.1   Dynamic Equivalents

In the following text the term "simulator" is used to denote a hardware subsystem integration device for use in a situation where all subsystem

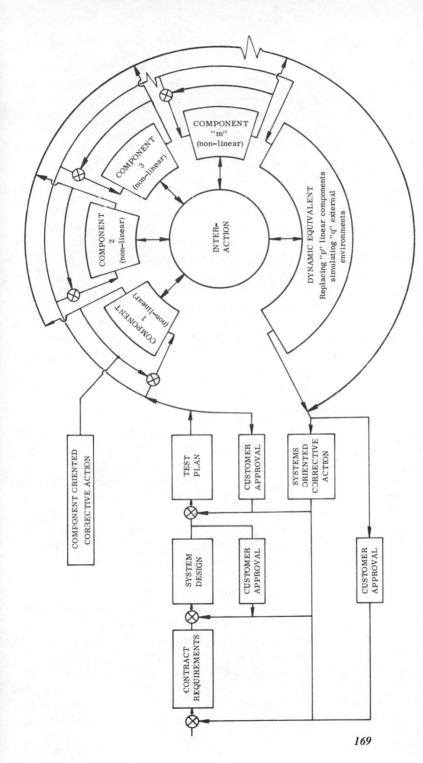

Figure 6.8. Simulation testing.

components are not available and a dynamic equivalent of the mathematical model is substituted for each missing component as shown in Figure 6.8. Such a simulator consists primarily of hardware and is often referred to as an "iron monster;" computers are used only to provide supporting functions.

In such a simulator the dynamic equivalent accepts the input signals from real interfacing components, processes them, and provides the same output signals as would the actual components under the same conditions. Such a dynamic equivalent can only be constructed when the function of the replaced component can be described by a mathematical model with sufficient accuracy. A dynamic equivalent is a piece of hardware, but many dynamic equivalents include a computer that calculates the values of the dependent variables in the model with the changes in input. If the entire subsystem consists of components whose performances can all be described adequately by models, the subsystem can be completely simulated on a computer, and there is no need to build a physical or hardware simulator. This extreme case defines the simulator preferred by programmers as mentioned above. A physical and dynamic simulator is required only when the subsystem includes components whose performances cannot be adequately predicted by a model.

6.9.2 Origins of Simulators

Simulator testing gained its initial wide acceptance in the aerospace industry with the introduction of power operated flight control systems, and its use has expanded since to other areas. The introduction of powered flight controls brought out into the open the absolute necessity for more adequate quantification of design and test requirements for stability and control.

The interactions between a powered flight control system and the airframe structure are extremely sensitive, hence the control system is designed to operate exclusively in a given airframe. To support the flight control system development, the airframe should be available to serve as a development bed for the flight controls. The airframe, however, does not exist at the beginning of the development. Hence a simulator having the expected structural characteristics of the airframe is used as the development bed for the flight control system. As airframe development proceeds and the expected structural characteristics undergo changes, the simulator is changed accordingly, and airframe and flight controls are developed simultaneously. At the end of the development period, the structural characteristics of the simulator will be the same as those of the airframe, and the flight control system can be transferred from the simulator to the airframe.

However, it should be noted that testing disciplines that existed in the aerospace industry before the advent of power operated flight controls are not considered simulation even though the same principles may be involved. Hence, static-test facilities, wind tunnels, hydraulic test stands, and many electronic test stands are not considered simulators, even though the environment and other operational conditions may be "simulated" by artificial means.

Except in flight control systems, the word "simulator" is not used whenever there are reliable and recognized analytical prediction procedures, such as in structural design, and tests are performed for either of the following functions:

1. To verify analytical results, as in the static test of a wing.
2. To determine the characteristics of a special element, as in the element test of a stiffened panel or the wind-tunnel test of a specified configuration.

The word "simulation test" is normally reserved for subsystems integration tests that include a number of different functional entities, and some of these entities cannot be sufficiently defined by existing models.

6.9.3   Need for a Simulator

The lack of mathematical models, or the lack of adequate predictive capability in existing models, is, therefore, the key factor in establishing the need for a physical and dynamic simulator or "development bed." In marginal cases there may be considerable difference of expert opinion on this question of whether an existing model can predict performance of a component with sufficient accuracy to eliminate the need for a physical simulator. Before a system is actually integrated, the way to prove that a physical simulator is really needed is to build one and demonstrate its usefulness. This is never an easy task to justify when funds are in short supply.

6.9.4   Linearity of Components

A mathematical model that predicts the performance of a component with sufficient accuracy for systems integration is known in simulation terminology as a "linear" model, regardless of the power of the variables or complexity of the equations. The corresponding component is referred to as a "linear" component (Figure 6.8). If the performance of a component cannot be predicted with sufficient accuracy by a mathematical model, the component is classified as "nonlinear." In this context, the term "linearity" is used to denote sufficient agreement between performance measured on an actual component and performance calculated or predicted on the basis of a model. This terminology is an extension of classical mathematical terminology. In classical mathematics, predictive models are classified as linear or nonlinear, depending on the power of the variables. In simulator technology, components are classified as nonlinear only when they are not subject to mathematical treatment. In any specific case the exact meaning of the word "linear" must, therefore, be deduced from whether the context refers to simulator or classical mathematical terminology.

Nonlinear components (Components 1, 2, and 3 in Figure 6.8) cannot, therefore, be replaced without special provisions; the actual components must be used in the simulator. For example, if man is a component of the subsystem and he performs nonlinear functions, he cannot be replaced but must be included in the simulator. The human is most sensitive to interactions, and his inclusion in a simulator converts it into an adaptive device of greatest potential. To replace a nonlinear component requires simplifying assumptions regarding the characteristics of the component. This is sometimes done because it may be the only feasible way to approach a problem, but, in interpreting the answers, proper consideration must be given to the effect of the assumptions.

6.9.5   Environmental Simulation

In a functional subsystem-block-diagram the effects of the external environment can also be represented as interacting functions and treated as a constituent block of the subsystem. Simulation of this block is particularly useful in providing a substitute for the external environment whenever there is a dynamic interaction between the environment and the subsystem characteristics.

6.9.6   Developing Design Requirements for Nonlinear Components

As an extension of subsystems integration testing, the purpose of simulation testing includes the verification of design requirements at subsystem and component interfaces at all levels of the system. As mentioned before, such requirements are derived by systems engineering techniques that rely on models based on simplifying assumptions. One of the usual major assumptions in such calculations is linearity over a restricted operating range, that is, it is assumed that the actual measured performance will coincide with the performance predicted by the model with sufficient accuracy to permit systems integration. However, many interactions between components in modern systems are nonlinear, and linear (predictive) models are of limited validity. Without the linear model it becomes very difficult to derive and establish subsystem and component requirements from the overall system objectives. Even in simple cases, computations involving nonlinear components, at the present state of the art, are not compatible with system development schedules.

In particular, a simulator is useful when insufficient evidence exists to confirm the validity of the assumed linear model for the subsystem, or when the program schedule does not provide sufficient time for the development of a model of acceptable validity.

In these circumstances the simulator provides a powerful new tool to support, supplement, and sometimes even substitute for analytical procedures in developing design requirements for subsystems and components. Design requirements developed in this manner sometimes defy attempts to be expressed analytically, and the most effective method of coping with this difficulty has been to use the simulator for the development of hardware. In cases involving nonlinearity and similar difficulties, there are many advantages to be gained by conducting the component development testing on the simulator, including the use of trial and error techniques.

6.9.7   The Human in the Loop

The inclusion of the human in the loop has been a significant technical development that has permitted simulation to open new technical frontiers. The development of systems by simulation now also includes the training of the operator simultaneously with component development, and the increasing skill of the operator has an important impact on the emerging component design requirements.

6.9.8   Dynamic Characteristics

In addition to involving the human in the loop, simulators have proved particularly helpful when the problems to be studied include dynamic characteristics, power amplifications, and other advances in the state of the art, where critical characteristics are unpredictable.

Dynamic characteristics, for example, are important in the design of airframe structures and flight control systems. The transfer of dynamic loads through an elastic structure depends on the exact local characteristics of the structure and its support. Classical structural dynamic models consist of lumped masses, massless springs, and dampers. Except for the simplest types of laboratory demonstration models, the relationships of these three parameters become so complex that only gross overall predictions can be made with the required degree of accuracy.

6.9.9 Trade-Offs Involving Design Tools

It is also worthwhile to state clearly that engineering trade-offs are made not only between components, subsystems, and performance, but also between available development tools and performance. A traditional analysis, for example, may be replaced by a complicated simulation on a particular design project to economize on weight, not simply because this design tool is new and better. Every design analysis has its accuracy limitations. When this limit is reached, "confidence margins," "safety margins," or even "ignorance factors" are added to compensate for worst-case conditions. A "better" analytical tool, even if more complicated and expensive to use, is one that improves the accuracy of prediction to the point where the "confidence" is adequate.

6.9.10 Simulator Applications

The power amplification requirements in flight control systems have increased dramatically during the past decade with advances in vehicle size, gross weight, and operating speeds, as well as increased requirements for precision flight at various speeds. Simultaneously, the effects of the operating environment have become more severe, such as thunderstorm, clear air and mountain-wave turbulence, and temperature shocks. Exact amounts of high forces are required at the precisely correct times. Since the physical strength of the pilot is limited, the performance of a control system depends on the practically instantaneous amplification and metering of high levels of power. The amplification factors, based on the pilot's input, may range between 1 and 250,000, and the metering problem involves bursts of hundreds of horse power for very short intervals of time. Coupled with these problems is the one of feedback of information to the pilot, both in terms of control system feel and vehicle response.

The use of simulators under these circumstances is a crutch often used for lack of theoretical knowledge. As soon as adequate theories are developed to predict the critical parameters with some acceptable degree of confidence; a few carefully selected flight tests should be sufficient to verify the analytical requirements used to establish the initial design. However, the difficulties involved in establishing new theories and constructing models where nonlinearities are concerned should not be underestimated; there is no validity in the customary extrapolation of related historical data. Any change in a component creates a new component in a new environment, and all previous experience can no longer be assumed to be valid. The changed component is now subject to unexpected performance or failure with respect to the new feature as well as with respect to any of its old characteristics. The impact of the change on all the other components of the system also must be questioned.

Additionally, there are many other mechanical, electrical, and electromechanical parameters that cannot be specified and predicted with the required degree of accuracy on the basis of an established valid model. The simulator has extended systems analysis beyond the limitations of linear or predictive models, and the field of simulation has developed into a new major engineering discipline.

Simulators have been particularly useful in establishing design requirements for aircraft flying qualities, as well as for developing related hardware. For example, the performance requirements for a military airplane include maneuvering requirements, such as rate of roll and pull-out load factors, and specify the specific maneuvers the airplane must perform in order to demonstrate that the requirements have been met. The satisfactory performance of the demonstration tests depends not only on the characteristics of the machine but also on the capability of the pilot.

6.9.11 Flight Simulators as Design Tools

In a military fighter aircraft, it is essential to provide a high degree of maneuverability in vertical plane maneuvers and in rolling pullouts; these are design characteristics that conflict with other performance characteristics because of the relationship between airframe weight and strength. In the past, predictive models in this area have been quite unreliable. In the struggle to meet stringent performance and weight requirements, some aircraft have turned out with marginal longitudinal stability. This is a particularly difficult characteristic to improve, because corrective action usually involves adding more surface (and weight) to the tail. The addition of weight so far aft of the airplane center of gravity (cg) changes the moment of inertia of the airplane and many of the associated flight characteristics. When trouble of this type occurs, corrective action is usually based on astute evaluation of pilot reaction.

As the contractor's test pilots develop more skill in a particular aircraft, pilot reaction usually becomes less negative. Based on an evaluation of the rate of change in adverse pilot reaction, the designer can decide how much additional stability is actually necessary, giving consideration to the fact that less experienced service pilots will be less skilled than the test pilots making the initial evaluation for the contractor. Although this process is less than scientific and usually requires several costly changes in both the aircraft primary structure and the control system, it still permits a skilled designer to achieve a near optimum compromise between maneuverability and stability.

Much progress has been made recently in developing more reliable predictive models in this area, and optimum trade-offs are easier to establish in a simulator using human pilots. Although our predictive models are getting better, they are not yet sufficiently reliable that all work can be done on paper and we can proceed to build hardware without a simulator. In fact, at the present time no new model of a military airplane may make its first flight before the flight control system has been ground tested in a simulator.

6.9.12   New Systems in New Environments

The development of new systems for new environments can be contrasted with the weight-control program where an engineering requirement can be established in advance and the development can be monitored, all on paper and in complete detail, to assure that the final product will have the desired qualities. This procedure simply cannot be followed to the same extent with respect to flying qualities, where the achievement of optimum trade-offs requires the fine touch of an artist rather than of an engineer. Furthermore, flying qualities rank high among those characteristics that spell success or doom for a project; they are critical in determining whether pilots trust and like to fly a machine. When problems arise in this area, they are usually given the highest priority, regardless of their impact on cost or schedules. Success requires thorough management understanding of the problems involved as well as knowledge of the specific capabilities of the people who are going to be assigned to the job of developing the solutions.

The simulator has become a powerful device for studying the behavior of complete new systems in new environments. Large simulators, for example, encompass the full range of the Apollo moon landing mission, including rendezvous and docking maneuvers in space, and lunar landings and takeoffs. In these lunar simulators, the astronauts were not only trained in their detailed tasks, but also while they were developing their skills, they influenced the design of the hardware so as to achieve near optimum system performance in an interaction of equipment with their human skills.

6.10 QUALIFICATION TESTING

The term "qualification test" has a variety of meanings, primarily dependent on the product to which it is applied. In this book it is used in its most comprehensive system-oriented sense to include all contractual tests designed to demonstrate that one or more representative specimens comply with the specified contractual requirements. It is, therefore, a generic term that includes tests known by a variety of specialized names. The expression applies to an operating model and not to a unit, since the purpose is to demonstrate the supplier's ability to comply with the terms of the contract.

Figure 6.9 identifies the four controlling items involved in a component qualification test as:

Component requirements
Qualification test plan
Qualification test requirements
Qualification test specimen

The customer approved qualification test plan must, in the opinion of the customer, adequately reflect the specified component requirements. In the test, itself, the qualification test specimen must comply with the qualification test requirements established in the test plan.

6.10.1 Applicability

The procedure applies at every level of procurement. In advanced systems, every component or item of materiel supplied under an individual contract is either subject to a qualification test, or is certified to be the equivalent of a component or of materiel qualified under a previous contract. The military services and NASA maintain an information exchange program that registers all qualification tests and makes this information available to all contractors (80).

Normally, qualification tests are contractual demonstration tests of the final configuration of a fully developed and deliverable product; therefore they do not provide any useful new design information, except as unforeseen failures occur or unsatisfactory conditions are uncovered and corrective actions indicated.

This statement is not completely true in the airframe and structural areas, where tests are sometimes extended beyond the contractual levels for design reliability to obtain information of engineering interest to the procuring activity.

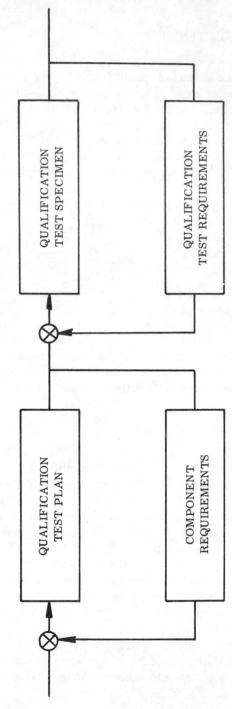

Figure 6.9. Qualification testing.

In complete aerospace systems the qualifying procedure may involve one or more specimens and a combination of ground and flight tests. Once a product has passed its qualification tests, the design may not be changed unless at least the changed feature is requalified. In any case, it may be necessary or desirable to obtain prior customer concurrence.

6.10.2   Type of Tests

The qualification test is a go/no-go type of test, and the product is not acceptable unless the test is completed satisfactorily. If a premature failure occurs during the test, the requirement must be waived or appropriate corrective action must be taken, including redesign if indicated, and it must be demonstrated with the corrected or redesigned item by retest that the failure will not recur. Normally, a reputable supplier will not inform the customer that he is ready to run qualification tests until he has performed equivalent tests in-house on an informal basis and he is satisfied that his specimens will pass the required tests.

6.10.3   Contractual Implications

Under this procurement philosophy, the customer assumes the risk that a product that passes the qualification tests will be satisfactory in service. The supplier assumes the risk that he can design and fabricate a product that will pass the qualification tests on the date specified.

6.10.4   Correlation of Tests and Models

The magnitude of these risks in advanced systems development, with contracts of more than a billion dollars and up to 10 years lead time, points out the need for contractual qualification test requirements that correlate with both analytical design control methods as well as with data feedback from actual field operations. Correlation with analytical methods assures the supplier that the test specimen, when constructed, will pass the qualifying test. When the simulated test conditions are correlated with the desired service life and operating conditions, the customer is confident that he is buying a product with satisfactory life, without having to wait until actual wearout occurs in the field. Furthermore, the supplier does not have a long-term warranty hanging over him, since his equipment is deemed to have met its service life requirements.

The emphasis on correlation of test results with analytical methods is equally important to both customer and contractor, yet it is something unique. In the commercial consumer industry much testing is devoted to determining the minimum quality levels at which the product will still perform satisfactorily; that is, how bad can a product be, and still sell. Consumer goods are often sold under warranties with the expectation that a small percentage of defective units will be returned. Design requirements are established by the producer, at his own risk, completely independent of the consumer. If more than, say, 5% are returned, the quality will be increased. If less than, say, 2% are returned, it usually is profitable to institute a cost reduction program to achieve the desired lower level of quality. This commercial type of testing is unknown in the advanced systems industry.

6.11   ACCEPTANCE TESTING

A successfully completed qualification test provides evidence that the supplier has the capability to comply with the requirements of his contract. Another testing procedure is used to assure that all follow-on units are at least as good as the specimen that survived the qualification test. The acceptance test procedure shown in Figure 6.10 involves:

Qualification test specimen
Acceptance test plan
Acceptance test requirements
Acceptance test specimen

The acceptance test plan is designed to provide assurance that the specimen offered for acceptance is equivalent to the qualification test specimen. This procedure is based on the fact that the qualification test has provided evidence that the qualification test specimen complies with the requirements of the contract. It is important to realize that this is all that has been demonstrated, and that, therefore, it is the qualification test specimen that has been established as the standard for acceptance, and nothing else. The acceptance test must show that the follow-on units offered for acceptance comply with the requirements specified in the customer approved acceptance test plan.

6.11.1   Production Monitoring

The acceptance procedure is hardly ever a single test, but consists of a number of measurements and observations taken throughout the fabrica-

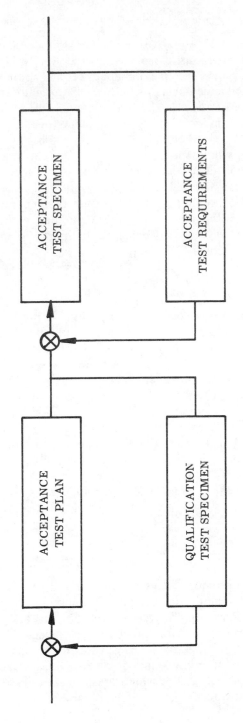

Figure 6.10. Acceptance testing.

tion cycle. To be most effective each check must be simple, easily administered, and taken as close as feasible to the operation being monitored. The design and maintenance of an effective system of monitoring in the advanced systems industry is normally the responsibility of the quality assurance department, which is made independent of both engineering and manufacturing departments of the contractor's organization. Military contracts usually require the implementation of MIL-Q-9858A (73), which specifies the establishment of a "quality program" to assure that all products and services conform to the requirements of the contract. For a detailed description of a quality program refer to Chapter 10. Although quality assurance is a separate discipline, a good design engineer will be thoroughly acquainted with all applicable procedures and will configure his design to facilitate the maintenance of the quality of his product.

6.11.2   Role of Quality Assurance in Design

In the same sense that it is good practice to involve production planning engineers in the initial design process (Section 5.9.3), it is equally desirable to assign quality control engineers to the design staff or to the design review procedure to assure that detail parts are inspectable, as discussed in Section 7.6.13. Quality control engineers can also assist in the preparation of the qualification test plans to assure that they include all the elements required to serve as a baseline for subsequent acceptance tests. The more that representatives of such downstream functions can be involved in the design reviews during the initial design process, the easier will be the transition from the prototype to the production phase. In addition, such individuals usually have the experience to make suggestions that result in cost savings in the prototype phase.

6.12   CHAPTER SUMMARY

System engineering provides a systematic method for analyzing needs, for identifying and quantifying pertinent values, for trading-off these values to arrive at near optimum system requirements, and for integrating analysis, design, and testing for effective system development. These procedures have proven effective in the advanced systems industry in the design and development of large and complex technically advanced, long lead-time systems of high value, which are produced in relatively small quantities. In the advanced systems industry the application of systems engineering has been responsible for some of the most spectacular new technical systems in the history of the world, such as the Apollo.

However, systems engineering is not a rigid formula the application of which automatically guarantees success. In this book the systems engineering process has been shown to consist of a series of closed-loop, iterative paths; as such the procedure cannot be applied piecemeal here and there—it's all or nothing.

Systems engineering is an approach that must be skillfully adapted to the problem at hand, requiring a professional understanding both of the problem and of the method. It should also be understood that the best solution that can be expected from the normal application of systems engineering technology is a workable solution, not an "optimum" or "best" solution. To achieve a superior solution, inspired, inventive, and creative technical genius is required, in addition to best systems engineering practices.

The systems approach, when applied to systems development, requires intimate and continuous interlocking cooperation between customer and contractor in a procedure in which each has very specific and different responsibilities. Figures 6.11 and 6.12 summarize the responsibilities of contractor and customer respectively. To provide visibility into the principles involved, the figures show only the major functions in their simplest form. Note that *contract definition* is shown as a joint responsibility on both figures.

Customer Responsibilities

The customer is an operational agency with a defined mission to perform. His creative challenge lies in the performance of that mission; the development and acquisition of equipment needed to support him in the performance of his mission is a necessary adjunct to his prime responsibility. For example, the prime responsibility of the military is the defense of the nation. The development and acquisition of weapon systems is necessary to support this prime responsibility.

In Figure 6.12 the key operation shown is the analysis of operational plans. This function is shown as a feedback control activity to indicate its key status. This analysis must include all factors involved in deployment: consideration of all past operational experience, evaluation of all operations with current equipment, and estimates of anticipated future missions that the customer agency may be called on to perform. This analysis then provides the knowledge and background for the detailed planning for a specific system and the establishment of the technical requirements necessary for its procurement. In the equipment acquisition process, the customer's responsibility is the definition of the technical requirements

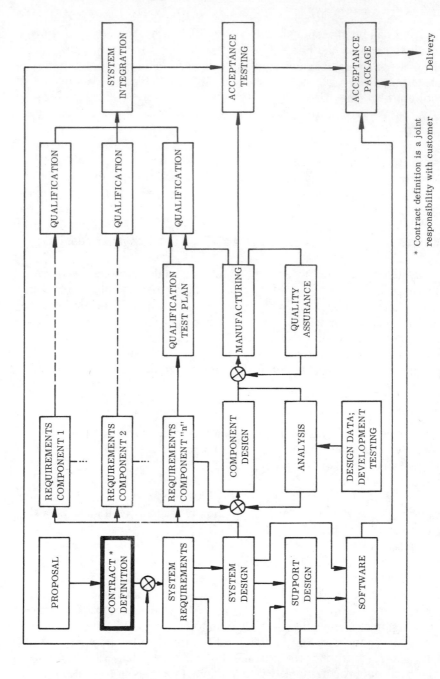

Figure 6.11. Major contractor responsibilities in systems development—development of equipment to pass required tests.

* Contract definition is a joint responsibility with customer

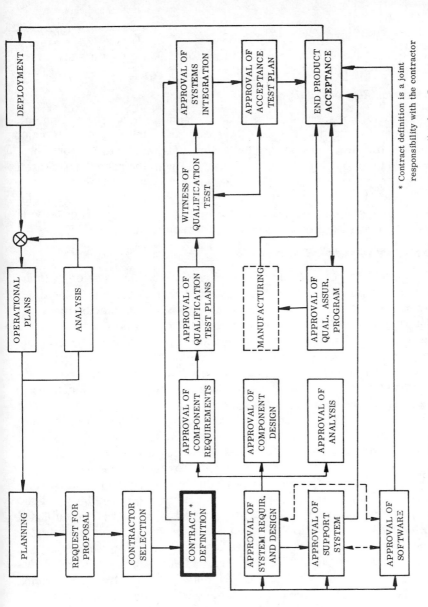

Figure 6.12. Major customer responsibilities in systems development—coordination of requirements and use.

* Contract definition is a joint responsibility with the contractor

185

for equipment that will enable him to perform his future missions in the most effective manner.

Contractor Responsibilities

The contractor's responsibility is to develop and produce the equipment, and to demonstrate that it meets the specified requirements. The major challenges for the contractor are the design and manufacturing functions. These two functions are shown in Figure 6.11 with their feedback control activities to indicate their critical status.

The other blocks of Figure 6.11 indicate some of the other major functions that support the design and manufacturing functions or provide for the demonstration of compliance with requirements.

Customer-Contractor Relationships

In Figure 6.12 there are a number of blocks indicating a customer interface with the contractor. The blocks generally call for the timely customer approval of a completed development phase; these blocks represent monitoring and evaluation points during system development. At these points "customer approval" indicates concurrence that a phase of the development has been completed in accordance with the specified requirements based on the analysis of operational plans.

Customer and contractor have different orientations and functions in the systems development process, and it is the orderly and systematic meshing of the different viewpoints that provides the basis for the eventual development of a satisfactory system. The intimate interlocking of customer and contractor professional personnel in the various feedback loops affords opportunities for product improvement that are unique.

There is no counterpart to this procedure in the development of commercial consumer products. There is no technical representative of the customer who can evaluate consumer product proposals on an unbiased professional basis and who then can make funds available for features that are advantageous primarily from the customers' point of view.

The interlocking but conflicting functions and responsibilities of customer and contractor engineering personnel in the advanced systems acquisition process is very difficult for the uninitiated to understand. As mentioned in Section 6.5.8, with some allowance for editorial license, it might be compared to the relationship of prosecuting and defense attorneys in important trials where mistakes cannot be tolerated. Both parties are primarily concerned with arriving at a correct decision, and their best guarantee of achieving this goal is by the rigorous adherence to established procedure. Meeting the estimated schedule and budget allocated to courtroom trials are secondary considerations. Furthermore, in important court cases there is always a high probability of appeal, and great emphasis is placed on accurate recordkeeping. The paperwork approach associated with courtroom procedures is certainly comparable with that in systems engineering.

However, systems engineering personnel are not as concerned with the appeal and review procedure as are the courts, and engineering paperwork will not be executed with comparable completeness. Engineering paperwork usually consists of memoranda, reports, and test data exchanged between professionals who are all thoroughly familiar with the project, rather than of a complete presentation to be interpreted by professionals not associated with the project.

7

Design Assurance

In previous discussions it has been pointed out that the contractor's responsibility is to design and construct systems that will pass the customer approved qualification and acceptance tests. The customer on his part must assure that the tests are designed so that systems that pass the prescribed tests will provide satisfactory service.

This chapter discusses the tools available to the contractor to assure, during the design and development phase, that his proposed paper designs, when fabricated, will pass the specified tests. The basic procedure is outlined in Figure 6.5. In this chapter some of the representative analytical techniques associated with this procedure are discussed and their effectiveness evaluated.

7.1 LIMITATION OF THE EXPERIMENTAL METHOD

The analytical techniques used in the feedback control loop to the design function in Figure 6.5 normally are based on the experimental method, that is, on predictive hardware or engineering models substantiated by reproducible experiments (see Section 1.3). A great part of the systems engineering effort is expended in analyzing operational situations and in identifying corresponding critical design and demonstration conditions by, for example, such established techniques as the velocity/load-factor (v-g) diagram in structural airframe design (17) and the Nyquisit diagram for determining stability criteria (8). Each of these experimental techniques is applicable to a specific but limited problem area susceptible to such analytical treatment. In their totality these techniques represent the application of the experimental method in engineering technology. The application of these techniques constitutes the basic subject matter of the normal engineering college curriculum.

Compromised Requirements

Although an impressive array of such procedures or models is available today, in the total design and test of a complex advanced system there remain significant areas that are not susceptible to such effective analytical control, that is, they are beyond the state of the art. There are also many practical considerations that simply do not lend themselves to exact analytical treatment. For example, a military airplane is designed and developed under contract in conformance with engineering obligations and procurement procedures that are keyed, to a marked degree, to military procurement policies, public statutes, and economic considerations that are essentially arbitrary. The actual contract requirements invariably represent extensive compromises evolved by groups that have diverse experience, views, and responsibilities. The airplanes resulting from such contracts are used for a decade or more by entities that have little, if any, influence over contract design requirements. Accordingly, during such usage, the aircraft are subjected to operations that often are quite different from those delineated for design and development.

Extending the State of the Art

In systems development, experimental procedures provide excellent control in many areas, while in some few other areas adequate control is not available. Evidence of this is provided by the fact that, no matter how carefully an advanced system or device is designed, built, and qualified, unexpected failures or unsatisfactory system characteristics still occur in tests and in service.

This situation often is encountered in a dramatic manner when a new military aircraft is introduced into service. All new military high performance aircraft extend the state of the art, that is, they extend the flight regimes of previously available aircraft. They seldom do so without surprises, that is, unexpected and uncontrollable flight maneuvers occur as, for the first time, the aircraft enters a new flight condition. An interesting example on a modern military airplane occurred during symmetrical pull-up maneuvers when it was suddenly subjected to unexpected violent alternating positive and negative accelerations. The total absolute value of the changes in load factor exceeded 14. There was no "failure" as such, but the flight characteristics of that airplane were unsatisfactory.

Development of New Design Control Tools

Partially as a result of these limitations of the experimental method, new design control techniques are continuously being developed. The use of the simulator is discussed in Section 6.9. Another new set of design control techniques is based on the use of statistical methods, in particular those classified as "operations research." Although these new procedures certainly do not close all of the existing gaps in technology, when properly applied they provide the designer with some powerful tools not previously available. These applications constitute a new developing and dynamic technology; the operations research based techniques sometimes are referred to as the assurance sciences. Among the best known control applications in design are reliability, maintainability, system safety, and integrated product support. A related procedure, quality assurance, applies primarily (but not exclusively) to manufacturing. Such statistical methods are never as intellectually satisfying as experimental methods, but they do offer a method of approach when the scope of a problem is beyond deterministic analyses by available experimental methods (32). A characteristic of the systems assurance viewpoint is that it encompasses the total system to assure that all areas are included, in particular those not adequately controlled by the experimental disciplines.

7.1.1. Some Brief Definitions

The various assurance sciences are to a great extent all dependent on the same or similar data collection systems, and there is a great deal of overlap between the various techniques. In a specific organization the scope, activities, and effectiveness of the various assurance operations will be more dependent on personalities, historical development, and customer preference, than on any other factors.

The official definitions of the terms used in the assurance sciences are listed in MIL-STD-721 (81). However, the concepts behind these official definitions are often expressed with different words; for example, the definitions of ref. 59 are worded differently. The few definitions in the following paragraphs either are not included in MIL-STD-721, or they differ slightly in wording to make them more suitable for inclusion in this particular discussion based on the concepts of a contractor's engineering and manufacturing responsibilities as outlined in Figure 6.11.

Engineering. Responsible for the definition of the product to be built, described in terms of drawings and specifications, and for the demonstration that the design complies with all customer and contractor requirements. Engineering is also responsible for the *completeness* of all product data, and for providing necessary technical support to other internal and external organizations.

Manufacturing. Manufacturing is responsible for the fabrication of the product in accordance with the definition (drawings and specifications) provided by engineering.

Within this idealized engineering-manufacturing framework, the systems assurance disciplines and techniques are defined as follows:

Quality Control. This discipline is responsible for the certification that the manufacturing facilities have the required technical capability to produce the specified product; and that the fabricated product conforms to the engineering requirements established in applicable documentation (drawings, specifications, and supplementary documents). Quality engineering advises design engineering on quality matters and assures that all critical product characteristics can be demonstrated.

Reliability. The term *reliability* refers to a product characteristic that is a measure of the ability of the product to perform specified missions without failure.

Safety. The term *safety* includes consideration of equipment, personnel, and facilities, and refers to the operation of the product without personnel injury, loss of life, or substantial (specified) economic loss. *Safety does not consider effectiveness of mission accomplishment.*

Maintainability. This is a measure of the effort required to keep the product in an operating condition (that is, so that it can satisfy mission requirements) as measured in units such as maintenance man-hours and/or costs of maintenance per hour of operation. *Maintainability engineering* is a before-the-fact design activity which assures that the product is configured so that it will meet its maintainability and integrated logistics requirements or criteria. *Maintenance engineering* is an after-the-fact activity which determines the procedures for keeping equipment in an operating condition.

Integrated logistic support. The planning and assurance of all those procedures, facilities, equipment, software, and personnel that are essential to maintain the entire system in a ready condition, capable of completing satisfactorily all missions it may be called upon to perform in an operational situation, is called integrated logistic support. This includes management, maintenance, check-out and troubleshooting, automatic test equipment support, furnishing and installing spares and repair parts, personnel and training, technical data and publications, facilities, support equipment, and status reporting.

Common Base – Different Objectives

The overlap and similarity between these various systems assurance techniques can be illustrated by a comparison between safety and reliability. A military aircraft, for example, may have been unable to accomplish its mission because of a failure of its radar, or because its guns would not fire or bombs would not release, but it made a safe return to base. Here, the reliability of the aircraft system is in question, but not its safety.

Example: Airline Safety

An example of the successful application of the assurance sciences in a most difficult area on a very large scale is provided by the commercial airlines of the world. This industry always has been deeply concerned with safety and reliability. Safety devices and practices are costly; however, even if cost is not a consideration, 100% safety cannot be achieved. The objective is to provide "adequate safety" in a responsible manner, compatible with economic reality. No other commercial industry has ever had to face such difficult problems involving the safety of life of such large groups of people as today's airline industry, and none has acquitted itself in such a responsible manner.

A basic question in this connection is, what is "adequate safety?" It is interesting to note how the airlines have made use of commercial insurance to cope with probabilities to quantify this problem. A recognized and defined hazard may be countered in two following ways:

1. By additional safety practices or devices.
2. By the purchase of additional commercial insurance.

If the cost of the additional safety practices exceeds the cost of the additional insurance, there is little incentive for the airlines to adopt the safety practices. If the hazard is in fact critical, the way to enforce adoption of the safety practices is by government regulation, which may result in a fare increase.

Consumer Index of Airline Safety

As far as the individual air traveler is concerned, "adequate safety" is provided when there is a lack of accidents and an atmosphere of safety that exists when the two following basic conditions are satisfied:

1. When good engineering practices are followed in design, construction, operations, and in quality assurance.
2. When commercial insurance is available to the air traveler at reasonable rates.

Although the words "good" and "reasonable" as used here are subjective, an excellent index of airline safety in terms generally understood by the public is the cost of airline travel insurance. Although the general public may not be able to judge the quality of aerospace engineering practices, the insurance companies engaged in the airline business are capable of doing so. As a result of their total evaluations, airline travel insurance is available today on a commercial basis at remarkably low rates at every airport serviced by regularly scheduled airlines. Insurance rates provide, therefore, even for the individual traveler, an overall measure of commercial airline safety.

Limits of Discussion

A review of the definitions listed in this section leads to the conclusion that the scope of systems assurance is very broad and will exceed the boundaries of a single book of reasonable size. This discussion is, therefore, for reasons of space limitations, restricted to a very brief description of a few examples of modern applications of systems assurance. Maintainability has already been discussed in Section 5.10; and logistics and support in Section 5.11. The following sections are concerned with reliability. Section 9.15 discusses risk management, and Chapter 10 covers quality control.

7.2 THE NEED FOR RELIABILITY CONTROL

With the tremendous growth of advanced systems in size and complexity during the last two decades, new reliability assurance techniques have been developed to broaden the basis for making design decisions. The need for reliability control of equipment is well established in the literature (123). Loss of life, economic losses, and the associated public scrutiny of technical operations have all combined to make unreliability, or the occurrence of failure, close to intolerable (124).

These are not local problems; they exist wherever the frontiers of technology are being extended. Polovko, a member of the USSR Academy of Sciences, states in his excellent book *Fundamentals of Reliability Theory* (18), that "Reliability is the central problem of modern engineering and is of national importance." All major U.S. weapons systems contracts today contain requirements for reliability control programs. An extensive bibliography of reliability literature has been published by the Army (82).

As mentioned before, the development of sophisticated advanced systems involves difficult trade-offs between technical performance, schedules, and funding. Reliability analyses can provide a rationale for clarifying many of these trade-offs. To be fully effective, reliability analyses must be prepared in a timely manner with a broad understanding of the impact of the controlling factors in a particular situation, especially of the economic forces involved.

7.2.1   Origins of Reliability Assurance

The initiative for the establishment of formal reliability control (164) originated after World War II in at least five different professional groups associated with the following:

1. Electronics, because of increasing equipment complexity.
2. Power operated primary flight control systems, owing to safety of flight considerations.
3. Maintenance engineering, because of the high cost of support and the scarcity of expert maintenance personnel.
4. Flight safety engineering, because of increasing cost of accidents.
5. Quality control, with their quality records and statistical capability to provide the analytical support.

Of these various groups, the statistically oriented quality control engineers in the electronics group were the first to produce a mathematical model for reliability and to obtain official recognition for it in the United States (83).

Current reliability methodology is founded on the statistical study of the occurrences of failures. Although the basic mathematical concepts involved in reliability analyses and predictions are rather simple, the effective application requires mature engineering judgment and may be a very complex task. This, of course, is true of many statistical applications.

7.2.2   Objectives of a Reliability Program

The objective of a reliability control program is to provide assurance to the customer during the design and development phase that the system, when eventually deployed, will perform with the reliability specified in the contract. The major tasks associated with a reliability control program are the following:

1. To determine in quantitative terms the reliability that the product or system must achieve if it is to perform its intended tasks in a satisfactory manner; and to prepare a set of contractual system reliability requirements.
2. To provide assurance to management during decision making in the design process that the proposed product or system will meet the reliability requirements.
3. To demonstrate, before deployment, that the product or system has actually achieved the reliability required at that phase of development (see Section 11.3.2).

The mathematical treatment of reliability predicts the probability of success for a device during a specified mission. As with all mathematical applications, a simplified model is required. Two of the usual major simplifications made in reliability analyses are

● Mission conditions are held constant.

● Mission time is treated as the variable.

Reliability technology is design oriented; that is, it is primarily concerned with formal requirements, design analysis, and formal demonstrations, in consonance with the feedback control loops depicted in Figure 6.5.

7.2.3 Cost and Public Scrutiny

An example of both the need for, and the progress made in, reliability control is provided by ref. 149 which states:

. . . Kennedy Space Center has developed and maintained over the past three years, a capability to conduct space launchings, on a schedule established up to a year in advance.

The launching schedule which has been maintained on the Apollo program is an achievement which, only a very few years ago, was not considered possible. Four successive Apollo launchings have been conducted on time, in full view of the entire world, and in an atmosphere created by the public press where slight technical errors are exaggerated and assigned a consequence completely out of proportion to the engineering responsibilities involved.

A measure of the magnitude and complexity of launch operations came to light when the scheduled launch of Apollo 9 was postponed from Friday to the following Monday because the three astronauts had contracted head colds. The cost of this delay was reported in the press to have exceeded $500,000. This high sum for a weekend delay is an indication of the magnitude of a space launching operation as well as of the importance of maintaining operations on schedule.

7.3 TECHNICAL CAUSES OF UNRELIABILITY

Reference 123 discusses some of the major technical causes of unreliability. This list of causes is reproduced here with some additional items.

1. Lack of accurate methods for predicting new operating environments.
2. Deficient customer requirements.
3. Incompatibility of requirements and the design state of the art.
4. Unit-to-unit variation in ability to resist failure—while components are all within specification limits.

5. Design errors.
6. Manufacturing errors (out of specification limits).
7. Maintenance errors.
8. Human errors, including communications errors.

Items 5, 6, 7, and 8 can be controlled by a number of practices such as good engineering; reliability and quality assurance; and maintainability, maintenance, and safety engineering procedures. Unfortunately the mere existence and formal application of control procedures does not completely prevent the occurrence of errors. However, intelligent and consistent application of these control techniques can minimize the occurrence of errors to where they can be held to a tolerable level.

Solutions to the problems posed by items 1, 2, 3, and 4 require extensions of present technology that tax the ingenuity and capability of all those engaged in these activities. Decisions involving extensions of the state of the art are always difficult to make; furthermore they are always of a most critical nature and provide the majority of the causes for schedule delays and cost overruns.

7.3.1 Traceability

Another requirement that is receiving a great deal of attention today is for traceability of materials and piece parts by batch or lot to the original manufacturer. Records must be organized so that, in case of a failure, the batch or lot from which the part came can be identified, and all other parts made from the same lot or batch can be located and tested and/or replaced as necessary by like parts from other lots certified as satisfactory. In recent years, for example, the public press has devoted considerable attention to the recall by the manufacturers of defective automobile parts that may involve safety.

7.3.2 Break-Out

In the military an unresolved problem associated with traceability and responsibility for material failures is provided by the so-called "break-out" procedures. To assure the continuing availability of essential supplies in cases of emergency such as an enemy attack on supplier facilities or strikes during a time of critical military operations, it is normal military policy to develop alternate sources of supply for as many critical items as possible. A military procurement agency is always uncomfortable when it must rely on a single source for an essential item. It is a standard requirement that all drawings and specifications for spares and replacement parts be prepared so that the items can be supplied by any alternate qualified source.

A problem arises when the parts procured from an alternate source are not of the same quality as the parts supplied by the original prime contractor. For example, one military agency procured critical helicopter parts for the power drive system directly from a competing supplier rather than from the prime helicopter manufacturer. These break-out parts carried the prime helicopter company's part number; they were true carbon copies in appearance, but did not necessarily comply with all material, processes, and workmanship requirements. The procurement agency did not have the inspection and test facilities or the technical know-how to determine the difference between parts supplied by the prime contractor and those procured from the alternate source.

When these parts failed in service, the prime contractor had difficulty in avoiding the financial responsibility associated with the consequences of the failure. The supplier claimed that the drawings and specifications prepared by the prime contractor did not contain all of the information required to produce a satisfactory part.

7.3.3 Proprietary Procedures

Break-out becomes especially complex when special skills or proprietary procedures are involved. In Figure 2.1 it is pointed out that "how to do the job" is the responsibility of the contractor and is the basis for competition. The "how to do the job" is often based on proprietary procedures that may be essential to the proper definition of the design characteristics of systems, subsystems, components, and elements, as well as for the construction of much of the specialized hardware.

7.3.4 Customer Visibility into Proprietary Processes

A modern system may include special parts produced by possibly 2000 suppliers organized in several tiers. The amount of proprietary information associated with such an industrial complex is so vast that it is physically impossible for any procurement agency to be aware of all the significant details. In some cases, specialized suppliers develop unusual skills but are not aware that these skills are unique to them, whereas other suppliers are not aware that their techniques do not correspond to the state of the art as practiced by their competitors. Furthermore, there are long-standing statutory restraints against the disclosure of proprietary information by government agencies to competitive bidders.

Military procurement agencies face a dilemma in advertising for the competitive procurement of critical spare and replacement parts. In many cases, they really have no way of knowing from the drawings and specifications available to them what special skills and proprietary information and procedures are required to produce satisfactory parts. Without such knowledge, they cannot evaluate properly the qualifications of competing bidders. On the other hand, there is a strong need for the military to develop alternate sources of supply of critical parts to assure a continuing flow of supplies in critical situations. In these circumstances, procuring agencies have little choice but to continue to take substantial risks in the break-out of critical parts.

7.4 RELIABILITY AS A PROBABILITY

A generally accepted definition of reliability appears in MIL-STD-721 (81) as follows:

Reliability is the probability that an item will perform its intended function for a specified interval under stated conditions.

7.4.1 Definition of Failure

The above definition is based on the concept of failure as a discrete event that is positively discernible from a nonevent, such as heads or tails occurring in the flip of a coin. Although this is a simplification necessary for the application of mathematical theory, in actual practice it is not always easy to identify a failure in absolute go or no-go terms.

Considerable care must be exercised in any given situation to develop a precise definition of failure and to make all legitimate failure data available for analytical purposes. Failure data are needed inputs to statistical reliability analyses, but properly designed advanced system components are inherently quite reliable, and failures of such components are relatively infrequent. The problem in advanced systems is that they are so large and consist of so many parts. Even if only a very small percentage of these parts fail occasionally, the total number of failure reports is large. However, few failure reports refer to the same part under the same conditions. Many of the failures that do occur or are reported are marginal or borderline cases.

Classification of Failures

Failures may be final or intermittent, instantaneous or catastrophic, or gradual. The term "catastrophic" is used here in its technical sense and is descriptive of a class of instantaneous and complete material breakdown. It has no implications regarding safety, disaster, or severity of the consequences. Sometimes an item will be reported as having failed when the value of one of its critical characteristics has gradually degraded in effectiveness to some percentage (such as 50%) of its initial value as manufactured. Failures may be independent, or they may be the result of previous failures (dependent). Failures may also be classified by their effect on system operation (consequence); that is, system loss, mission abort, subsystem breakdown, repair required before next operational cycle, and nuisance failures (see discussion in Section 7.6.3).

Ref. 125 contains the following definitions:

Defect. An instance of failure to meet a requirement imposed on a unit of product with respect to a single quality characteristic.

Failure. The termination of the ability of an item to perform its required function.

These definitions are designed for use within the framework of ref. 125, which assumes the existence of documented quantitative requirements in a relatively stable situation. A defect is then easily identified as a characteristic that does not conform to a requirement.

7.4.2 Dynamic Factors in Failure Definition

Sometimes a definition of failure is based on a particular conceptual framework which in turn may result from two basic and interrelated viewpoints: technical and business. The preponderance of one or the other viewpoint in a particular situation will be the significant factor in any given case.

Figure 7.1 shows the dynamic relationship of a number of such factors that affect the definition of a failure. In the center of the figure it is indicated that a failure exists at any point if performance does not meet the requirement. Performance may be determined during inspections or tests, or under operational conditions.

Performance is a variable that can be degraded by product defects or by operational errors; these two items in turn are the results of one or more of the causes listed in Section 7.3. It should be noted that not every product defect or operational error results in a failure; a failure occurs only when the effect of a product defect or operational error is sufficiently severe to cause the performance to degrade below the minimum acceptable level or to increase to an unacceptably high level.

The requirement also varies with time as a function of expectations. Expectations are formulated by a very complex process that balances

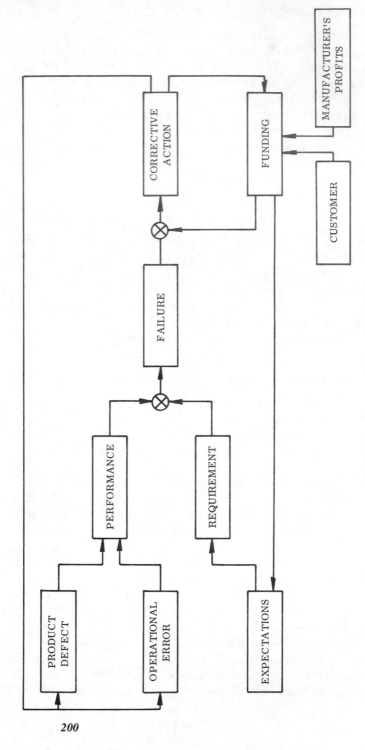

Figure 7.1. Factors affecting the definition of a failure.

200

technical factors against business considerations, such as state of the art and the availability of funding.

Under normal conditions a failure is a signal for the initiation of corrective action. On critical programs, such as the space program, a vehicle may not be launched unless all failure reports have been "closed" by appropriate corrective action. In practice, this requirement is met when the cognizant design engineer approves the implemented corrective action report. Approval signifies that, in his opinion, the cause of the failure has been eliminated to the extent that the failure will not recur, that is, the product defect or the operational error has been removed.

Unfortunately, depth of corrective action is a function of available funding. Failures are unexpected events, except in those cases where tests are deliberately continued to failure. Adequate funding for corrective actions is never provided in advance. Such funding must, therefore, be provided on an emergency basis, either by the customer by "embezzling" from other phases of the program or from other programs, or the funds must come from the manufacturer's profits. Lack of funds for implementing forceful corrective action will result in a revision of the expectation, a corresponding degradation of the requirements, or a failure to comply with the contractual requirements. If the requirement is degraded sufficiently, performance may be adequate for an appropriately modified contract obligation in which the observed condition now complies with the requirements. At the inception of a design and development project, however, a manufacturer cannot depend on renegotiating agreements for modified requirements as a means of coping with future failures or defects.

7.5 STATISTICAL TREATMENT OF RELIABILITY

In the application of the mathematical theory it is necessary to keep in mind that the term "failure" is a complex concept that is a function of a particular balance of technical and business factors in force in a given set of circumstances. In complex advance systems, what may be considered a failure under one set of conditions may not be a failure in another, and the definition may change with time. This condition might be compared with illness in a human, where the problem is to measure how well the sick individual can perform tasks of varying difficulty. The distinction between success and failure in advanced systems, therefore, cannot be compared with the simple statistical concept of "heads or tails" of a flipped coin.

The statistical treatment of reliability has gained wide acceptance in the past two decades and is well recorded in the literature, so that there is little to be gained in discussing the technical aspects in this book. References 18, 19, and 21 represent some of the excellent text books available. References 123 and 164–168 record some of the milestones in the development of the art, while refs. 83–92 are only a few of the many government publications that have supported the acceptance and development of the process.

7.6. ENGINEERING TECHNIQUES OF DESIGN ASSURANCE

In addition to the mathematical models, a number of engineering design oriented techniques have been developed to provide analytical design support within the constraints of the feedback loops shown in Figure 6.5. These are all effective to some degree in assisting the designer in achieving reliability, but not in measuring achieved product reliability. Without such measurement there is no truly objective means of determining the effectiveness of these techniques at the system deployment point of Figure 2.3.

This section describes these procedures as briefly as possible since more complete descriptions are available in the references.

7.6.1   Reliability Requirements and Apportionments

The establishment of the reliability requirement for a proposed system is a systems engineering task that must be performed in coordination with the analysis and definition of the proposed missions. Reliability is one of the many factors that must be traded off against one another to arrive at a set of feasible system requirements (see Figure 5.2). The reliability requirement impacts strongly on cost, performance, and schedules (21). See also "Conflicting Requirements," Section 3.1.4.

Reliability Requirements

Normally system reliability requirements are established on the basis of experience with past and existing systems. However, occasionally a new system will be subject to hazards that have never been encountered before, and the establishment of reasonable requirements may involve some creative thinking.

The Apollo Program provides a dramatic example. In manned space flights, system reliability requirements must be compatible with the risks the astronauts are asked to take. On the Apollo Program the establishment of these risks provided a unique challenge, since there was no precedent.

To obtain guidance a study was made of the probability of survival of Naval pilots during carrier landings in peacetime and in war. The results showed no significant difference in the probability of survival between war and peacetime operations. This was considered surprising until it was realized that 85 to 95% of all wartime flight operations are training flights, identical to peacetime operations, to maintain ground and flight personnel at top proficiency.

This study led to two further investigations: one of the probability of survival of racing car drivers at the Indianapolis Track, the other of bull fighters in the Mexico City Ring. These two subjects were selected because of the high risks involved in both activities, and because of the completely different cultural environment and nature of the risks. Again the results of the study were surprising in that the risks at Indianapolis and at Mexico City were practically the same; however they were of a higher order of magnitude than for carrier pilots.

Although the scope of these studies was limited, a number of tentative conclusions were formulated. At Indianapolis and at Mexico City the risks are self-determined; the individuals themselves control the risks by their own decisions in a continuous trade-off between risk and glory. The glory for carrier pilots is not as great as for the winners at Indianapolis or Mexico City; they cannot be asked to assume risks that are as high.

On the other hand, it was also concluded that volunteer astronauts on an historic first trip to the moon are the potential recipients of glory much in excess of that of winners of earthbound events. They can, therefore, be asked to assume greater risks, and they knowingly accepted these risks.

These quantitative calculations had a significant impact on the formulation of the original Apollo mission reliability requirements. Now that the flights of Apollo have been successful, the Apollo requirements will serve without question as the baseline for future space travel developments.

Reliability Apportionment

Given the system reliability requirement, a model of the system should be constructed showing the various levels of assembly, subassemblies, components, and elements and defining the relationship between all units, both vertically and horizontally. The system reliability requirement is then apportioned down the line among all units, based on the estimated relative reliability of the individual units. The apportionment must be such that when the reliability requirements for the various components are combined up the line in accordance with the product rule, the system reliability requirement will be satisfied.

The process of apportioning reliability requirements from the top system level to lower levels is not an automatic procedure. Available textbooks contain a number of principles, but there is no substitute for judgment and experience coupled with a sound knowledge of the system and of all the components involved. Estimates must be made of the expected reliability of all the components in the system. Each component has its own expected value. The reliability requirement for a header element at any level is successively apportioned to the trailer elements on the next lower level in proportion to the ratio of the expected reliabilities of the various trailer elements on that level.

The initially apportioned reliability requirement normally does not coincide with the expected reliability for any given unit. It may take considerable trial-and-error adjustments of several parameters to arrive at a reasonable and consistent apportionment acceptable to all those involved.

The apportionment should be revised periodically as the design progresses and better estimates become available for the various components. It should be noted, however, that once a reliability requirement is contractually established for a subcontracted or purchased item, it is not advisable to make changes unless absolutely necessary, since a financial or schedule adjustment may be required. Apportionment techniques are discussed in many textbooks, including Lloyd and Lipow (21).

7.6.2   Configuration Analysis

Configuration analysis is a trade-off technique useful in assisting the designer in selecting the best design configuration. The technique is widely used and known by many names; the terminology used here is from ref. 117, which explains the technique as follows: The purpose of configuration analysis is to provide the designer with a systematic, objective, and quantitative technique which will assist him in making a better decision than he could make intuitively. The method prescribes a systematic procedure designed to disclose significant factors which otherwise might be overlooked.

It is assumed in this discussion that the reliability objectives have already been established as described in Section 7.6.1. The next step is to record systematically all possible system arrangements, or at least all of the possible workable arrangements that the designer can think of in the available amount of time. Each possible arrangement is shown in a schematic diagram or by other suitable means.

At this stage it is usually possible in a rough screening process to determine that certain system configurations are obviously inferior; this is part of the design control process of Section 6.5.2. When these inferior configurations are eliminated the designer is left with a small number of competing configurations from which he cannot make an intuitive selection.

At this point it is advisable for an analyst other than the designer to review the design, since he will not share the designer's biases. The analyst attempts to determine all significant or critical factors that must be considered in making a final selection from the remaining competing configurations. The parameters will vary with type of system and with environment; each case must be carefully analysed to assure inclusion of all factors bearing on the problem.

Quantification of Critical Factors

Each critical factor or design parameter must be defined by a quantitative expression. A number of factors such as weight and cost can be estimated readily with sufficient accuracy in numerical terms based on a paper design proposal, but there are many other pertinent parameters for which accepted definitions do not exist, or that cannot be accurately estimated in the paper proposal stage. Reliability, availability, and maintainability are such considerations. However, since the purpose of this analysis is to establish comparative rather than absolute values, it is not necessary to have rigorous definitions as long as the following conditions are met:

1. All competing systems are treated in the same manner.
2. The definitions are not biased in favor of a particular configuration.

Nonetheless, the best way to obtain valid relationships is to compute them with accurate absolute values; every reasonable effort should be made to obtain valid definitions.

Each pertinent factor is computed for each competing configuration. The results are then presented in such a manner that the advantages and disadvantages of each configuration are apparent and can be balanced one against the other, helping the designer arrive at an objective decision. See ref. 117 for an example.

7.6.3 Failure Effect and Mode Analysis

This type of analysis is a design review technique and consists of an independent critical review of the proposed system design coupled with a systematic examination of all conceivable hypothetical failures and an evaluation of the effects of probable assumed failures on the mission capability of the system (116).

Functional Block and Sequencing Diagrams

The system is described in a functional block diagram defining the critical functions that constitute the system. A critical function is one that must be performed if the system is to complete the missions stated in the specification. A separate diagram is required for each operating mode.

In this diagram the functional blocks are all interconnected by lines that represent the inputs to an upstream function as the outputs from each related downstream function.

The analysis is best performed in a set of tables organized by functional blocks or groups of associated blocks. For each functional block all significant design requirements are listed, those derived from the system specification as well as those established by the designer. Compensating (fail-safe) provisions for possible failures are noted.

In preparing the functional block diagram for a complete system, appropriate blocks should be included to depict the functions performed by the operators—the flight crews and ground maintenance personnel. Failure rates of personnel are not as well established as those for equipment, but work is being done in this field, and some failure rates are available.

In defining the various possible operating modes, consideration should be given to conditions where the system is changing from one operating mode to another. These transitional states may constitute operating modes, which often warrant separate analyses.

Complementary to the functional block diagram is the sequencing diagram, which shows the order and times at which events occur (158).

Hardware Descriptions

Next, a description of the hardware constituting each functional block is prepared. A systematic review by the analyst establishes that the hardware in each block is capable of satisfying the requirements established for the block.

Assumed Potential Failures

A systematic sequence is followed, where each functional block is assumed to fail in turn. A failure of a block is now defined as a break in an output line, with the output signal in its most critical position or adverse condition. If each output line is, in turn, systematically cut, either alone or in critical combinations with other failures, all possible hypothetical functional failures in the system will be identified. Any condition where the system output does not meet the design requirements should be considered a failure. Input lines do not have to be considered, since an input to one function is an output from a preceding function.

Any change in the order of events shown on the sequencing diagram is also listed as a potential failure.

Possible Causes of Failures

For each hypothetical failure so identified, the hardware description is examined and the possible causes are determined and described. The hardware components that must malfunction in order to cause the hypothetical failure are listed. The failure descriptions are expressed in terms of breakdown of specific parts (such as, shorted item, open circuited item, structural failure) or of a configuration or design deficiency. It is especially important to recognize the latter type of failure early in the design stage, because corrective action at a later time may be particularly troublesome. Common examples of configuration deficiencies include lack of provisions for sufficient fuel to accomplish long-range missions and inadequate aerodynamic control to perform rolling maneuvers. In one case, permanent wing tip tanks had to be added to a fighter aircraft to enable it to meet the combat range established in the specification. In some early models of jet aircraft unanticipated gun-gas concentration in the cockpit was disclosed during trials. This contributed to weapon system unreliability since the crews had to wear oxygen masks whenever the guns were fired.

Consequence and Criticality of Potential Failures

In the analysis the symptoms and consequences of each assumed failure are listed as they affect pertinent higher levels of assembly and the mission capability of the system. The compensating provisions inherent in the design or alternate operating modes and fail-safe features are reviewed. This examination covers any corrective action needed, either automatic or provided by an operator; evaluates the results of such corrective action; and indicates the resulting degree of equipment degradation. The analysis is reviewed to establish the most likely hypothetical failure modes and the probability of occurrence under operational environments. This aspect of the analysis requires a knowledge of the failure rates of the various components under the expected environmental conditions.

The assumed failures and failure modes can be classified in categories which, together with the probability of occurrence, provide a measure of the criticality of the failure. For each assumed failure an estimate is prepared of the percentage of system degradation that occurs as a result of the failure. Failure classes, for example, might include the following categories.

Class I. An equipment or component breakdown or degradation that prevents the advanced system from accomplishing its mission.

Class II. The same as Class I, except that automatic provisions are available to compensate for the failure, or the failure can be detected and corrective action can be taken by the crew during operations. Corrective action may be by adjustment of equipment, where provisions for such adjustments have been made; replacement of failed items, where spares are feasible and provided; or by complete bypass of the failed-item function by the crew, using other modes of operation.

Class III. Equipment or component performance is degraded; the system will function and perform its intended purpose, but not within specified limits. These failures do not constitute system breakdown.

Class IV. Equipment or system operation is not noticeably affected; these are nuisance type failures.

Designer Versus Independent Analyst

It is emphasized that the real purpose of this kind of analysis is to have an independent analyst work closely with the designer and furnish his inputs during the formative stages of the design, in the feedback design control loop of Figure 6.5. A failure effect analysis cannot be performed effectively by a designer who is struggling to define a design that he believes will work; it must be performed by an independent analyst. In the construction and analysis of the block diagrams and hardware descriptions, the failure-effect-analysis procedure provides the analyst with the means for developing a "red-team" or "adversary" viewpoint, quite different from that of the designer. The real benefits of the analysis derive from the convergence of the two viewpoints.

7.6.4 Fault-Tree Analysis

Fault-tree analysis is a systematic troubleshooting procedure performed on paper. It is a useful supplement to other reliability assurance procedures during the design stage. In conjunction with the failure-effect-and-mode analysis, it becomes an excellent device to help the designer understand how the elements of his system operate and provides him with a basis for trade-offs.

For each potential failure mode identified in the system oriented, failure-effect-and-mode analysis, a fault-tree analysis can be constructed for the proposed hardware. The detail with which this is done depends on the probability of occurrence of the failure and its criticality. It is unrealistic to apply a rigorous treatment to a problem requiring only an approximate solution.

The fault tree is literally a tree-like diagram with a selected potential failure (or undesired event) at the top and the various causes, or combinations of causes, listed as a sequence of events making up the various branches below. Essentially, the tree is developed during an analysis from the top down, but in the case of an actual failure, the logical sequence of events flows from the bottom to the undesired event at the top. The diagram provides an easily followed, orderly, and concise description of the combinations of occurrences within a system that can cause the undesired event (22, 169, 170, 201).

At any given level the fault tree defines the inputs required to obtain the indicated output. Inputs pass through "gates" which are of two types, customarily designated as "OR" gates and "AND" gates. If any one of a number of inputs will permit the sequence of events to continue through the gate, it is an OR gate. If the coexistence of all inputs is required to produce the output event, the gate is an AND gate.

Automatic Checkout Equipment

The most important application of the fault-tree analysis has been in the design of automatic checkout equipment (ACE) or automatic test equipment (ATE) for electronic equipment. A piece of automatic checkout equipment is a mechanization of a fault tree; each branch of the tree is followed systematically, and each gate is checked, until a failure is detected and the failed component isolated. Corrective action usually is performed manually.

However, more and more ATE is being designed for mechanical applications, especially in conjunction with trend analysis, discussed in Section 7.6.11 (171). Such equipment is being designed from a systems viewpoint to monitor the operational integrity of mechanical, hydraulic, and electrical subsystems and to predict incipient breakdowns of critical parts in costly advanced systems. Typical "signatures" for each measured parameter are developed, and predictions regarding equipment life can be made as these signatures change with accumulated operating life. The use of ATE helps to reduce operational costs by eliminating two longstanding maintenance problems:

1. Failed part identification: In breakdowns of complex systems it is sometimes impossible to immediately isolate the failed component. Troubleshooting is performed by replacing one suspected part after another until the system is restored to an operating condition. In this process, many serviceable components are removed until finally the failed unit is found.

2. Predicting part wear-out: Preventive maintenance on many sophisticated advanced systems is performed on the basis of replacing parts before wear-out, in accordance with an established time between overhaul (TBO). A TBO provides for the replacement of a critical part before wear-out takes place on a statistical basis. However, the longevity of some parts of a given design may be considerably longer than that of others, and the severity of the service environment is also a factor. TBOs are based on safe statistical values, and many serviceable parts are removed before they begin to wear out.

ATE in conjunction with trend analysis provides a capability for monitoring the remaining operational life in critical parts on an individual basis; a positive indication is given when an individual part must be replaced. Hence parts may be replaced near the actual end of their service lives when they are about to fail, and no parts that are still serviceable need to be removed.

The design of automatic checkout equipment has developed into a major new engineering specialty which has a significant impact on improving the reliability and maintainability of electronic systems. There are two basic approaches.

1. To develop the automatic checkout equipment concurrently with the basic system.
2. To design the basic system so that existing automatic checkout equipment can be used.

An extensive literature has been developed (90), and the field has become the subject for specialized university courses (202).

7.6.5 Static Strength Testing

This method is attractive because it is simple and independent of time. It is based on a deterministic simplified abstract model constructed on a base of assumed ideal design conditions. The fact that the assumed ideal conditions may never occur in actual operation bothers critics of the method, but does not detract from the usefulness of the model. Structures designed in accordance with this model perform satisfactorily within a limited range of operating conditions. The model, therefore, is a useful analytical tool when applied within its range of validity, which encompasses most structural design. To be successful, the analyst must possess precise knowledge of the limits of this range.

The static strength requirements for an airframe are given in terms of "limit" and "ultimate" loads (or corresponding allowable accelerations), which are established on the basis of past experience and in cases of new designs are verified by wind-tunnel tests. The procedure is described here in some detail because it is different from the procedures used by structural engineers in fields other than airframes. Generally, structural engineers are constrained by cost considerations, which still allows them a degree of freedom in configuring their designs. The airframe procedures are designed to assure specified airframe strength at minimum weight —one of the most important considerations in aircraft design. These procedures provide an illustration of design control that could possibly be adapted with advantage to other engineering disciplines where similar constraints exist.

The term "static strength" refers to the capability of a structure to perform satisfactorily under a specified constant load condition. In a static test, a given load condition is applied and held constant for a specified length of time, usually three minutes, before readings are taken.

Margin of Safety

For design purposes the static loads and resulting load distributions throughout the structure are derived from specified static and dynamic conditions: take-off, flight maneuvering, landing, and other operational environments. A mathematical model is constructed to represent the structure, permitting the determination of the load distribution in each structural element.The construction of such models is discussed in detail in Bruhn (17). Given the design configuration and material of each member, a static margin of safety can be computed. The margin of safety in each member must be equal to or slightly greater than zero, otherwise the structure will be overweight. Such a structure is considered statically safe, and its reliability under static loading will be 1.0 from an engineering viewpoint. The static margin of safety (M.S.) is given by

$$\text{M.S.} = \frac{F_u}{f_u} - 1 \qquad (7.1)$$

where F_u = allowable ultimate stress for the material of construction
f_u = design ultimate stress

The allowable ultimate stress for a given material is taken from an approved materials handbook such as MIL-HDBK-5 (91). Such documents may not contain allowable stress values for some of the new materials, particularly for nonmetallic materials or for composites (which in effect are "materials" created by aerospace manufacturers rather than by customary material producers). Under such circumstances the systems manufacturer must make such investigations and tests as are necessary to establish allowable stress values, and must document the results for customer approval.

Safety Factors

The maximum load anticipated at any particular region of the structure during an operational event is called the design limit load. The design ultimate load is then the design limit load times a factor of safety, usually 1.5 for an aircraft airframe. The factor of safety provides not only for inadvertent overload conditions, but also for any imperfections that can reasonably be expected in material properties and in manufacturing and quality assurance procedures. Many aerospace contracts contain provisions modifying this factor, making it more severe at critical points such as at the wing-fuselage fittings, by multiplying the design ultimate load by a "fitting factor" greater than unity; and reducing its value for low probability flight conditions, such as continued flight with a shut-down engine resulting from an operational failure.

The major purpose of increasing the safety factors at important fittings is to prevent a failure from occurring at these expensive fittings. By adjusting the safety factor at different parts of the airframe, it is possible to contain failures to easily repairable elements, were a failure to occur. Safety factors are selected so that the probability of their being exceeded is negligible.

Static Tests

The validity of the analysis is demonstrated in a static load test, variously called a failing load or static strength test. The specimen is subjected to loads and load distributions in the laboratory consistent with those assumed in the analysis, reproducing the conditions for which the Margins of Safety are computed. The stress distribution in the structure under test is measured and compared with the distribution indicated in the analysis. The failing strength (failing load; failure mode) will then indicate the achieved margin of safety, which also can be compared directly with the value predicted in the analysis. If the measured load distribution and the failing load are in sufficiently close agreement with the analytical values, the analytical values (and not the measured ones) will be recognized as the strength characteristics of the aircraft. The test specimen is considered a special case incorporating possible material and manufacturing imperfections, and the analysis is recognized as the standard.

Flight Tests

Subsequent to the satisfactory completion of the required static tests, a series of instrumented take-off, flight, and landing tests is performed to determine the loads in critical members over the allowable range of take-off, flight, and landing conditions. Critical measured loads are compared with the corresponding critical load requirements used in the design analysis. These comparisons validate the original design requirements or indicate design deficiencies.

Validity of the Method

Despite the fact that some of the assumptions described above are questioned by statisticians, this classical static strength analysis procedure has been one of the most successful techniques in engineering practice. However, its success is based on the fact that there is only one variable, load; and a direct relationship between load and stress can be established.

Another critical factor contributing to the success of the approach has been the integrity and discipline with which it has been applied by engineers and managers who have been acutely aware of their responsibility for safety of life. This responsibility, more than any other tangible factor, has been the driving force in establishing the quality of airframe engineering. The analyses, test plans, and test results have invariably been recorded in written reports; traceability of requirements has been maintained, and accountability of responsible designers, analysts, and test engineers established. These reports have been filed in central depositories from which they are retrievable, so that if unexpected failures occur later in service, possibly under unanticipated conditions, it has usually been possible to retrace the design history of the failed element and establish improved design procedures to prevent the same failure from recurring.

7.6.6   Safety Factors as a Means of Design Control

An interesting aspect of Equation 7.1 and the factor of safety concept as a means of design control is that it provides management with a means for imposing an uniform standard of safety throughout a large structure being designed by a large number of individual creative engineers, often geographically dispersed, each with his own ideas of what constitutes adequate safety.

An individual engineer does not want that portion of the structure he has designed to fail. However, both specified operational environments and specified factors of safety are beyond his control; they are stipulated by the customer's design specification. The designer must maintain a zero or slightly positive Margin of Safety (Equation 7.1) and must also remain within the allotted weight, or compromise system performance. Between these limits, there is little that he can do to add his own personal margins of safety. Since all designers must comply with the same discipline, even if dispersed geographically, the resulting structure will be of approximately equal strength throughout; no portions of the airframe will be noticeably stronger than others, except in those regions where the management has deliberately assigned a higher factor of safety or where extra strength results from the effects of overlapping design requirements.

Any experienced structural designer is fully aware of the harassing consequences that result from a premature or unexpected failure. He normally does everything in his power to assure that his portion of the system will be as trouble free as is possible. The concepts of Equation 7.1 and the factor of safety provide effective management tools for keeping these desirable personal objectives of individual designers under control.

Conflicting Requirements

One of the problems in establishing aircraft design requirements is the very large number of critical operating conditions that must be considered (take-off, in-flight maneuvering, landing, etc.). Because of the magnitude and complexity of this problem, it has not been possible to derive as set of structural design requirements that satisfies all known critical flight conditions, but does not overlap in some areas. Since each structural element must be designed for a zero or positive margin of safety for every stipulated design condition, there will always be some portions of the airframe that have excess strength because of the overlap. Nevertheless, this excess strength is the result of the way the design requirements are written, and is independent of the discretion of the designer.

Design Control in Nonstructural Areas

In other areas of aircraft design, that is, nonstructural areas, reliability requirements and the maximum weight allowances provide the two limits that theoretically box in the designer and force him in meeting performance requirements to conform to a level of reliability and safety that can be held reasonably uniform throughout the system, or can be controlled locally at the discretion of the management. When these controls are not exercised in an effective manner, it is not uncommon to find designers compromising system performance by adding personal factors of safety, and hence weight, based on their own subjective evaluation of the risks involved.

7.6.7 Dynamic Life Testing

In aircraft design a distinction is made between static and dynamic loading conditions. The actual critical operational loads on airframes generally result from dynamic conditions: take-off, flight maneuvers, or landing operations. Dynamic loads change with time; they can resemble hammer blows. As discussed in "Margin of Safety," Section 7.6.5, most of the elements of an aircraft structure are designed and tested for static loads derived from dynamic operating conditions. However, there are always some critical elements in an aircraft structure that must be designed and tested under dynamic loading conditions. Normally, all of the dynamic elements of an aircraft engine or helicopter transmission are designed and tested under dynamic conditions. This increases the design difficulties in these subsystems by orders of magnitude. In particular, the performance of a complete aircraft structure is demonstrated under dynamic conditions. For example, landing loads are imposed in the laboratory in a "drop test." The entire airplane is dropped vertically repeatedly from various heights onto a solid platform.

The capability of an airframe or other structure to perform satisfactorily under actual repeated dynamic loadings resulting from a variety of operating situations is referred to as "fatigue strength" or "fatigue life." To establish the spectrum of loadings required for the determination of fatigue strength, analytical consideration must be given to all possible load distributions that are likely to occur in all planned missions for which the system is being designed and developed.

Selection of Elements for Fatigue Analysis and Test

The determination of dynamic loading spectra and of fatigue strength is a complicated analytical process. The demonstration of fatigue strength is costly and time consuming. To develop a safe aircraft, this determination and demonstration must be performed for critical structural elements and assemblies of the airframe. The responsible analyst must know what portion of the airframe or other machine can be justified by simple static methods, and when he must resort to much more complex dynamic analyses and testing. A skilled professional analyst can be expected to examine a structural or mechanical layout (preliminary drawing) and recognize what techniques are required in particular situations, just as a trained physician can listen to a man's heart and be expected to recognize whether the man is sick and what medicine to prescribe.

Multiple Loading Conditions

As soon as more than one load or environment is present in combination with other variables, such as transient loads, extreme temperatures, shock, vibration, humidity, transients in electrical or fluid power, cavitation, and the like, the problem of assuring reliability becomes extremely complex or even indeterminate. It is difficult to establish how much stress is caused by which variable or combinations of variables. Measurement of reliability requires the establishment of standard spectra of combined environmental loadings. The loading spectra are applied gradually until the specimen fails, and the failure mode, location, and loadings at failure noted. The failing load level then can be compared with the specified loading level in an equation similar to Equation 7.1 to indicate a margin of safety. If a number of such tests are performed on several specimens, and the margin of safety is consistently the same for all specimens tested, the data can be used to calculate reliability.

Loading Spectra

Unfortunately, because of a number of factors such as the lack of validated operational data, the lack of analysis of whatever data does exist, and the high cost of validating proposed loading spectra, standard spectra of combined operational and environmental loadings are very difficult to obtain. In a systematic study of this problem, Navy structural design engineers have collected and collated a vast amount of data during training, combat, and simulated-combat operations. These data have been used to derive the design load spectra of ref. 75, which include magnitudes and frequencies of loads less than, equal to, and greater than design limit loads. They represent the probable loading environments for planned uses and lives of aircraft. The modes and locations of failures of aircraft test articles during laboratory life tests, using these derived load spectra, have been compared with reported modes and locations of failures during routine service operations. These comparisons have furnished evidence for confidence in these design load spectra and have also confirmed the necessity for substantiating the reliability and safety of aircraft structures by appropriate combinations of analyses and tests, both for dynamic and static strength.

Except for the spectrum of ref. 75, other standard spectra do not exist for structural reliability testing. There are some standard qualification testing spectra for combined environments (33), but this literature is limited. Today, there still is no recognized general method for the laboratory measurement of structural reliability under combined environments.

Application of Statistical Methods

Because of the high cost of the consequences of failure, more attention is being focused on the reliability of mechanical components under dynamic loadings. Reference 172 is an example of some data analysis conducted on automotive components. A great deal of work is being done on the application of statistical methods in the analysis of mechanical failure modes of critical parts, such as fatigue, brittle fractures, stress concentrations, creep, crack propagation, and the behavior of metals under shock loads and in the plastic range. A representative sample of this work is given in refs. 126, and 173–181.

7.6.8 Overstress Testing

Overstress testing consists of testing at higher than known or anticipated operational levels. In such tests, a combination of a number of simultaneous operational and environmental conditions is imposed on the specimen, and selected test conditions are deliberately increased until a failure occurs in the device. Such testing provides an excellent tool for identifying the weaknesses in a device.

Generally the goal of an overstress testing program is to demonstrate that the specimen is capable of surviving some assumed loading level, such as 1.5 to 2.0 times the anticipated combined operational and environmental conditions, for a specified length of time without failure. Quite often it is discovered that the specimen will not survive all of the required or anticipated combined operational and environmental conditions without failure when these conditions are applied simultaneously in the laboratory. When a failure occurs, the failed part should be redesigned so that the failure will not recur.

Application of the method normally requires a number of arbitrary assumptions on the part of the test engineer. This procedure is often not acceptable to many scientifically oriented individuals. It is questioned if failure modes forced under overstress conditions can be correlated with those occurring under actual operational environments. Methods of artificial aging generally are unsatisfactory. Nevertheless, there are reported cases in the electronic industry where the skillful application of overstress testing techniques has repeatedly proven to be a useful method for achieving reliability by indicating to the designer where to take timely corrective action, even though the quantitative value of achieved reliability may not have been actually determined. On the lower level of detail design of electronic equipment, safety margins are related closely to the common practice of derating electrical components (87). Overstress testing is much more of an art than a science.

Reference 127 presents a status report on overstress testing in which a fivefold improvement in reliability was achieved on some 15 pieces of airborne electronic equipment with less than 200 hours of testing per unit. Reference 182 contains a remarkable report of a successful missile program that was based on a test to failure program, closely integrated with design.

Overstress testing to failure was also used extensively in the develop-

ment program of the Lunar Module of Apollo (LM), utilizing the procedures described in refs. 33 and 185. The reliability requirements were so
high that there was no way of incorporating standard reliability life testing
into the development schedules.

7.6.9 Reliability Life Testing

Classical reliability demonstration procedures require life testing to failure of several specimens in their final configuration under representative
operational conditions, and the cumulative test duration must be a multiple of the MTTF or the MTBF to be demonstrated.

To demonstrate achievement of reliability, several specimens of the
final configuration are needed, and the tests may require appreciable
calendar time. The tests normally continue until failure, so that the
specimens are lost. Very little engineering knowledge is gained if no
failures develop. Meaningful requirements must be expressed in terms
measurable during testing as shown in Figure 2.3. MIL-STD-781B (92) is
widely used for the design of formal reliability demonstrations.

7.6.10 Field Trouble Data Collection

A field trouble data collection system is needed to satisfy the technical
requirements discussed in Sections 5.10.3 and 5.11.10.

The parameters required for the measurement of field reliability are:
part and serial numbers of the failed and replacement units, system
identification number, type of mission, descriptions of system symptoms
and component failures, environmental and operating conditions under
which the failure occurred, mission time, component and system operating times, number and classification of failures per component type, date
and location of failure, results of failure diagnosis and accident investigations, time to troubleshoot and to repair, availability of spares, administrative time, down time, turn-around time, and other pertinent information.

Some of this information is readily available from available field records; the careful recording of system utilization and operations management has always been a normal procedure in a well-run operation. Additional effort is required to record component operating time and the
number of component failures and to collect, process, and analyze the
necessary data. The analyst should be able to retrieve the operational
history of the failed system, especially data on unusual environmental or
operational conditions to which the system may have been exposed in

previous missions. To obtain all of the required data normally requires routine data collection from several different and unrelated sources such as field operating, repair, and logistics facilities as well as central headquarters planning, purchasing, and monitoring offices and the contractor's design engineering data and diagnostic laboratories. Of particular value is the fact that all such data refer to actual operational conditions. The analyst must also have available to him all qualification test data and failure mode and effects analyses to serve as a baseline for failure diagnosis.

Identification of Failed Components

The identification of the failed component can sometimes be difficult. System repairs are sometimes made by the successive replacement of parts until the system failure or complaint disappears. Many of the replaced parts will not have failed, but they will be returned for overhaul anyway. The increasing use of Automatic Test Equipment (ATE) will assist in isolating failed components, but at present the areas monitored by ATE are still limited.

Various contracts for spares may provide different warranty conditions, and this should be known to the analyst. Parts acquired in the break-out process from competing suppliers will have different characteristics. Hence, in addition to the identification and serial number of both the failed and replacement parts, it is often useful to obtain the contract numbers of both parts, whenever this information is available. The contract number usually appears on the packaging and sometimes on the name plate. In some companies complete traceability information is coded into the part number.

7.6.11 Trend Analysis

Polovko (18) points out that a failure is usually preceded by adverse internal variations in a system, just as death in a living organism usually is preceded by its sickness. Trend analysis is a technique for monitoring operating mechanical equipment that identifies incipient failures before they can be detected by other methods (such as performance degradation or breakdown in service). The procedure consists of measuring certain significant parameters on a regular, periodic basis. Maximum and minimum limits are established by extensive testing for each parameter; the acceptable limits are set wide enough to encompass the individual variances that occur from unit to unit. Within these limits, a careful plot for an individual unit of a parameter of interest establishes the "personality" of a unit. This plot indicates the profile of normalcy for each individual unit. An observed trend away from the norm is the signal that a specific failure or class of failure is imminent.

Bruneau (183) gives an early report of the application of trend analysis to jet engines. His report contains a list of specific incipient failures that can be identified on the basis of this procedure. Another successful operational program is the spectrometric oil analysis program reported by Bond and Ward (93, 94). This program monitors the wear of oil-wetted parts and is used extensively on jet engines and helicopter transmissions. A review of various other approaches appears in ref. 184.

7.6.12   Design Reviews

Design reviews are conducted on various management and technical levels for a multitude of purposes. Some of the more important types of reviews are as follows:

Contractual milestone reviews

Management visibility reviews

Technical assurance reviews

Analytical reviews

Each of these reviews is discussed briefly in a cursory manner in the following sections. For a more detailed discussion see NASA SP-6502 (95).

The success of a design review depends to a great extent on how well the participants are prepared. It should therefore be required that all pertinent documentation be distributed to all participants in a timely manner before the review, that is at least 10 days, to permit them to become thoroughly familiar with the subject matter to be discussed.

Contractual Milestone Reviews

Most contracts for advanced systems development provide for formal design reviews at prescribed milestones for the purpose of determining the extent to which the contractor is complying with contractual requirements. The contract will specify that specific data items be delivered or certain hardware items be completed by specific milestone dates.

In Section 3.1.4 there is a discussion of the difficulties in resolving conflicts in requirements, in balancing requirements against the state of the art, and the tendency to overspecify requirements, in particular in relation to available funding. As these problems impact on the development process, they will surface at the design reviews. The design review attempts to resolve as many of these problems as possible; objectives may be redefined and waivers may be granted (see Figure 7.1). The new agreements are recorded in the minutes of the design review meeting. Unresolved problems are written up as "action items" for further consideration by either customer or contractor. It has been noted that the contract is a dynamic instrument, subject to change. The design review board normally does not have the authority to change contractual commitments; when action items require contract changes, they are submitted to customer and contractor managements where they are reviewed and contractual changes negotiated as may be indicated.

Reference 95 lists and discusses the following milestone reviews:

Preliminary design reviews at system, subsystem, and component levels.

Prepackaging reviews for electrical, electromechanical, and similar critical components.

Prerelease review for all components prior to manufacturing, also at subsystem level.

Acceptance reviews for all levels.

Special purpose reviews and buy-off (acceptance data package).

This reference also lists the detailed objectives of a design review, as may be appropriate to a particular milestone, as follows:

To evaluate the capability of the design to meet the total system requirements.

To determine the effects of procurement, assembly, test, shipping, storage, human factors, and maintenance on the achievement of the design goals.

To identify problems of process control, production, and parts and material procurement.

To consider the effects on performance of proposed configuration changes.

To evaluate the operational functions of the design with respect to the current known state of the art.

To evaluate the adequacy of the specification to meet the intended operational use.

To determine whether the design conforms with the specification requirements.

To provide for optimization of design within the functional performance and reliability requirements, schedules, and funding.

Management Visibility Reviews

Another class of design reviews is organized to provide management visibility. A briefing is prepared and presented to customer or contractor management, or to both, to acquaint them with the progress being made on the program. Such briefings should be based on complete staff work. Any technical problems and the associated risks should be clearly, completely, and honestly represented, in as much detail as necessary to develop a full management understanding. Proposed solutions, in particular with respect to reallocation of resources or the need for new resources should be clearly indicated.

The briefing should leave management with an accurate concept of program status, of the probability of success, of what the problems are, and a clear set of recommendations as may be needed to solve existing problems.

Normally no management decisions are made at such briefings, but they should provide management with the background and required data needed for top management to take corrective or supportive action as may be indicated.

Technical Assurance Reviews

Technical assurance reviews are popular devices for assuring management that the technical aspects of a project are being conducted in a competent manner. Reference 95 states that to satisfy this objective the review team must have a collective technical competency greater than that of the designers of the project being reviewed, and must have the confidence of the management group to whom it reports.

Given an experienced and competent design team, it is very difficult to organize a design review team that has greater technical competence than the design team responsible for the project to be reviewed. In fact, a good experienced design team can always favorably impress any design review committee to any extent desired. Generally, an alert management is well aware of this situation and this type of design review is kept to a minimum, if held at all. The major requirement for a design review team in this situation is that it enjoy management confidence.

However, in some areas of advanced system design, the engineering disciplines involved are too new, and most contractors have not had the time to develop mature and experienced design teams. In these cases, independent design review committees ("red teams") are indispensable for providing management with the assurance that the project is progressing in a competent manner.

Members of the design review committee should be technical experts not connected with the project being reviewed. Reference 95 suggests that the committee be made up of various combinations of the following:

Permanent reviewers (to provide continuity).

Recruited specialists.

Staff or consultant authorities.

Analytical Reviews

For providing technical assurance, by far the most indispensible, powerful, and effective design reviews are the contractually required analytical reviews provided by the systems engineering staff assigned to a design project. These analytical reviews are prepared and coordinated under the supervision of the engineering project management; are documented in technical report format; and are normally subject to customer review and approval, usually at or prior to contractually specified milestones. The effectiveness of this system depends to a great extent on the customer's in-house technical capability to properly review the technical reports submitted and to influence design trade-offs in a timely manner to the customer's best advantage. Emphasis is placed on "in a timely manner" because of the excessive costs always associated with customer required changes after the contractor has proceeded beyond a certain point.

The review process consists of the analyst reviewing a proposed design as soon as it is formulated by applying engineering or statistical models as provided for in approved handbooks or manuals to predict the performance of the proposed design (see Figure 6.5). The engineering or statistical models used must be of the type that yield quick and timely results. The time available to formulate the design is limited; any analysis that requires the designer to wait for results will simply be ignored. The analytical predictions are compared with the corresponding requirements. If the requirements are met, the design will proceed as planned. If the requirements are not met, the proposed design is modified or the methods of analysis are refined until the predictions agree with the requirements.

If the prediction cannot be made to agree with the corresponding requirement within the time limits and with the resources available, the project engineering management is alerted and must decide either:

To allocate more resources and time to the design and analytical effort.

To direct that another design or analytical approach be found.

To ignore the analytical prediction and accept the risk that the design will pass the qualification test anyway.

To get the customer to modify the requirement or grant a waiver.

To accept the penalty for not meeting the requirement.

A simple illustration of this type of analytical design review is given for the instance of weight control in Section 6.6.

These analytical reports provide evidence, before the prototype is built and tested, that the proposed design will comply with requirements. To assure that the customer will agree with the analytical methods employed and will approve the reports in a timely manner when they are submitted for approval, the analysts are usually in continuous contact with their counterparts in the customer's organization. In a well-run engineering design organization, every attempt is made to resolve all possible conflicts with the customer before a technical report is submitted for approval. If a contractor is having trouble meeting a requirement, it is normally to his advantage to attempt to resolve the difficulty at the lowest possible level as soon as possible. The customer also wants the project to succeed, and the sooner he becomes aware of potential trouble, the better he can cope with the problem and attempt to find a solution within his own house.

7.6.13 Quality Assurance in Design and Development

Quality assurance is normally thought of as an after-the-fact activity and is discussed in greater detail in Chapter 10. However, there are benefits to be gained by including representatives of the quality assurance organization in the design and development cycle and making them part of the design review activities (128, 129).

The success of this operation depends to a great extent on the selection of the proper individual. He must have had extensive experience in production, and he must be able to communicate easily with design engineers, gain their confidence, and become accepted as a member of the team.

Provide User Viewpoint

The prime product of the engineering department is documentation: drawings and specifications. As a routine matter, the quality engineer should review this output from the "user" viewpoint to assure that drawings and specifications are understandable and not subject to misinterpretation, that the product can be measured and evaluated as anticipated by the design engineer, and that the accept/reject criteria are clear and enforceable.

This activity will associate the quality assurance representative with his major tasks in advanced systems development. He can inform designers of pitfalls that can occur in production, which they may have overlooked. In Section 2.4 it is stated that acceptance of the prototype is contingent on its passing the qualification test. The accepted prototype is then the standard for follow-on production items, which must be equal or superior to the prototype with respect to all specified characteristics (Section 6.10).

Acceptance Test Coordination

In practice, production parts are built from and inspected to engineering drawings and specifications; they are not compared to the physical prototype parts. The actual dimensions of the prototype parts, however, will vary from the nominal drawing values within the allowable tolerances.

On critical parts, it may be necessary to determine the allowable differences between the production parts and the part that passed the qualification test. Tolerances must be specified to assure that the production parts are sufficiently similar to the prototype parts so that system performance will not be degraded. Acceptance tests must be designed to specify a minimum of measurements that must be taken to assure this similarity. In cases of complex equipment, it may be necessary to repeat the complete qualification test on the first production article. Sometimes, one unit will be chosen at random for a full or partial qualification test from each successive lot of possibly 10, 25, 100, or more units. In cases where destruction tests are required, statistical acceptance test plans must be designed. Statistical plans are also required for parts or elements produced in large quantities.

Quality Versus Cost

Since the design cannot be changed once the prototype is qualified, except at high cost and at least partial requalification, many of the tolerance and acceptance test determinations should be made during the development stage. To minimize his risks, the designer normally tends to specify quality characteristics (precision, tolerances, finishes, materials, processes, tests) that may be more costly than necessary. The quality assurance representative should be prepared to advise the designer on two following specific items:

1. The minimum quality levels required to achieve the specified or desired performance.
2. The cost associated with achieving various quality levels.

Much of this information on minimum quality versus performance for routine design should be available to designers in the form of company handbooks; however, in complex advanced systems there will be new conditions arising where the designer will require expert assistance. Furthermore, the quality assurance representative should assure that the available handbooks are up-to-date, containing the latest information developed by the quality assurance organization.

Review of Manufacturing Plans

Section 5.7 discusses the manufacturing plan and its impact on design. The quality assurance representative should interface with the production planning engineers in advising designers on achieving adequate quality at minimum cost. In particular, he should review the manufacturing plan to assure that the planned facilities, especially those of proposed subcontractors and vendors, are adequate to provide the required quality.

7.7 EFFECTIVENESS OF CURRENT DESIGN ASSURANCE TECHNIQUES

Current methods of reliability control have certain characteristics that limit their applicability to certain specific problem areas. This leaves other phases of advanced systems development without effective reliability control. In particular, no convenient, quick method exists for measuring the achieved reliability of large complex advanced systems and subsystems before initial delivery within the dynamic demonstration loop of Figure 2.3.

7.7.1 Application of Mathematical Techniques

From an engineering viewpoint, mathematical techniques are useful only to the extent that their predictions coincide with the subsequent actual behavior of the equipment under service conditions. Mathematical models provide the basis for design control to assure that a proposed design, when built, will pass the prescribed tests to demonstrate that it meets the requirements.

Technical Integrity

Mathematical models also supply the key to an understanding of the functional characteristics of equipment as necessary to assure *technical integrity*. Given an operational system, the various parameters representing the operating conditions can be plotted on one or more related diagrams. There are many innovative and sophisticated methods for presenting operating conditions in a graphical display. The extreme or limiting values can be identified for each anticipated operating condition. These limiting values can be interconnected, thus forming an envelope or tent that encloses all operating conditions.

In any large system it is not economically feasible, or technically practical, to test the product with any degree of statistical confidence in all of the anticipated critical operating conditions defined in this manner. Hence, the evidence that the equipment will operate satisfactorily in all conditions in the envelope or tent is furnished by engineering analyses, sometimes supported by simulation studies. These analyses refer to the laws of physics and chemistry as applied with a technical understanding of how and why the equipment works, and they provide the rationale and justification for all design decisions.

Given these engineering analyses and a knowledge of his resources, facilities, and of the criticality of the product, the test engineer selects a number of test points (conditions) that in his judgment are most representative of the operational envelope or tent. Most of these selected points represent limit conditions, but some may lie well within the boundaries of the envelope or tent.

The major objective of these tests, and this is the hidden key underlying the engineering approach, is to demonstrate the validity of the engineering analyses. If the test results confirm the analytical predictions at the selected test points, it is then assumed with some degree of confidence that the analysis at all other untested points is equally valid. The confidence in this assumption depends on the quality of the analysis, the number of test points, and the degree of agreement between test results and analytical predictions.

In some cases it is not feasible or possible to demonstrate the analytical predictions and the test engineer must back off; he may design tests to verify critical assumptions or, as a last resort, to provide evidence that the analysis is not incorrect.

The term "confidence" in this context is based to a large extent on professional engineering judgment; it represents *technical integrity* and is quite a different concept from statistical confidence.

Control of Unit-to-Unit Variation

Section 7.3 lists unit-to-unit variation as a major contributing factor to product unreliability.

Whenever existing analytical techniques are capable of predicting the performance of a product, unit-to-unit variation need not be a problem, provided that equally effective procedures are used in material procurement, manufacturing control, and inspection. Since requirements, analyses, and test correlations exist, standard quality control procedures, intelligently applied, are adequate to control unit-to-unit variation.

However, in the case of components subject to random failures, acceptance criteria cannot always be established to take full advantage of quality control procedures. No adequate models exist from which all these criteria can be derived.

Furthermore, the classical statistical reliability theory, based on the exponential distribution, does not take unit-to-unit variation into consideration. In reliability demonstration testing, a test without failure of 10 units for 100 hours each, or 1000 unit hours, is the equivalent of 100 units at 10 hours each, also 1000 unit hours. This is not a realistic measurement of the parameters of interest in many special components.

7.7.2 Application of Engineering Techniques

Engineering procedures apply experience, judgment, and intuition to supplement formal reasoning to provide approximate working solutions to immediate problems.

Within the dynamic control loop concept of Figure 6.5, the skilled application of the engineering assurance techniques of this chapter can be quite effective in assisting the designer in formulating the component and system configuration to minimize the effects of random failures when they occur in field service.

These techniques include configuration analysis, the use of redundancy and fail-safe designs, failure effect and mode analysis, fault-tree analysis, multimodal analysis, and similar procedures. The associated mathematical models, which refer to the quantitative reliability and other systems assurance requirements and which utilize applicable, valid failure rates, serve as useful guides in trade-off studies to determine how much redundancy, margins of safety, maintainability, and other features to provide and where to include safety provisions.

Several space programs have demonstrated capability to design and produce complex new advanced systems of very high reliability. The most intensive reliability programs have been conducted on our space programs up to and including Apollo, and they have achieved a degree of success. It is doubtful that a reliability program on any other project can yield better results. It is therefore instructive to examine the Apollo program and regard it as a baseline representing the best that can be done. Now that Apollo has demonstrated the feasibility of space flight, it is doubtful that equivalent resources will ever be made available again for such a complete reliability program.

Achieved Reliability

In actual space flight, the Apollo Spacecraft exceeded expectations, as has the Orbiting Astronomical Observatory (OAO) (130). However, without the capability to measure the achieved reliability during development, these systems were built as well as it was possible to build them with available resources; they were then launched with a prayer for success. At the time of the first launch, a scientific assessment of the probability of success was not available.

The Shillelagh Missile is another example of good reliability, achieved through the judicious use of engineering design techniques and tests during the development cycle (182).

An overall evaluation of the various procedures discussed above is provided by the review of the Mercury project reliability program by French and Bailey (96), which includes the following conclusions, and which apply equally well to the Apollo project:

An effort was initiated in the Mercury Project to make a quantitative reliability assessment and obtain an overall estimate of mission success and flight safety based on test time and failures that took place during the ground test program. The estimate of the reliability of the Mercury spacecraft utilized mathematical models of the subsystems together with failure rate data derived from actual test experience on the system parts and components.

In general, the results were not satisfactory because the applicability of the failure rate data was always highly debatable. It was a basic ground rule of the approach to manned space flight that a failure during development and preflight tests always resulted in a corrective action designed to eliminate all possibility of repetition of that particular type of failure. Hence, past failure data never applied directly to the then-current articles.

The first Mercury space flights with new systems could not be delayed pending statistically rigorous reliability tests to assure demonstration of reliability goals. The problem was, therefore, to decide, by a combination of engineering judgment, common sense, experience, and intuition, just when the last serious "early development" types of failure had been eliminated . . .

As a result of the experience in the Mercury Project, the role of numerical reliability assessment in manned space programs may be summarized as follows:

1. It is desirable to *specify an overall numerical reliability goal* to insure that adequate attention is directed to reliability in the design stage. This goal should be apportioned or budgeted through a mathematical model down to the various subsystems and their components. The subsystem designer should be required to show that his subsystem is capable of absorbing the expected number of random or statistical type failures of parts without serious consequences or without exceeding his reliability budget.

2. *The logic flow diagrams* which show functionally the systems sequence of action were especially useful since they represented primary and critical abort paths, crew inputs, and principal events. They reflected the basic ground rules relative to choice of alternate modes of operation and aborts. From these diagrams the effect of a component failure could readily be determined.

3. Beyond this point, the usefulness of formal quantitative reliability assessment procedures is debatable: the most effective approach from here on is to concentrate on establishing *a testing program* and *quality assurance program* that will assure detection and correction on all the unproven design and induced sources of system failure before flight.

7.8 CHAPTER SUMMARY

Figure 7.2 shows in simplified form the relationships of the various methods discussed in this and previous chapters with respect to the systems development process. The achievement of quality, reliability, maintainability, integrated logistic support, and safety is influenced by many activities throughout the development process.

Starting on the left side of Figure 7.2, coordinated *reliability and maintainability requirements* are developed from the *system requirements*. Based on the system-layout and design, the system reliability and maintainability *requirements are apportioned* among all of the components of the system.

In parallel, and coordinated with any system design activity, running *reliability predictions* are prepared based on the mathematical methods to indicate the degree to which the proposed scheme will satisfy the established requirements. Configuration analysis (that is, trade-off techniques) is employed to assist in the selection of the best configuration.

Throughout the entire equipment design and installation phase, reliability predictions are prepared for every proposed component to assure that the components selected for use in the system satisfy the apportioned requirements. In the failure mode and effect analysis, the details of the design are reviewed to assure its operational integrity, that is, that good engineering practice is followed and that the right kinds and amounts of redundancy are included. All cases where a component failure is likely to cause a system failure are specifically indicated.

Test plans are developed to indicate the procedures to be used in demonstrating that the developed hardware conforms to the analytical predictions and meets all requirements.

Figure 7.2 associates *maintainability* considerations with the equipment design to call special attention to the growing importance of these requirements. In addition to the usual design considerations, maintainability

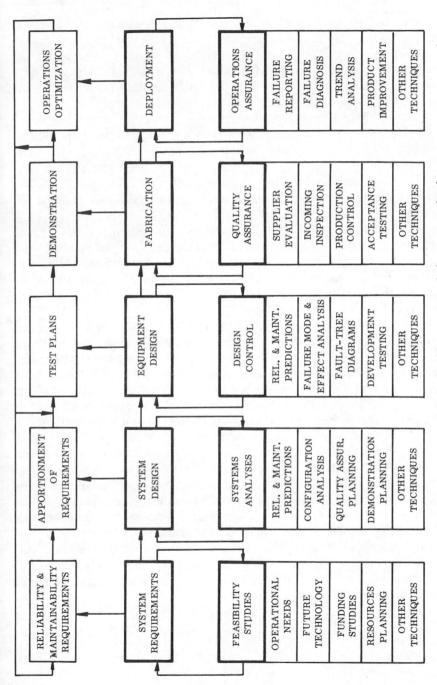

Figure 7.2. Reliability and maintainability assurance during systems development.

231

requires that the operational, support, and logistics plans be considered to assure that the system can be operated, supported, and maintained when deployed as required under the anticipated operating environments. The procedures associated with unscheduled as well as regular maintenance must be outlined, and the interface established with regular ground and special support equipment. Fault-tree diagrams are indicated in Figure 7.2 as an example of the analytical tools available in this area.

Quality assurance procedures are shown associated with the production phase; these procedures provide assurance that the selected manufacturing facilities have the capability of producing a product that will satisfy all requirements (before-the-fact), and that all elements of construction, components, and higher assemblies offered for acceptance will conform with requirements.

Demonstration consists of *qualification and acceptance testing*. Qualification assures that the manufacturer has the capability of meeting all of the requirements of his contract (after-the-fact). Acceptance testing provides assurance that the production components and systems offered for acceptance are at least as good as the specimen that survived the qualification tests.

When complex systems enter the *operational test phase* they are closely monitored, and every attempt is made to control the operating conditions. Operations take place in accordance with established procedures by trained operators, maintenance, and support crews. Trend analysis and ATE provide for the monitoring of certain operational points to detect incipient failures and to provide the operator with greater confidence that his equipment is in satisfactory operating condition.

A *continuous product-improvement program* is maintained on many advanced systems. For example, a spacecraft may not be launched until all previous failures have been adequately diagnosed and corrected to provide assurance that the failure will not recur. Elaborate failure reporting systems, especially during the initial operational phase, assure that all failures are reported and diagnosed properly, and they provide a measure of reliability. Much advanced equipment is monitored by such means throughout its entire service life. *Failure causes* are determined, and this information is *fed back into the design and manufacturing processes,* both to achieve correction of existing equipment and to develop better design guidelines for new equipment.

8

Organization of the Technical Development Staff

The successful development of new advanced systems takes place in an unique environment that determines the organization and character of the technical development staff. In Figure 2.1 the manufacturer of commercial consumer products is shown to assume the entire risk for the development, based on his market analysis. This business environment determines how he organizes his technical development staff to best achieve his objectives.

Design to Meet Customer Requirements

This book is concerned with the development of new advanced systems where the financial risk is too great to be assumed by a manufacturer. The customer assumes the financial risk. The requirements are set by the customer, and the system is developed to meet the customer's requirements. The customer monitors the development to protect his investment. Both customer and contractor must subscribe to the same objective of developing a satisfactory system; they are both part of the same development team, although each has different responsibilities. This relationship presents an environment for the operation of a technical staff different from the one that exists in the development of commercial products.

This advanced systems development environment is not easy to understand for those who have not been closely associated with the process. Manufacturers of consumer products who sell their products on an open market have a constant incentive to reduce their costs and increase their profits. For them it may be a matter of survival to make their product as inexpensive as possible.

Commercial Procedures

Commercial consumer products are sold with a warranty, and the process is closely monitored by the marketing department. If too many units are returned in an unsatisfactory condition, the quality must be improved. If none is returned the quality is excessive, and the product is redesigned to reduce the quality. Hence mature commercial products are designed, often over a long period of time by trial and error, to survive marginally in their operating environment. The manufacturer of consumer products lays the greatest value on achieving the minimum quality required to meet competitive market expectations (which may change with time), and his technical staff has expert knowledge regarding the costs involved in attaining various quality levels.

Cost Impact of Design Objectives

This major concern with cost naturally results in products less expensive than those designed by engineers whose major concern is meeting new advanced requirements. Military authorities occasionally express concern that the cost of military products is higher than that of comparable commercial equipment and forget that they should think in terms of life cycle costs rather than initial costs. From time to time military officials succumb to the temptation and buy commercial products on the open market for military field use. Normally they insert "requirements" in such purchase contracts. Manufacturers of commercial equipment generally do not understand the meaning of the term "requirement" as it is used in this book. The military field environment is invariably more severe than the commercial environment, and the commercial equipment procured in this manner seldom survives as expected in the military field environment.

To produce a satisfactory advanced system, a contractor must have a set of objectives different from his commercial counterpart, and his technical staff will be organized to support his objectives.

As an example, in the following sections the objectives and technical staff organization in the military aerospace industry are reviewed and presented as a baseline, since this industry has been a leader in developing advance systems in partnership with its government customers.

8.1 AEROSPACE INDUSTRY OBJECTIVES

To be effective in any organization, management personnel must support the objectives of the organization. Because of its traditions and the nature of its products, the aerospace industry has developed a set of objectives that are uniquely characteristic of itself and which are relatively uniform throughout the industry.

The leadership of the industry has its roots in a common heritage created by a small band of dedicated aviation pioneers who only a generation ago were all struggling with the same problems. The industry has been in a good position to maintain its traditions because of the relatively small number of companies involved. The Aerospace Industries Association consists of less than 100 member companies. Although these companies employed at one time more manpower than any other industry group in the country, the top management group is small.

It is also characteristic of the aerospace industry that most technical advancements are first developed for use on military products or under government contract; only after the advancements have demonstrated their merits in military service or under government sponsorship are they adapted for commercial use. This practice results from the fact that commercial operators cannot assume the financial risks inherent in introducing improved, advanced products into commercial service, except on a very limited basis.

The military, on the other hand, is responsive to a different environment established by potential threats; it cannot afford to lag technically and must strive constantly for the most advanced weapons systems, since the security of the nation is at stake.

In this environment, the average prime aerospace contractor subscribes to a set of corporate objectives similar to those described in the following sections, listed in four categories: general, technical, business, and social:

8.1.1 General

To participate prominently, as a developer and supplier of top-performance, aerospace systems and related products, in the defense of the Western World, in the exploration of space, the geophysical aspects of the earth, the deep sea, and in similar pioneering endeavors requiring the application of the highest technical skill.

To supply the best available products or services within the scope of corporate activities.

8.1.2 Technical

To provide first-line, top-performance air-weapons systems to the U.S. Armed Services.

To provide first-line, top-performance air-weapons systems to allied military services, in conformance with national policy.

To provide first-line, top-performance spacecraft systems to NASA.

To provide superior commercial and general aviation systems to sophisticated customers (normally concerned with life cycle costs).

To provide top-performance marine systems based on advanced ocean engineering developed as an outgrowth of aerospace systems technology.

To supply on a commercial basis other products and services developed as a fallout of the special skills and technology originating as a result of aerospace systems technology.

To provide research facilities available on a contract basis in such areas as may further the development of advanced aerospace systems and related products.

8.1.3 Business

To be a low-cost producer.

To be in the top profit bracket compared to similar companies.

To maintain broadly diversified product lines.

To maintain a strong financial position to permit the assumption of substantial risks in the development of business opportunities requiring the application of advanced technology.

8.1.4 Social

To provide a good place for people to work.

To be a good neighbor in the communities in which facilities are located.

To participate in programs utilizing the handicapped and under-privileged.

To contribute to the socioeconomic stability of the communities in which facilities are located.

To promote broad utilization of technical advancements.

8.2 STABILITY OF THE TECHNICAL STAFF

One of the key resources of an advanced systems contractor is his technical staff. A forward-looking contractor, therefore, will consider it one of his prime management responsibilities to provide for the continuing development and stability of his technical staff. The financing of this technical staff constitutes a significant problem.

Figure 8.1 shows the ratio of income versus development costs as a function of production quantities. On contracts calling for little or no production, the income will cover costs plus profits. However, if the number of production units is large, the ratio of income versus develop-

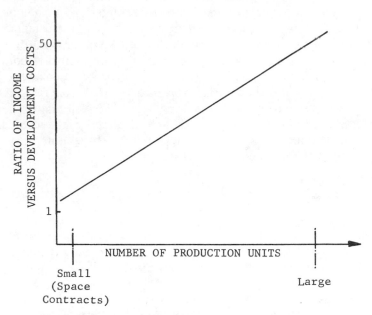

Figure 8.1.   Ratio of income versus development costs as a function of production quantities.

ment costs also will be large, or inversely, the cost of development will be small compared to total income. As an extreme case, consider commercial mass-produced items. The number of production copies is so large, that the development cost per unit is practically insignificant, as, for example, with razor blades.

8.2.1   Unpredictable Workloads

Defense development contracts are not awarded at regular intervals in a manner to foster level employment for the staff of defense contractors. Large production contracts result in a larger ratio of income versus development costs and thereby provide a contractor with the financial resources for maintaining and improving his technical capability until the next contract is won. However, the smaller the production volume, the less will be a contractor's ability to offer the stable employment necessary to maintain his technical staff.

The most severe problems of this kind are associated with space contracts, which involve only very small production lots. Unless a contractor has other suitable work on hand, with the completion of a major space contract he has no choice but to disband his development team.

Some procuring agencies do not look with favor on awarding three successive development contracts to the same contractor. A contractor may do an excellent job on his first contract and be awarded a second. By this time he will know the internal workings and the strengths and weaknesses of the personnel of the procuring agency so well that on the third contract he may be in a position to take full advantage of the technical and management weaknesses of the agency and thereby counterbalance the advantages of the competitive process.

8.2.2 Diversification

The major solution to the problem of stable employment for the technical staff has been "diversification," with contractors soliciting contracts from a variety of government agencies. However, most companies are careful to assure that such contracts always contribute to the technical growth of their staffs.

Diversification has not been sufficient to provide for stable employment of technical personnel, and there are large numbers of engineers who follow the contracts, moving to whatever company is awarded the latest contract. This practice is not too inconvenient in the Los Angeles area where there is a concentration of a number of companies; however, in the rest of the country the industry is geographically widely dispersed, and not everyone will put up with the inconvenience of moving from company to company. As a result, many capable engineers, after an initial trial period, leave the defense field for employment in less glamorous but more stable industries.

8.3 MANAGEMENT OF THE TECHNICAL DEVELOPMENT STAFF

With the great growth in the number of engineers in a contractor's organization and the contractor's need for diversification, the problem of organizing the technical staff has become a matter of top management concern. In a small company organizational deficiencies are compensated for by informal personal communications. Each man knows what the others are doing, and, as a responsible member of the team, he easily coordinates his own efforts to best support the efforts of others.

As the team becomes larger, however, these informal methods of communication gradually break down, and organizational channels of communication must be provided. The effectiveness of the organization now depends on the proper and efficient design of these communications channels. Organizational boundaries, even within a single company, represent communications barriers—in effect, invisible "walls" between groups of people. To permit communications to flow easily back and forth between organized groups, "gates" must be provided in these walls. The number and location of these gates is of critical importance.

8.3.1 Managing Large Numbers of Creative Individuals

It is essential to note that managing creative technical talent is a new problem in our society. Individual major aerospace contractors employ thousands of engineers. Never before in the history of mankind have such tremendous numbers of creative professionals been concentrated in one organization. This is an entirely different problem from organizing the same number of noncreative workers to achieve an objective broken down into coordinated detail tasks, such as building a pyramid, manufacturing an automobile, or operating the postal service, difficult as these tasks may be.

8.3.2 Customer In-House Staffing

The organization of a technical development staff in military or governmental procurement agencies is an equally major problem, although the character of the problem is different. In contrast with the variable size and character of an industrial development staff, the corresponding technical staff of a military or governmental procurement activity remains more or less stable over extended time insofar as its total authorized complement of personnel is concerned. However, the officer rotation system within the military service causes a continuous periodic turnover in military personnel, and there is also considerable unscheduled turnover in civilian employees. Hence, the almost continuous training of some personnel at all levels of the organization is an ever present problem.

The primary technical work of the staffs of the procuring activities falls naturally into five main categories of work as follows:

1. Translating anticipated military needs into technical requirements, preparation and promulgation of invitations to bid, evaluation of proposals, and the preparation of recommendations regarding contractor selections.
2. Obligating manufacturers to realistic, up-to-date requirements.
3. Monitoring and enforcing compliance with technical and program requirements.
4. Assessing the quality, systems effectiveness, reliability, maintainability, life cycle cost, and safety characteristics of delivered systems, and issuance of instructions on usage.
5. Maintenance of acceptable levels of systems effectiveness, reliability, and safety characteristics of systems throughout their periods of use.

So long as the rates of procurement do not change rapidly, as usual in peacetime, the size of the staff of a procurement activity need not increase rapidly to maintain proper awareness of the state of the art. However, the total authorized complement will have its ups and downs in response to the changes in political orientation in the executive and legislative branches of the government. These factors influence the allocations of funds for salaries, for research and development, for personnel training and improvement, and for procurements and operations of defense systems.

The technical specialists on the staff of a procuring activity are exposed over the years to the results of the efforts of their many technical counterparts in the industry, involving many types and models of a variety of advanced systems. This diversified exposure provides a technical specialist with an unique opportunity for growth of expertise. Inevitably, from time to time, a procuring activity will lose some of its most effective and best trained employees to other agencies or to industrial establishments.

8.3.3   Management Operating Mode

The question, therefore, both in industry and in government, of how to organize and manage a large group of creative engineers for maximum effectiveness, is a formidable challenge. In desperation many managements today fall back on practices such as "management by crisis" or "management by indirection." Once assignments are made and resources allotted to a project, the technical managers sit in their offices between design review meetings until subordinates bring them problems. Then, in a crisis atmosphere, they assist in the solution of the problem. Unless something goes wrong, competent group leaders who are within their budget and schedule allowance deal directly with their counterparts in the customer's organization and receive little if any guidance from their own project supervisors. To come to grips with the problem, some basic principles are explored in the following discussion.

8.4 RESPONSIBILITIES OF THE TECHNICAL DEVELOPMENT STAFF

The industry objectives of Section 8.1 establish technical excellence as the keystone of all industry operations. In support of these objectives, the responsibility of the technical development staff might be listed as follows:

1. To design, develop, qualify and supervise the fabrication of products and systems that satisfy corporate objectives and that can be produced on schedule and sold at a profit.
2. To provide technical support to other corporate departments, to subcontractors, and to suppliers, in the discharge of their responsibilities and obligations.
3. To develop and maintain the superior technical excellence required in a competitive environment to comply with corporate expectations.
4. To maintain regular contact with technical customer agencies to promote their forward-looking thinking, to assure that their requirements are feasible and adequate, and that they are aware of the contractor's technical superiority.

8.4.1 Project Versus Functional Units

An advanced systems contractor's need to diversify has been established; as a result his organization will be working simultaneously on a number of projects.

Each customer is concerned that his project receive its share of management attention and be staffed by the best available personnel. Customers encourage contractors to establish projects as independent and self-sufficient organizational units, with all of the skills required to complete the project. Under this plan, an advanced systems corporation would merely provide a holding company umbrella management for a number of independent projects that would have little to do with one another, except that they all report to the same top management and share in their joint successes.

School Health Care

Simon (23) contains considerable discussion of the advantages and disadvantages of the project versus functional relationship. In one typical example the problem of providing health services for school children by the two methods is discussed. In the project method, the school board hires its own physician. On the other hand, utilizing the functional department, the school board requests the health department to provide the necessary services. The problem is to determine which plan provides the best health services for the children.

The school board is educationally oriented; working for the school board, the physician is under the direction of people who have little professional understanding of his work. When a health department provides the services, the physician is professionally associated and supervised by individuals whose primary interest is health and medical matters.

Neither of these arrangements in themselves can be shown to assure better quality of health services. If the school board is financially strong, it will probably hire its own physician. If it does not have the money, or does not have enough children to keep a physician busy for the time available, it will request the health department to provide the necessary health services. The organization, by itself, usually does not affect the quality of the professional services being offered. Whether the physician is supervised by the school board or by the health department will not affect the professional quality of the services he provides for the children.

This example, which is typical of the discussions in the literature, does not come to grips with the basic problems of an advanced systems contractor in providing technical excellence in a dynamic, advancing technological field. These problems involve extending the state of the art as shown by the dashed line of Figure 2.2; this is a function that is not typical of other industries. This point is illustrated by means of the two following extreme examples.

Company Carpenter Shops

Consider, as one extreme case, a discipline that seeks no appreciable growth in the state of the art. In a company with a number of plants it may be feasible to have an individual carpenter shop in each plant, provided that each plant generates enough work to keep the shop busy. The rate of advances in the state of the art in carpentry is so slow, that all shops will have the latest technology available to them. The individual carpenter shops need not even be aware of each other's existence in the same organization, yet they can all provide equally good service. Economic considerations will be the basis for organizing carpenter service and determining the best degree of centralization. A certain amount of competition between the various shops may be healthy and inspire all shops to exceed themselves in quality and quantity of their production. The development of new technology is not involved.

Intelligence Operations

At the other extreme, consider the intelligence operation of a front-line military command. Success and survival may depend on the quality (accuracy and completeness) of the intelligence. Here the cost of gathering critical elements of intelligence immediately becomes secondary. A strong centralized center is indispensible. Unrelated bits and pieces of individually uninterpretable information are gathered from many sources. They are collated with all other available information in a systematic search for new meanings. Much of the information gathered will be redundant, misleading, or useless; but the intelligence agency cannot afford to take a chance and restrict its information gathering activities. There must be assurance that every human effort has been made to collect all available information, and that all of this has been processed. The emphasis is on completeness of information. If two or more centers existed in competition under the same command, the decision maker would have no assurance that one center did not have more information than the other. In these circumstances competition should not exist between centers—only between opposing military commands.

This situation is recognized in some commercial areas where there is competition on advanced concepts. In the field of high fashion, each house has its central intelligence service. Automobile companies act much the same way with respect to the styling of their annual models.

Functional Design Units

A similar condition exists in the design and development of a new advanced systems involving extensions of the state of the art. To beat the competition, a contractor's capability in key technical specialties must be better than that of his competitors. There is a basic difference between a satisfactory carpentry job and the development of an advanced system that extends the state of the art further than any competitor is capable of doing. It is essential that the manager of an advanced systems corporation understand the difference and that he adopt a policy consistent with the business that he is in. An advanced systems contractor must be vitally concerned that the capabilities of his functional units in each key technical specialty are superior to those of his competitors.

In this environment, there must be only one functional unit for each key technical specialty in a contractor's organization, and each functional unit must have the full and active support of top management in aggressively maintaining itself as the most advanced center of technology in its specific area. Within one corporation there should be no distracting competition between centers. All energies should be concentrated on improving this technical capability of the functional unit for genuine competition with comparable functional units in competing firms. This is a prime responsibility of a functional unit. Every piece of new technical information must be carefully acquired and integrated with existing technology so that it can be exploited to the maximum benefit of the corporation and its customers. Such new pieces of information are scarce, and a contractor cannot afford to let his functional units miss any opportunities to acquire such new information.

8.4.2   Operations of a Functional Unit

Engineering is a predictive science and has certain features in common with all other such predictive sciences, including biology and economics.

The work of the technical specialist consists of collecting, collating, and analyzing data on past experiences; organizing this material for the purpose of generating models, generalizations, or procedures from which predictions can be made; and testing the accuracy of these predictions in suitable applications. The measured accuracy of the prediction is an indication of the usefulness of the model and the success of the specialist.

The fun, the intellectual challenge, lies in the ability to make an accurate prediction, as in the analysis loop of Figure 6.5. The development of this capability is the thing that inspires the specialist; he wants above all to improve his predictive capability. The technical specialist always has one foot in the past, one in the present, and one in the future. He knows that to do the kind of a job he would like to do he also needs three hands.

In support of this effort of developing an even better predictive capability, the specialist needs to check himself continuously; he needs some means to measure his progress. He needs problems to work on to test his ever improving capabilities. For this purpose, the corporation supplies him with "current projects."

Current Projects

Current projects are welcomed by functional units as opportunities for the application of advanced theories that have been or are being developed, in order to test their effectiveness. The procedure is illustrated in Figure 8.2. The model represented by the center block is designed on the basis of past experience. Given project requirement(s) and the necessary resources, the model controls the development of the configuration of the project to

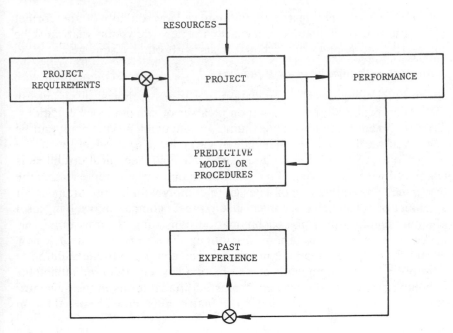

Figure 8.2. Project design.

achieve the required performance. The achieved performance is compared with the requirements; this comparison adds to the treasury of past experience and provides a basis for improving the model.

The output of interest to the technical specialist is the accuracy with which performance has been predicted. The viewpoint of technical personnel is fundamentally different from that of business personnel. It is essential that the manager understand what constitutes the intellectual challenge that inspires and motivates his technical staff.

8.4.3 Number of Functional Units

A critical management problem is the determination of the number of functional units in a contractor's organization.

A functional unit should represent a basic technical discipline constituting a unique viewpoint that affects all phases of design. For example, an airplane can be designed on the basis of aerodynamic, structural, and functional considerations only, without any conscious thought being given to weight control. If weight control is then introduced as a new formal discipline, it will enter as a new consideration into every design decision. This is not just an additional review on top of the former application of previous disciplines; it is a new viewpoint that must be integrated and applied on an equal basis with other disciplines.

Analogy with Simultaneous Equations

Breen (203) compares the application of various technical disciplines in the design process with the solution of a set of simultaneous equations. Each technical discipline representing an independent viewpoint can be compared with a single independent variable. The configuring of the design to comply with the requirements of a single technical discipline is compared with the solving of an equation in one variable representing the discipline. The configuring of a design to comply with the requirements of a number of independent technical disciplines is compared to solving a set of simultaneous equations, each equation consisting of terms in all variables representing the various disciplines. The introduction of a new discipline to the development process can be compared to the addition to a set of existing simultaneous equations of a new equation for solving for an unknown previously neglected, plus additional terms in the new variable to all the previous equations. The matrix must now be solved for an additional unknown.

Adding a New Function

This comparison is useful for calculating the impact of adding a new functional unit to a technical development staff. A single equation can be solved in some minimum number of operations. The solution of two equations with two unknowns requires more than twice the number of operations required to solve a single equation.

Consider a matrix of 10 simultaneous, linear equations with 10 unknowns. To solve this matrix requires some minimum number of operations. Add an eleventh equation with an eleventh unknown. The number of operations needed to solve this new matrix increases by at least 30%, although the change in requirements is only 10%.

Consider a technical development staff consisting of 10 functional units representing basic disciplines. Add a new functional unit representing a new basic discipline, such as reliability or maintainability. This analogy indicates that the engineering labor required in the technical development staff to take effective advantage of the contributions of the new discipline increases by a minimum of 30%.

Minimizing the Number of Functional Units

The obvious conclusion is that the number of functional units in a contractor's organization should be kept to an absolute minimum. No new functional units should be added except for basic technical reasons to cope with essential, new requirements such as the addition of a unit specializing in the delivery of atomic weapons or in escaping from the associated blast and radiation effects. The interactions between functional units is a subject that requires a great deal of management thought. In most companies there is considerable overlap between existing functional units. Sometimes this overlap is cleverly disguised. Nevertheless, in most engineering organizations the functional units do not always truly represent independent technical viewpoints. Efficiency could be increased by combining such functional units. Overlapping jurisdictions establish barriers and impose artificial constraints on the members of a technical staff. Management should be able to pick the brains of its technical people; to do this effectively, only those constraints may be imposed that are necessary to synchronize outputs which must go together in an orderly way.

The author, for example, sees no reason why all design oriented techniques based on the statistical analysis of operational data cannot be embraced by one functional unit, regardless of whether the techniques are called reliability, systems safety, maintainability, systems assurance, or what-not.

8.5   PROJECT RESPONSIBILITIES

The contractor's project manager *is the contractor* to the customer. He receives the specifications and funds from the customer, and in turn he is responsible for the timely delivery to the customer of a satisfactory product, associated services and other software, within the allotted resources.

To support him in fulfilling this responsibility, the resources of the corporation and its associated subcontractors and suppliers are at his disposal, subject to the needs of other project managers in the corporation. The contractor's program manager must organize these resources for maximum effectiveness.

Customers Prefer Project Organization

Many procuring agencies encourage corporation project managers to establish independent and self-sufficient organizations, shown in simplified form in Figure 8.3. All necessary functional units are under the exclusive direction of the project manager. The manpower in these functional units may be supplied by corporate parent units, which then often deteriorate into job-shopping operations, or personnel may be hired directly from outside of the corporation.

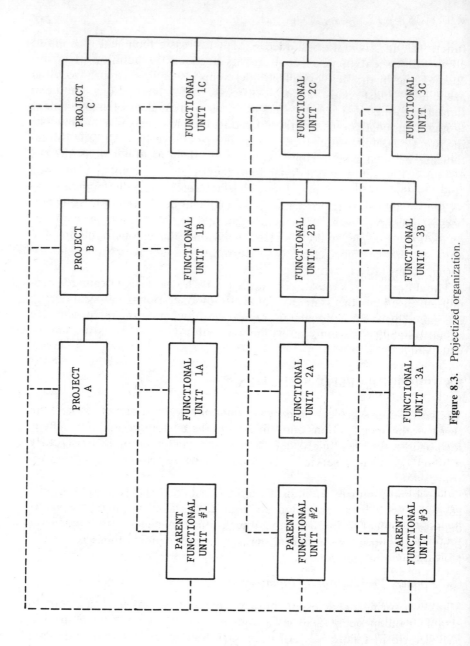

Figure 8.3. Projectized organization.

Customers often feel that in this type of organization they can better hold top men on the project. In some cases government customers will not permit the release of advanced techniques for use on projects of other agencies because of competition between agencies.

In the author's experience, top men are denied a full opportunity to grow in this isolation, and often they become frustrated. Project loyalties are developed that are detrimental to the customer and the corporation. The author has found that in strongly projectized companies, project personnel in need of technical information and data on advanced subjects find it easier to obtain this information from professional associates in competing corporations than from other projects in the same company where the information originated. The poor exhange of information between projects invariably results in each project "reinventing the wheel" over and over again.

Use of Functional Organization

The author has observed the operation of strongly projectized development programs both in government and in industry, and for the development of advanced systems, recommends consideration of the functional organization of Figure 8.4. Project managers assign the responsibility in various technical areas to the head of the parent functional unit. It then is the responsibility of the head of the parent functional unit to provide each project with the necessary services to satisfy the contractual requirements and also to coordinate his activities with those of other functional units. The project manager maintains cognizance over all activities on his project, but his authority is exercised through the heads of the functional units.

In this scheme, technical and business responsibilities are in better balance than in the scheme of Figure 8.3, and this permits the exercise of greater creativity on the part of the two groups involved. Specifically, the head of a functional unit is technically oriented and is responsible for technical quality. The project manager is business oriented and is responsible for technical adequacy and for profit. The project is a profit center while the functional unit is a cost center.

In Figure 8.2 it is pointed out that the models used in the design and analytical development loop of Figure 6.5 are based on past experience. In technically advanced developments in which the state of the art is being extended, it is essential that the models be based on all available information. The author prefers the organizational scheme of Figure 8.4 over that of Figure 8.3, because it facilitates the exchange of current experiences through the parent functional unit. The scope of work is

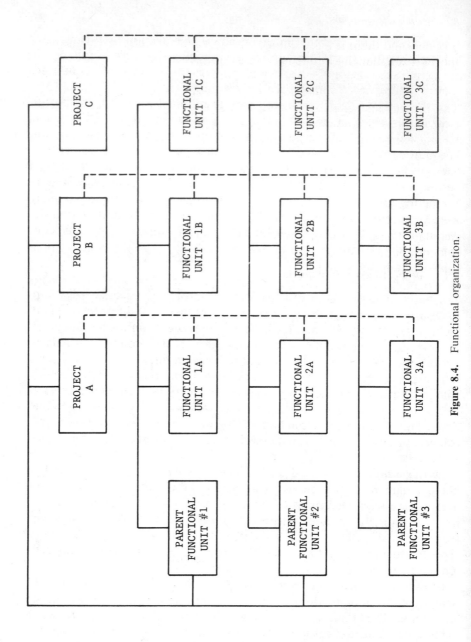

Figure 8.4. Functional organization.

broader, and there is more opportunity to keep top designers working on the most challenging problems across project lines.

Evaluating Functional Units and Subcontractors

The project manager also has a basic responsibility for evaluating the contributions of the various functional units and subcontractors. In any large organization, some units will be stronger than others. However, the strength of the total organization can be hampered by its weak links. Therefore, project managers must assist technical management in establishing a standard of technical excellence that each individual functional unit must strive to maintain and exceed; and they must monitor the performance of the functional units on their projects to assure that they do maintain these standards. The standards must be clear and universally understood, and those functional units exceeding the standards should be appropriately rewarded.

Where functional units are weak, effective action must be taken to reinforce their technical capability and help them meet the minimum standards of excellence.

8.5.1   Technical Versus Business Conflicts

The balance between technical quality, on one hand, and technical adequacy and profit, on the other hand, is most visible in the process that takes place on the lowest level design board.

Responsibility for profit on a development program is an exceedingly complex subject. The performance of certain tasks early in the development phase by high-priced professionals can pay off handsomely downstream in savings of lower-priced labor. For example, early, well thought-out design decisions can have a significant effect on such characteristics as producibility in downstream manufacturing operations.

Creative Design Engineers

The creative design engineer is dedicated to developing the best possible product with all of the resources that he has at his disposal. On a new job a creative design engineer never has enough time to do the kind of job he would like to do, because as soon as he formulates a concept, he gets a better idea. A definition of a good creative design engineer might be one who never runs out of new ideas.

There are few men in this category of creative engineer. Nevertheless, these few creative engineers are the ones who are responsible for progress and for the development of products that are superior to those of the contractor's competitors. The manner in which the creative designer is managed is, therefore, of critical importance to the success of the project and the future of the corporation. His creativeness must be stimulated and not stifled. This is difficult to do in a larger organization where the special treatment of individuals runs counter to administrative procedures which, for convenience, are designed to deal with classes of employees—where each man is a number.

Business Oriented Project Managers

In contrast to the design engineer, the job of the business oriented project manager is to determine needs, to allocate resources, and to motivate people in a manner that will result in the highest overall profit to the company. This function may be executed as follows.

Say that a good creative design engineer is allotted 10,000 hours to produce a new design. At 7000 hours he may have developed a design that appears to be quite satisfactory to the manager, but as a true, creative design engineer, he will not be satisfied with it. He will, therefore, downgrade his accomplishments in his reports, or even attempt to suppress them, and will attempt to use his remaining 3000 hours to evolve either an improved concept or a new and better design. It is most important to realize that the design engineer is not trying to promote his personal advantage. He is not trying to deliver an inferior product and pocket the difference. The facts are just the opposite: all of the resources that he can acquire, even if he has to steal them, are used to improve the quality of the product.

The job of the project manager in this case is to recognize that at 7000 hours his first-line designer already has developed a perfectly acceptable design, and that the company should not spend any more funds trying to improve it. He must recognize this fact regardless of the contrary information that the design group is furnishing him. His proper action, therefore, is to strike a bargain with the head of the functional unit and arrange for the transfer of his first-line designer to a different suitable job, leaving a less creative engineer to complete the odds and ends, and supervise construction.

It is essential that the transfer be arranged by the head of the functional unit in accordance with the scheme of Figure 8.4. The fact must be faced that a creative design engineer has little respect for a business oriented project manager and is much more willing to accept a decision from the technically oriented head of his functional unit.

A project manager also must be alert to the opposite situation. It may turn out to be impossible for the best available creative designer to develop an acceptable design within the time allocated to him. Again, the responsibility of the manager is to recognize this situation and come to the rescue of the designer. He must make revisions in the budget in a timely manner so that the schedule will be met.

Adversary Viewpoints

The point made here is that the prime responsibility of the creative designer is and must remain one of unrestricted commitment to quality of the design. To assign him to collateral management responsibilities is to inhibit him in his basic talent, which can result only in a decrease of his creative potential and productivity—the most precious resource in the company. The creative designer cannot be made responsible for assuring company profit.

The project manager is the one who is committed to making a profit for the company. He must develop a keen sense of judgment for adequacy of design and for estimating the resources necessary to achieve an adequate or superior design. His is a different personality, and he has different talents than the creative designer.

Distractions

Advanced systems houses are in business to develop and produce the best possible technically advanced systems. This is a very difficult technical job which they want to perform on schedule and within allotted resources. Any activity that detracts them from this goal decreases their capability to meet their basic objectives. The one distraction that advanced systems contractors can do without is legal involvements of any kind. Project managers must therefore be alert in all activities to keep the company safely within those bounds so that investigations and renegotiations will be minimized and other legal distractive side activities avoided. This applies both to company relations with its customers and with its subcontractors and suppliers.

In fact, relationships with subcontractors and suppliers normally are conducted on a highly ethical plane. In the process of source selection, and in evaluating and monitoring subcontractor and supplier performance, contractors learn a great deal about the proprietary processes of individual subcontractors and suppliers. These trade secrets are treated in a confidential manner by the contractors and are carefully protected in compliance with statutory requirements.

Project Organization at the Shop Level

As the work becomes less creative, the advantages and disadvantages of project versus functional units change. For example, on the lowest manufacturing level, a project shop would be wholly dedicated to the success of the project. When special tooling, tolerances, or higher levels of workmanship are required, as is often the case with space hardware, the project shop organizes itself as a matter of course to satisfy the requirements as necessary.

A common corporate shop is motivated by production quotas; all projects are treated alike. A standard of workmanship is established compatible with maximum production capability, and this standard is maintained on all output. Special standards only reduce the production output that is the measure of the shop's efficiency. If special requirements are requested, project personnel are told that that is not the way to do things.

Real World Practices

It is interesting to note that the general tendency is to projectize mainly the creative engineering effort but then to make use of common facilities at the lower organizational levels. In both cases this seems to be the most disadvantageous form of organization for the development of advanced systems.

8.6 MANAGEMENT OF A FUNCTIONAL UNIT

The functional technical unit, as discussed here, is the single center of technology in the corporation for a specific technical specialty. For effective management the number of such functional units should be held to a minimum, and all such units should be of approximately the same size. The technical scope of a unit should cover a recognizable basic discipline founded on a body of historical experience, represented by at least one national professional society, and identified with a specialized set of analytical tools or design and testing skills.

The prime responsibility of a functional unit is to be the best in the business. In line with Figure 8.2, the functional unit collects pertinent data, constructs predictive models on the basis of past experience, and applies these models to the company's projects. The ability to predict accurately the performance of new proposed systems while the system is still in the design stage is the measure of the quality of their service.

It has been mentioned that a contractor is normally required to subcontract a portion of the work on a project to distribute the economic benefits over a greater geographical area. The functional units in a contractor's organization normally do not cover all the skills required to develop all aspects of the system. As a result, the project draws on a number of subcontractors to supplement the contractor's own organization. The contractor must decide what portion of the project will be done in-house and what portion he will subcontract.

8.6.1 Outside Competition

In some large multidivisional (commercial and defense) organizations, the parent company maintains a central engineering department that stands at the disposal of the various divisions. These divisions (profit centers) may be free to call on the central engineering department or on outside consultants for services, or to organize their own divisional engineering departments to provide the needed services. In several cases the central engineering department, offering its services in a competitive environment, has been an outstanding success because it can provide superior services at competitive cost. The major reason for this is its broader experience base acquired by servicing several divisions.

This approach seems to provide a workable model in an advanced system contractor's organization. In determining what portions of a project to subcontract, the project manager should be able to compare the services available from a subcontractor with those available at the same cost from a corporate functional unit. In general, he should be required to make use of his own functional unit only when it is shown that this arrangement results in specific benefits, such as better service for the same or less cost. A businesslike customer-client relationship between project management and functional unit appears to be a healthy and most productive arrangement. Competition between functional units and outside consultants or subcontractors, where feasible, appears to be an effective and stable means for assuring that the functional units maintain their competence and remain abreast of the state of the art.

8.6.2 Model Builders

The contribution of a functional unit is based on application of its predictive models. These models normally cannot be developed while working on a project; project work generally makes use of those models available at the beginning of the project. However, in parallel with a project, the experience being gained should be analyzed and used to improve current models, that is, to extend the state of the art, for use on the next project.

8.6.3 Customer Relations

In maintaining his posture with respect to his customers, that is, the projects, the head of a functional unit must keep himself informed of the requirements that procuring agencies intend to place upon current and future projects. He must be particularly alert and take all necessary action to assure that these requirements make sense. Part of his continuing sales effort must consist of convincing his counterparts in the procuring agencies that he does have the best technical center in his specialty, and in educating them so that they will establish requirements that will result in the best product. Simultaneously, he must maintain or build up his functional unit's technical capability so that it will be able to comply with the requirements better than corresponding units in competing contractor organizations.

8.6.4   Training

Training of personnel is another important task of a functional unit. Experience in a large contractor's organization should not be a personal matter. The experience and know-how should rest in the unit's organization. Experience and policies must be recorded and incorporated into models, written procedures, and manuals, so that a newly hired college graduate (or a man from another company) can receive appropriate guidance to become part of the team in a reasonable time. These models, procedures, and manuals must be flexible to be able to absorb new knowledge and experience as it becomes available. The functional unit must be sensitive to new needs and requirements and must respond quickly and in a technically competent manner.

The strength of a functional unit, nevertheless, rests on its individual technical specialists: designers, analysts, and model builders. There must be management awareness that these men must be provided continuously with challenging work and professional opportunities to foster their continuing growth. It takes decades to develop such individuals, and, once they are available, it is easy to lose them overnight through administrative blunders.

There is also a great deal of administrative and public relations work associated in running a first-line functional unit. Personnel training in this area may become more necessary as a company becomes larger.

8.7   EVALUATION OF PROJECT AND FUNCTIONAL LEADERSHIP

The responsibilities of projects and functional units as profit and cost centers, respectively, has been discussed. The following discussion points out some of the difficulties involved in evaluating performance.

8.7.1   Contractor Project Managers

Project managers are normally evaluated on their ability to get work done on time and within allotted resources. However, there are other factors involved, some of them far-reaching and extremely difficult to evaluate.

Associated with the ability to get work done on time and within allotted resources is the manager's ability to define the job properly in the first place and then get an adequate resources budget to do the job. In a climate of austerity, initial estimates are always optimistic and are prepared on the basis of a "success philosophy." That is, it is assumed that all tasks will be performed correctly the first time and will be on schedule. On an advanced systems development program, this will never be true. The project manager must recognize what additional work is essential to maintain the minimum required technical quality of the project, must sell this concept to the customer, and must find the necessary additional resources to get the work performed in an adequate manner.

Part of the job also includes contract definition. Often contracts are negotiated by a group of people different from those who administer them. After several years of a development program, it often is not clear whether a particular needed analysis is a contractual requirement or not, and whether or not adequate funding is provided for it. The case for clear and complete contract definition cannot be overstated; all work should be completely and accurately defined, and the funding identified, to provide as good a base as possible for further negotiations.

Customer Acceptance

An important measure of project quality is customer acceptance of the system, both in terms of follow-on orders and the customer's inclination to award new contracts to the company. Also, the length of the production cycle and the service-life cycle of individual units are significant parameters of project success.

Company Profits

The project manager's responsibility for company net income has been discussed. In many cases, however, his decisions result in company profits through opportunities for cost reductions offered to downstream activities. For example, the project manager may make certain decisions about expensive hard tooling that may be very unpopular with both company and customer management. Yet these decisions may, several years later when the project manager is off on another job, provide downstream manufacturing departments with the flexibility and opportunity they need to redesign the production procedures to achieve very substantial cost reductions.

Public Relations

In view of the fact that most advanced systems are developed with public funds, it is desirable to promote favorable public interest in the system, although due care must be taken to protect classified defense information, where applicable. The number of favorable articles in the technical and public press and the sales of toy model kits are all measures of the project manager's effectiveness in this area.

8.7.2 Functional Managers

Research Contracts

The responsibility of the head of a functional unit has been described in previous discussions as focusing on the development of the leading center of a specialized area of technology in the industry. In developing new technology, outside research contracts provide an indication of the standing of a functional unit. Normally, most government agencies prefer to grant research funds to universities or research institutes; a contractor's functional unit, therefore, is handicapped in this competition. Nevertheless, to maintain standing as an outstanding center of specialized technology, it is necessary to win a portion of these contracts.

Technical Papers

An associated activity is the presentation and publication of technical papers in the professional press. Technical papers often report on the results of new research or the development of new models, but a contractor's personnel can fulfill a unique function in reporting on the application, usefulness, and accuracy of prediction of new models in the real world of Figure 8.2. The presentation of technical papers provides the leaders of a functional unit with a national forum where they can test their ideas and advanced concepts. It places the contributions of the functional unit on the public record. It informs potential customers of the capabilities of the functional unit. Customers develop greater confidence in the professional opinions of technical personnel when they have observed these men engaged in public debate in a professional forum with their counterparts in industry, government, and education.

Such presentations also have an effect on attracting superior personnel. Top technical men sometimes make up their minds about the desirability of working for a contractor on the basis of the quality of technical papers presented by the contractor's personnel. as well as the general impression made by such personnel during presentations. Technical papers and presentations also have an impact on educators when they evaluate contractors and advise their top students on career opportunities in industry.

Customer Relations

Of prime importance is the professional prestige, standing, and influence of the head of a functional unit in the eyes of the procuring agencies.

Through informal contacts and professional society committees, contractor's personnel can be of great assistance to procuring agencies in preparing forward-looking specification requirements. This is a continuing sales task and is most effective when the customer agency is not thinking about the issuance of a request for proposal. The time to make friends is before they are needed.

The most decisive evaluation of the contributions of a functional unit is provided by customer ratings of proposals. Even though a contractor does not win a proposal, the ratings given the various technical specialties provide guidance concerning where improvements in technical capability are needed or where brochuremanship for later proposals should be improved.

8.8 CUSTOMER IN-HOUSE TECHNICAL CAPABILITY

The customer-contractor interaction in contractual and technical matters has been repeatedly emphasized in several places throughout previous chapters, especially in Figure 6.5. To play his role in an effective manner, the customer must command a certain quality and type of technical talent. This requirement is discussed below for two specific areas: the evaluation of proposals and the monitoring of system development programs.

8.8.1 The Evaluation of Proposals

A proposal is prepared for the purpose of winning a contract, and it is designed to be responsive to those features that government evaluators, by their past actions, have shown that they consider important (186). Past decisions of government evaluators are carefully analyzed to determine the features of a proposal that will most influence the award. If it is brochuremanship that wins a contract, then the competition will be in brochuremanship. If the government evaluators have shown in the past that they are technically competent, the competition will center on technical aspects. The nature and quality of proposals is, in general, a reflection of what is known of the quality of the government evaluation team.

In view of the importance of the proposal-evaluation function, serious thought must be given to the organization and motivation of an evaluation group. The proposal evaluation is a technical function, separate and distinct from management. There must be a clean-cut delegation of responsibility and accountability. If evaluations are not conducted on the highest professional level and are not fair to all concerned, the government is not likely to receive full value for its money, and a lowering of ethics in the advanced systems industry will be inevitable.

Professional Evaluators

To provide leadership over a period of time, government evaluators must draw inspiration from their work. It must be apparent to them that they are making an important contribution, that the government is relying on them, and that their efforts are recognized and appreciated. They should have a feeling of belonging to an important, permanent team that is held responsible and accountable for its decisions. Their work should not end with the evaluation of an individual proposal; the same men should follow a project throughout its service lifetime, monitoring design and performance. They should provide a continuity of effort, with emphasis on experience retention. This continuous feedback of service information is essential to determine the effectiveness of the evaluators, and is also necessary for the discipline of the group. A man's reputation and standing should depend on the success of the projects that he has selected. (The reader should realize that the procedures suggested in this paragraph are not being followed at the time of this writing by any known U.S. Government procurement agency. Nevertheless, this procedure would be an effective way to improve the systems acquisition process. Many other procedures suggested in this book are in the same class.)

Government procuring agencies should encourage the development and retention of groups of such full-time professional evaluators by academic and practical training, by salary incentives, and proper organization, which could foster a strong esprit de corps. A customer will make out much better if he puts his trust in a thoroughly competent and experienced team of professional government-employed evaluators who can be trusted to shoulder the risk involved in initially evaluating the system and then monitoring its progress through the development years, rather than in trying to protect himself with volumes of rules and paperwork.

Competence of Engineering Groups

In every large engineering organization, there will be specialist groups that are less competent than others and who cannot or will not stretch their brand of technology to the ultimate limit. In the aerospace industry, for example, one indirect indicator of the level of technical competence of a specialist group is provided by its capability to estimate the weight of a proposed design, to obtain a realistic weight allowance, and subsequently to comply with the established weight requirement.

It is easy to determine how the weight of various subsystems have grown for past system designs during their development times since contract award. If no basic changes are made in the method of design control, it must be expected that the weight of new designs will grow with time in a similar manner. Given such an analysis on any given subsystem, it is possible to make a relatively realistic weight estimate at the beginning of the design.

A competent specialist group will make effective use of such information. Yet every major aerospace system in the past decades has been overweight. In each project there must have been a few subsystem groups that performed in a substandard manner; otherwise, it must be assumed that the estimated weight of the system was purposely misrepresented in design proposals on which the awards of contracts were based.

The aerodynamic and structural characteristics of aerospace airframes, for example, are critical functions of weight. It follows the flagrant failure to predict properly the total weight of proposed designs, at the time of contract award for the design and construction of the prototype vehicles, can lead to substantial degradation of systems effectiveness, quality, and safety characteristics of production vehicles.

Airframe design personnel, in some instances, have been able to offset the effects of improperly planned or improperly controlled weight growth by superior accomplishments in their technical areas of cognizance. There are limits, however, to the extent to which aerodynamic and structural design characteristics can be stretched to compensate for poorly planned weight requirements and weight growths.

So the evaluation of these advanced systems is not a simple matter. It requires a depth of understanding of the highly specialized technical fields involved, a knowledge of past experience, and an unbiased objective mind. The technical competence required is not to be interpreted in the ivory-tower sense. It includes a full awareness of the cost and time involved in development, production, training, and introduction into service operations. It requires courage based on knowlege and experience to exercise judgment in areas of high technical risk.

Evaluators versus Creators

In this connection it is well to make a sharp distinction between evaluators and creators. Both functions are worthy and significant technical professional activities, and they should not be mixed. A creator is normally biased in favor of his own design and can no longer act as an impartial evaluator. The evaluator requires a different type of cultivated dedication, similar to that of the project manager. The government, as a customer, is interested in evaluation; while creation in a competitive environment is the natural business of industry.

Government Laboratories

A government laboratory, for example, may be either one. It may perform creative work in competition with industry, or it may act as an evaluator. Towl (186) suggests that it should not do both. There are many areas where necessary creative research must be done in government laboratories, or it would never get done; but a government laboratory that is in active competition with industry may find it difficult to be an unbiased evaluator.

8.8.2 Monitoring the Development Process

The monitoring process assumes particular significance as soon as a development deviates in any way from the established plan. This deviation is characteristic of any advanced system development program and normally involves extension of the state of the art and associated design decisions that could not have been foreseen at the time of contract award. Each inclusion of such extensions of the state of the art requires either a contractual amendment or at least some other change in the development plan requiring customer approval. Such contractual changes can be quite extensive. There have been contracts where every single original provision was amended one or more times by the time the prototype was ready for demonstration, even though contractor and customer were highly competent in all respects.

These amendments are invariably associated with changes in funding and delivery schedules. As discussed in Section 2.2.6, a good contractor always has a greater technical capability to do more than the customer can afford; he is constantly preparing proposals for improvements that appear to be desirable. The customer must decide which of these improvements are meritorious and/or essential. In actual fact, the customer often does not have sufficient funds to do those things that are essential to the success of the program, and it is up to the customer's specialist to decide that the item is essential and then find the required funds elsewhere.

Design Information Feedback

First-hand knowledge of other proposals in a functional organization keeps the customer's specialist informed of the latest state of the art and what other contractors hope to achieve on future developments. Visibility into other development programs indicates how competing contractors and suppliers are coping with similar problems. Data from field operations indicate how well operators cope with deficiencies in operational equipment and provide a baseline for deciding whether current requirements can be compromised to any extent.

In fact, service troubles have a particular relevance to design monitoring, as indicated in the outside loop of Figure 6.5. For example, when a structural failure occurs in an operational situation the resulting investigation will include a review of the structural design analyses and test reports that were subject to the customer-approval cycle many years before the service failure occurred. Any lessons resulting from this investigation should be immediately brought to bear both in all the designs currently underway and in the proposals being reviewed. Unfortunately, it is most difficult to introduce changes based on such investigations to existing contracts, especially when additional funding is involved.

Customer Personnel Turnover

In Section 6.2.8, "Customer Monitoring," it is stated that the best monitoring system relies on a formal interface with the least amount of customer penetration into the contractor's organization. In some procuring agencies there is appreciable turnover of personnel. Customer personnel providing detail direction in the design phase generally are not in the same positions when the system becomes operational. Another group of customer representatives will be performing the monitoring function. A contractor who wishes to stay in business must not let his personnel be distracted by working-level customer contacts to the point where they lose sight of the contractor's basic responsibility for turning out a superior system. These objectives must not be compromised to please a customer's representative on an informal basis.

8.9 CHAPTER SUMMARY

The development of an advanced system requires a technical staff that has a different outlook and responds to different motivational objectives than does the engineering staff of a manufacturer of commercial consumer goods.

The emphasis in consumer products is to achieve the lowest initial cost, or life cycle cost in such industrial products as airliners, while maintaining acceptable quality in a competitive, dynamic open marketplace. The financial risk for the development is borne by the manufacturer; if he cannot retain a certain percentage of the market he goes out of business.

Advanced systems are designed to meet minimum requirements established by a technically sophisticated customer who is normally more interested in meeting requirements at minimum life cycle cost than in initial cost. The financial risk of the development is borne by the customer, who monitors the development to protect his investment. Customer and contractor, although playing different roles, are both part of the same development team. To perform in this environment of designing to meet advanced requirements, contractors must be dedicated to uncompromising technical excellence and must organize and develop their technical staffs to achieve this objective. The importance of this task is emphasized by making it a separate chapter in this book.

The engineering staff with its technical competence is one of the most important and most vulnerable assets of a contractor of advanced systems. To protect and cultivate their staffs, and to minimize personnel turnover, contractors attempt to diversify their product lines among several customers. It is suggested that a functional rather than a project organization is most suitable for this type of operation.

The customer faces many of the same problems in developing an in-house technical capability to properly discharge his responsibilities during systems development. However, the customer's engineering expertise is different from that of the contractor's. While the contractor needs creative designers, the customer must have expert evaluators to perform two basic functions: evaluation of proposals and monitoring the development process.

9

Systems Development Management

The term "systems development management" is used here to apply to the management of the advanced systems procurement and advanced systems engineering functions as defined in Chapters 2 and 6. It covers the conceptual, development, and activation phase of an advanced system up to the point where the customer assumes the full responsibility for its operation.

There are many textbooks and university courses on management, and it is not the author's intention to duplicate any of that material here. In particular, ref. 34 includes a section on management of a reliability program. It is the author's purpose to present some of his experiences in a format to help project management personnel understand the issues involved; not to produce a definitive textbook.

9.1 MANAGEMENT RESPONSIBILITIES

9.1.1 Project Manager

The responsibility of the project manager is to direct the successful development of his system within the framework of corporate policy. This job requires a thorough comprehension of the procedures outlined in the preceding chapters. Management must, for example, understand the division of responsibilities between customer and contractor as amplified by his contract, as well as the responsibilities of his various departments, such as purchasing, engineering, manufacturing, and quality control. He must assure that each entity in his organization discharges its responsibilities effectively. He must also assure that a replacement is available for himself when he is transferred to a new job (131, 132).

9.1.2 Intracompany Communications

In a description of the evolution of the project manager in Chapter 6 it is stated that systems engineering represents a new layer of management between the program manager and the design engineers who actually configure the product.

The personal contact between engineering supervisors and top management has been lost to a large extent, replaced by paperwork and layers of middle management with little discretionary powers. Engineers making important design decisions sometimes lose their frame of reference; it becomes difficult for them to identify themselves with top management, with the company, or even the system as a whole.

Within this framework it has become a major problem for top management to maintain visibility into the key developments that take place on the lowest level layout boards. It is significant that in spite of all the red tape and other communications barriers, in successful profitmaking organizations there is always a cadre of key engineering supervisors who find ways of providing top managers with the needed visibility so that they can spend their valuable time in resolving those problems that are most critical.

9.1.3 Need for Uniform Accounting Procedures

With the growth of modern advanced systems, the value of contracts has become so large that they have a significant economic impact on the geographical area in which the contractor's plant is located. Considerations other than technical become more significant in contractor selection. Defense contracts have become a means to support and further such other government objectives as the economic development of certain parts of the country.

As a result of these developments, corporate as well as government managers need better methods of planning, control, and accounting. The customer needs to compare the performance of his various contractors; he also is spending much more money, and there are more agencies looking over his shoulder. He has to account to Congress in a standardized manner for funds expended.

A number of new standard administrative procedures have been introduced across the industry. A functionally executed work breakdown structure, MIL-STD-881 (52), has been established, consisting of work-packages and detail tasks. Some companies have had to change their internal operations to be able to conform to the new reporting procedures.

9.1.4 Improved Performance at Lower Cost

Today the entire system of government defense and space procurement is being questioned by a Congress that cannot raise sufficient resources to do all the things that seem to be needed in a dynamically advancing technological age. The Congress is concerned because of high cost-plus overruns and examples of overmanagement and strangulation by red tape. The Apollo Program is rated as an excellent technical accomplishment, but much too costly. The industry is being challenged to do at least as well technically in the immediate future, but at much less cost.

This coming era presents a new and exciting challenge to the management of the advanced systems industry. Means must be found to remove the shackles of red tape from the necks of engineering supervisors, and better methods must be developed for encouraging them to use their talents to create better technical performance at less cost.

9.2 THEORY X VERSUS THEORY Y

In view of the large numbers of creative individuals required to develop an advanced system, management attitudes towards its personnel are of critical importance. This subject is discussed at the beginning of this chapter since all management actions discussed in the following sections are influenced by personnel attitudes.

In McGregor (24) two different approaches to work breakdown are presented, Theory X and Theory Y. The letters X and Y are selected to not show preference of one over the other. Theory X is based on the premise that the most effective work organization for production is the one that requires the least training in the worker. It assumes that the worker has definite productive skills and can be taught others, but only with considerable difficulty and expense. If the job is properly designed, the worker could be replaced by a machine, and it would be desirable to do so, if a machine could be built to perform the function at a lower total cost. The worker is assumed to be disinterested in his work and will perform his function properly only under constant monitoring.

Theory Y, on the other hand, assumes that people are interested in their work, want to do a good job, are willing to accept responsibility, want the company to be successful, and are motivated by the satisfaction they derive from their work.

9.2.1 Illustration—Theory X

Theory X, on its lowest level, can be illustrated by an imaginary encounter between an assembly-line worker and his supervisor. The worker's task is to install two bolts as the assembly passes his station. He has been installing these same two bolts for three years; he has no career prospects beyond installing these same two bolts for the next 30 years. Recently he was caught at two mistakes: one bolt was not properly tightened, on another the lockwasher was omitted. His supervisor rebukes him for lack of interest in his work.

The worker defends his attitude by reference to his complete lack of job satisfaction. He may compare his job with the simple chores he does around the house. How much more satisfaction he derives from warming the baby's bottle or taking out the garbage, jobs for which he is responsible. The supervisor points out the advantages of working for the best company in the industry: the highest hourly pay, the best cafeteria, the best hospitalization and pension plan, the best sports program, the longest vacations, and so on. If the worker does not like it, he can quit; there are plenty of others who would be happy to have his job.

The worker realizes he is trapped; if he quits he can only get another job in a similar company. He would be installing a different set of bolts on another assembly line.

Theory X exploits only the mechanical ability of people, assigning to each individual a minimum of responsibility. People respond with a minimum of interest and minimum identification with the job, the company, or the product. Theory X does not take advantage of the potential in the human worker. It establishes conditions that degrade rather than ennoble the worker. In particular, it does not provide the motivation considered necessary to produce high quality advanced technology products involving safety of life.

9.2.2 Theory Y in the Aerospace Industry

In Section 10.6, the importance of *making the individual feel responsible for quality and safety* is emphasized. This is the basic approach of MIL-A-9858A and of the discussion in Section 5.12.2. As a general (but not universal) rule, the tradition in the advanced technology industry has been to break down the work into individual packages that can be completed by one worker or by a small group of workers, and to hold them responsible for the quality of their work. In theory, management has emphasized the importance of craftsmanship and of doing the job right the first time. Workers are given as much leeway as practicable in establishing their own procedures. Incentive programs (awards, etc.) encourage them to develop better methods of production. As mentioned previously, this traditional trend is being reversed in some cases in the aerospace industries by the early introduction of hard tooling.

9.2.3 Potential Applications

The use of Theory X is not restricted to assembly lines; it has many applications, always with detrimental effects on human dignity and the quality of the work. When entering an office, it is interesting to observe what is the informal subject of conversation. In an engineering office where people are engaged in highly creative work, the informal conversation will be mainly concerned with technical matters associated with the work. Men are excited about what they are doing; they will live their jobs and put their whole selves into their creations.

If the conversation in an office mainly concerns hobbies, weekend and vacation activities, sports, and the like, it is very likely that the work is not organized in a manner to provide adequate job satisfaction to the individuals involved. This is not a matter of menial versus creative tasks, but of assignment of responsibility and authority. Not everyone is equally creative. People vary greatly in capability; but generally each individual will respond to a challenge on his own level.

If people are not challenged in proportion to their potential, they must find other outlets for their mental energies; they turn to outside activities to make life worth living. The employer is then not making the best use of his people. Creative people find themselves trapped and lead counter-productive, unsatisfying lives.

9.2.4 Implications of Theory X versus Theory Y

In an age of so many issues it seems unjust to single out Theory X as the culprit. Yet the author feels that it is a contributing factor to current unrest. In its concern for safety of life, the aerospace industry has in general rejected Theory X as an operating principle and has relied on the integrity of the individual to turn out a quality product. In view of the success of this policy, this insight might serve as a valuable lesson for the portion of industry that believes that it cannot achieve financial success without Theory X.

9.3 POLICY MANUALS

One of the problems with contractor objectives is the difficulty in making them known throughout the organization. This usually is done in a large organization by well-written policy manuals. (See Section 1.3.2.)

9.3.1   Decision Making in a Small Company

Top policy and objectives normally reside in the personalities of the top supervisors. In a small organization, subordinates make decisions based on their personal knowledge of the way their supervisors think. Subordinate supervisors must know what matters to bring to the attention of their own bosses and what decisions they can make themselves. An effective intermediate supervisor is one who enjoys the personal trust of his boss, and who through long years of service has demonstrated that he can be relied upon to make decisions the way the boss wants him to.

When this system works, it provides the most responsive and flexible method of management control it is possible to achieve. It is excellent in smaller, stable organizations. The major disadvantage of the system is that one supervisor can communicate effectively with only a limited number of subordinates, ideally not more than five or six.

9.3.2   Characteristics of the Work Force in a Large Company

As a work force grows, the number of intermediate supervisors cannot expand in the proper proportion. In the advanced systems industry, not only are the companies today very large, but project organizations have many temporary characteristics as they undergo several reorganizations as the project progresses from contract definition to design, to systems integration, to operational test, and to production and deployment. In a large company, key personnel keep moving from one project to another, and there is also considerable turnover of personnel between companies. As a result, people do not get to know each other as well as they do in a stable supervisor-subordinate relationship.

In these circumstances, unless good policy manuals are provided, lower-level supervisors feel themselves cut off from management; they feel themselves adrift and tend to identify with their own subordinates rather than with top management. They have no place to get guidance on matters requiring many decisions, some of which may be quite important.

For example, many "make or buy" decisions involve both technical and business management considerations. Many such decisions, involving in their totality considerable sums of money, are made by lower-level supervisors. It is essential that these supervisors be provided with explicit instructions regarding what is best for the contractor and the project from an overall viewpoint.

9.3.3   Rules for Decision Making

The purpose of policy manuals is to put the rules for decision making in writing. By studying the rules carefully, a lower-level supervisor will be able to make a decision himself, just the way the top supervisor wants it made. If he cannot make the decision to the satisfaction of his supervisor, the fault is with the manual, and the written rule must be improved until the top supervisor gets the results he wants on a continuous basis. In this manner, the top supervisor can make his desires known to a great many additional intermediate and lower-level supervisors. People want to do a good job; if given good guidelines and they know they are doing what the company wants them to do, they will feel more a part of management.

An important danger with written policy is that it is less flexible than a system based on personal contact. It therefore requires constant management attention to keep it meaningful; otherwise, it stagnates and becomes a barrier to improving productivity. This monitoring is best provided by the cadre of key supervisors mentioned in Section 9.1.2.

9.4. PLANNING, SCHEDULING, AND CONTROL

For the purposes of the discussion in this section, a *plan* is a list of tasks to be done, arranged in the sequence in which they must be accomplished to achieve a given objective. *Budgeting* refers to the allocation to each task of the resources necessary for accomplishment. For proper *control* each task must be small enough so that, at any time during the progress of the work, the supervisor can estimate the percentage of the work completed with sufficient accuracy.

Scheduling is the establishment and coordination of start and finish dates for every task. Scheduling assures that tasks are accomplished in proper order and that resources are used in the most economical manner.

The control function assures that the plan is carried out on schedule and within the allotted resources.

The terms "planning," "scheduling," and "control" in the advanced systems industry are normally understood in a narrow sense to apply to the day-to-day administration of a system development project. "Long-range" in this context normally refers to the next fiscal-appropriation cycle.

9.4.1 Objectives

The general objectives of a systems planning, scheduling, and control system may be stated as follows:

1. Establishment of manpower and facilities budgets for new jobs.
2. Running check to assure that current jobs are within established budgets and schedules.

3. Running check to assure that all contractual commitments are being satisfied. Special emphasis must be placed on control of contractual and engineering changes, and close-out of corrective actions arising out of deficiencies and failures uncovered or occurring during test or field operations.
4. Status report of available and potential manpower and facilities to assure adequacy for current and forthcoming requirements.

The planning, scheduling, and control system also provides the basis for financial accounting and must comply with all customer and corporate policies and procedures. All government customers, for example, are required to account for their expenditures to the Congress, and they must, therefore, keep their accounts on a basis consistent with their Congressional appropriations. Furthermore, the customer is required to maintain his accounts so that he can compare costs and performance of his various contractors. Therefore, all government contractors are required to submit cost information in the same format and use standard terminology (97).

In a company with several active projects, corporate officials want to compare costs and performance of items, such as testing, common to their various projects.

Collateral Requirements for the Accounting System

For systems management, the accounting system must provide visibility into the detail elements of cost and profit generation and must provide quantitative measures for evaluating performance and exercising control. The accounting system must be designed to provide maximum economic incentive to the project managers.

A few additional requirements, which are sometimes neglected, but which are believed to be of value, are the following:

1. Ease of conversion of manpower budgets to dollar budgets.
2. Provision of indices for evaluating the performance of the various organizational units contributing to system development.
3. Provision of indices for the establishment of requirements for quality of engineering services based on:

    (a) Minimum overall end-product costs.
    (b) Life cycle costs (if applicable).
    (c) Anticipated production quantities.

The impact of these objectives and requirements on the various aspects of the planning, scheduling, and control system is discussed in the following sections.

In view of the overwhelming amount of data and paperwork involved in the development of a large advanced system, maximum use should be made of computer applications (see Section 9.10).

Some of the administrative techniques associated with planning, scheduling, and control in the advanced systems industry are quite sophisticated and have been discussed extensively in the literature and the trade press and are the subject of many university and commercial courses. These techniques, therefore, are only described here in the briefest terms as necessary to provide an appreciation of their significance.

9.4.2 Long-Range Planning

An advanced systems contractor cannot control what contracts he will win on a competitive basis in future years. "Long-range" planning to the systems contractor does not have the same meaning that it has in a commercial organization that develops markets and services them over a long period of time. The systems contractor requires such a large concentration of specialized assets, he definitely considers himself in business to stay. He manages these tools of production in much the same way that any large manufacturer of commercial products does, and further discussion of this aspect of planning is unnecessary. Therefore this discussion is restricted to the unique aspects of systems development management and excludes those longer-range "total" integrated planning considerations that are primarily a corporate rather than a systems project responsibility, namely: facilities and capital assets; new technology and future systems; personnel skills, employment policies, salary planning, continuing education, professional development, professional activities, student relations, scholarships, and summer help.

9.5 PLANNING

The orderly execution of a systems development project requires three basic items of information.

Job definitions.
Available resources.
Completion schedules.

Resources and schedules are established with reference to units of work. Hence the task of defining the various jobs has to be accomplished first.

On systems development projects involving advances in state of the art (Figure 2.2), the detail work descriptions change throughout the entire development process. However, distinct work phases are identified. To simplify this presentation, only two phases are considered here, preliminary design and the development stage.

Advanced system contractors are fully aware that creative design engineers do not know when to stop working. Every contractor is well aware that his principal concern as manager is to distinguish between a perfect job and an acceptable one, and to terminate activities when they reach the latter state. The manager looks upon his planning, scheduling, and control system as his principal means of coping with this situation. Yet he is aware that this system only presents his engineering staff with an obvious conflict between exaggerated technical expectations and marginal resources, which they must resolve for themselves. While the manager can readily monitor cost and schedule performance, he is uncomfortably aware that the third side of the triangle—assessment of technical progress—is beyond his grasp. Present attempts to augment planning, scheduling, and control systems in this direction have done little more than describe the deficiency eloquently.

Managers themselves recognize that the trend toward more sophisticated planning, scheduling, and control systems is self-defeating, since these efforts divert contract funds from productive activities. "How much planning do we need?" is a central management question in the advanced systems industry. Obviously zero planning could lead to chaos, yet 100% planning leaves no resources for doing the job.

How Much Planning?

Current planning, scheduling, and control systems are based on the assumption that engineers do not know when to stop working, and also on the assumption that there is an optimum place to stop. Therefore one sees systems engineering management papers that advocate this recipe:

1. Define the total job.
2. Break down requirements into manageable tasks.
3. Define tests for each of the tasks to demonstrate that the requirements have been met.
4. Stop when test objectives have been met.

Unfortunately, this recipe applies more to a cook than it does to an engineer. The point at which a steak is "done" is so well defined and bounded by such distinguishable states (raw and burned) that the chef can be readily monitored for technical performance. The engineer's output is more like the educator's: just when does a student become "educated?" Although objective evaluation is a worthwhile goal we always strive for, it is unrealistic to expect it in every case.

In contrast to the overemphasis on identifying an adequate technical job, there seems to be an underexploitation of the very deep motives of the individual engineers. Engineers, like all employees, are traditionally motivated to perform tasks on the job, that they otherwise would not do, for money or the prospect of money. Yet in addition to this basic motivation, here is a group of employees applying countermeasures against management to enable them to do the job the way they think it should be done. One wonders what such drive could accomplish, if properly unleashed. For further discussion of this problem see Roberts (25).

9.5.1 Preliminary Design—Advanced Development

This phase covers the contractor's work completed before the award of a development contract. It normally consists of several stages, such as conceptual design (CD) and project definition phase (PDP).

All advanced systems contractors do a great deal of advanced development work, both with respect to future systems and to the new technology required to produce these new systems (such as new high-temperature materials or new high-strength joining techniques, etc.).They maintain close contact with potential customers to evaluate their future needs, so that when a customer issues an official request for proposal (RFP), all interested contractors are well prepared to submit a proposal in a minimum of time.

Detailed preliminary planning begins in the proposal stage in response to the RFP and must be done by responsible design and equipment engineers. Given the customer's needs, a proposal is prepared, not just in explicit response to the RFP, but one that must be superior to that of competitors who are also responding to the same RFP and who also want the business.

Proposals

A proposal defines the major characteristics of the proposed advanced system and its equipment. It indicates how well the proposed system complies with the requirements of the RFP and also satisfies other customer needs. It also outlines how the project will be organized and managed, as well as the construction, subcontracting, and delivery schedules. Anticipated costs are indicated. Reliability, maintainability, system safety, quality control, and support program plans may be included.

During this proposal preparation stage, block diagrams, configuration drawings, and schematics are established for the system and the various subsystems needed to meet the performance requirements. The equipment required to perform the various functions defined by these system drawings is also tentatively established. A number of potential subcontractors and vendors is contacted for advice in defining equipment requirements. All interested groups in the contractor's organization are drawn into the discussions and contribute their ideas.

Out of these discussions evolve preliminary specifications and drawings that define the individual subsystems and equipments, their demonstration and performance requirements, delivery schedules, and cost estimates. The engineering and manufacturing plans are outlined briefly. The engineering plans include preliminary schedules and outlines of what each engineering group will contribute. Each group prepares an estimate of the required man-hours to do its share of the job outlined in the proposal. These estimates are reviewed and adjusted by management, and finally each group receives an allotment of man-hours which will be used as the basis of pricing the proposal.

Sometimes a proposal suggests that the customer has asked for the wrong system in his RFP, and proposes a different concept which the contractor believes will better satisfy the customer's needs. Another technique is to submit a well-planned alternate proposal along with a proposal that is strictly responsive to the RFP.

9.5.2 Parametric Estimates

All contractors rely to some extent on so-called parametric estimates to establish the price of proposed systems. These estimating procedures predict cost and schedules based on certain key system performance characteristics such as weight, speed, range, and payload. They are based on careful analyses of previous bids of the contractor and all that is known of bids of his competitors, and how the bids relate to the then existing economic conditions. Although price is not necessarily the critical item in winning a contract, it is easier to sell a superior technical proposal if it is coupled with the lowest price. Therefore, the objective is to establish a price that is just under that of other competitors.

Since on any given proposal, the price is a closely guarded secret until bids are opened, the estimating procedures must be based on analysis of what competitors have done in the past, and guesses regarding their reactions to current economic conditions.

In preparing such price estimates, consideration is also given to the contractor's previous actual cost experience and to the cost estimates provided by the various groups in preliminary design who have prepared the proposal. However, in trying to win a contract, beating out the competition is the deciding factor. What the competition does is critical, and cost estimates prepared by individual in-house groups may be cut accordingly.

Affordability

If a contract is won but the final negotiated price is lower than that predicted by in-house estimates, a contractor will attempt to reorganize his facilities to cut his costs so that he can live up to his promises. To an extent it is possible to do this. Generally speaking, the customer's RFP is deliberately vague on quality levels. It normally implies that at least a "Rolls Royce" level is desired and may even be necessary to satisfy the requirements. The contractor's engineers ordinarily are disposed professionally to attempt to exceed all known standards in this regard, with the result that the cost of their proposed effort is invariably high. As a practical matter the customer's funds are limited, and someone must reconcile what is technically possible with what can be afforded.

Price Versus Quality

As soon as the dimension of price is added to the RFP, it begins to become clear to both contractor and customer just where in the price-range spectrum between Rolls Royce and Volkswagen this system must inevitably fall. This emerging concept of a high, medium, or low price-range system can only be arrived at by establishing a common technical baseline from which to make alterations. The combination of RFP plus proposal(s) establishes the common baseline, although not an easy one to interpret.

Once the quality level, or price range, for the new system has become clear to the contractor, he recognizes that he must take steps to modify the image of the project in the minds of the engineers who wrote his technical proposal. The contractor's planning, scheduling, and control system is his principle means of conveying the changes that must be made in system quality level. The contractor allocates budgets to each of the work task items described in the proposal. These budgets will be lower than the estimated costs, by amounts that on the average reflect the reduction in total system price. Each engineer is expected to reconcile for himself the conflict between the technical aspirations of RFP and of the proposal and the firm schedule and firm budget of the contract. The planning, scheduling, and control system is an aid in enforcing this reconciliation. It does not always work. The record of the advanced systems industry has been that the quality levels normally exceed expectations, but that costs and schedules generally overrun target values (1).

It is not at all clear why this inevitably seems to recur, although some of the best management theoretists have researched the phenomenon extensively (25, 133, 134). It is clear that the customer introduced the notion of the Rolls-Royce, but he made it abundantly clear that he could only afford a VW. Then he is offered a Porsche by the contractor and is expected to pay for it. Major deficiencies in customer requirements certainly invite contractors to underestimate costs to assure getting the contract, and this procedure guarantees that cost overruns will occur.

9.5.3   The Development Phase

Once a contractor has been selected, the details of the contract must be negotiated. Major problems revolve around unspoken questions of how much money the customer has and how well the proposed system will support the customer's operational plans.

A new project development team will be organized and assigned the task of executing the system development. This team will have inherited a substantial set of requirements, coupled with allocated resources. All of these can be changed as the design progresses, but not without good and sufficient technical justification.

Familiarization

The first task of the project development team is familiarization with the contractual requirements and the details of the proposal indicating how the requirements are going to be met, specifically: the performance requirements, the proposed hardware subsystems and their distinguishing features, the schedule and manpower allocations, and the subcontracting and manufacturing plans. Some of the project management personnel will have worked on the proposal, but their number will not be adequate to fill all key positions. The speed with which subcontracts can be let is a major problem at this time. Time lost at this point can never be regained.

Engineering Department Organization

The engineering design organization generally is based on equipment breakdown. The various assemblies and equipment are assigned to various levels of the work breakdown structure (52) in accordance with their "family" relationships, the size of the tasks, and the technical difficulties involved. A job may be small physically but, if the design is technically difficult and contributes a critical problem, it may be placed higher in the work breakdown structure so that engineering management can maintain better visibility over progress.

There is a great deal of leeway in how many man-hours can be legitimately consumed in doing a job in a competent manner. The major trade-off generally affects quality levels. This applies also to many commercial products, as can be demonstrated by the price differential between the economical Volkswagen and the most expensive Cadillac. Competent people do a better job in less time than a greater number of people who are less competent. When funding is tight, there is no choice but to utilize a small team of the very best people who work well together.

Job Definitions

Each specific detail design job is defined by discussions between project engineer and cognizant design group leader. Both are experienced, both know what is involved. The key factor is competence and mutual trust. Both know that no matter how carefully they plan and define the scope of the task, since the system is a new one extending the state of the art, unforeseen difficulties will arise, and the actual job will not be executed as planned. No matter how experienced both men are, it is not possible for them to foresee all of the problems. The key here is that the design group leader accepts the responsibility for doing the job as negotiated with him. In view of the imperfect planning, the important thing is that as soon as something new or unexpected comes up, as it will, everyone involved must be notified properly and quickly.

The exact definition of job expectations under these circumstances, although specific at any point in time, is flexible and is being continuously modified through mutual understanding. Actual written task descriptions can be only very sketchy; they list only the names of the jobs to be done, and they must be general enough not to impede the necessary flexibility. The scope of the job is defined by the manpower allocation.

Management Visibility

As a basic principle, a supervisor must always learn of impending trouble from the man directly involved. For a supervisor to learn of impending trouble from anyone else, worst of all from his own boss, destroys the relationship of trust between supervisor and individual designer. Open channels of unrestricted communication, therefore, are of critical significance. Sudden surprises from outside these channels of communication are intolerable. The management system must assure the identification and monitoring of troubles in a timely manner before they become serious.

The mutual understanding between project engineering and design group leaders concerns primarily determining what to do next to accomplish the best possible job with the remaining resources. If the available resources turn out to be inadequate and there is an impending overrun, the project engineer will have known about the developing situation for some time.

9.5.4 Work Breakdown Structure (WBS)

Cost Information Reports

Department of Defense customers require contractors to provide financial reports, cost information reports (CIR), in accordance with ref. 97. The purpose of the CIR is to require all suppliers of specified product categories to provide financial data in a standard manner, so that within product categories the equivalent cost figures from different contractors can be compared directly in a meaningful manner with assurance that the comparisons are valid. The CIR specifies the manner in which costs are accumulated and reported, and requires the use of a work breakdown structure (WBS). Cost items must be identified properly at a low level, and this identification must be maintained in reporting upward through the various higher levels of the WBS; the identification must not be lost by combining different items of information at intermediate levels.

The CIR defines the six top levels of the WBS. A contractor is not obligated to organize his company or conduct the job in accordance with the WBS, but he must accumulate costs and report them in this manner.

The CIR has a number of unique peculiarities. For example, no provisions are made for management reserves, and it is required that recurring and nonrecurring costs be recorded separately. There are provisions to assure that the customer does not pay for the same task more than once.

Furthermore, government customers require that financial reports be provided by contract line items, since they obtain their own funds in accordance with this breakdown. These requirements are important to the customer, but they need have little meaning to a contractor. The government accounting system is merely one of many possible systems; it is, however, the one that the government uses, and therefore all contractors must comply with it for financial reporting purposes.

Work Breakdown

The detail design of the WBS normally proceeds in accordance with MIL-STD 881 (52) which establishes for certain categories of defense material items uniform element terminology, definition, and placement in a family tree structure. It defines the upper three levels of the WBS for the following seven system categories: aircraft, electronics, missiles, ordnance, ships, space, and surface vehicles.

Level 1 consists of the entire item (system) as identified in the Department of Defense programming and budget system, either as an integral program element or a project within a program element.

Level 2 elements are major items within the total system.

Level 3 elements are subordinate to Level 2 elements.

The nomenclature associated with these levels is given in the MIL-STD 881 (52) for the seven system categories covered. It is assumed that all of the work required by a contract can be so broken down into individual lowest-level elements of work that each element is the responsibility of a single unit of the company. This assumption is generally not true, since even the lowest-level work element of a WBS generally involves a number of organizational units. Nevertheless, for bookkeeping purposes, all contributions by the various groups can be charged to the work element. All of these individual lowest-level elements can then be assembled in a WBS, which is a product oriented family tree composed of hardware, software, services, and other work tasks: it displays and defines the products to be developed or produced, and relates the elements of work to be accomplished to each other and to the end product. A number of similar or associated elements on one level are collected into a "header unit" on the next higher level. This combination of several lower level "trailer units" into a header unit on the next higher level can be repeated through successive upward levels until the single top-level unit is reached. A characteristic of the WBS is that the sum of all items on any given horizontal level adds up to the total system.

MIL-STD 881 (52) states that the WBS must be employed as a framework for technical and management activities and as a coordinating medium for systems engineering, resource allocation, cost estimates, contract revisions, and work execution. It requires that the WBS be used for scheduling and budgeting. The reporting of progress, performance, engineering evaluations, as well as financial data, must be based on the WBS. Furthermore, MIL-STD 881 (52) specifies that the WBS must serve as the framework for management control and provide traceable summarizations of internal data.

PERT Diagrams

Each lowest-level work element listed in a WBS can be used as the basis for the preparation of a schedule and for the allocation of resources. Costs can be accumulated against the work element, and the start and completion dates of work elements can be designated as "events" and included in a PERT (performance, evaluation, and review technique) diagram (98).

A contractor will himself have a number of requirements that may affect the design of the WBS; these requirements arise out of the way he wants to do the job and how he wants to account for it. He may, for example, wish to keep costs separately on each major subsystem, or break out the test program by test location.

Work and Control Packages

In practice, items on levels 5, 6, or 7 are sometimes known as "work packages." A work package is a code-numbered document that lists all tasks to be done, establishes the completion dates for each task, and indicates the manpower and resources allocated for the completion of the tasks. The manpower allocation is distributed over the development period. Most work packages cover such specific hardware items as a radar or a major structural assembly. One large development project reviewed by the author contained some 60 engineering work packages; two were management work packages, four were for computer programming, and the others all pertained to hardware subsystems.

The work-package concept was introduced because of the customer's desire to identify clearly a single individual as completely responsible for a unit of work—to eliminate "buckpassing" in large corporations. The work package is, therefore, normally under the control of a single individual, sometimes known as the "cognizant engineer." Generally, the numbered work package is the lowest-level "account" for budgeting, manpower control, and cost control. However, sometimes the lowest-level control is exerted on the next higher level. In this case, the unit over which control is exercised is called "control package," and each control package consists of several "trailing" work packages.

Work packages and control packages also provide a convenient breakdown for the acceptance procedures of Figure 9.1. A control or work package can be processed individually by treating it as an end product with its own acceptance data package. The alignment of the technical acceptance procedure with the financial accounting procedure not only provides greater management status visibility, but also saves considerable paperwork.

Cognizant Engineers: Responsibility without Authority

Unfortunately, many of the basic deficiencies in the operation of large corporations, which the customer is attempting to correct by the use of these concepts, simply will not go away just by designating a "cognizant engineer" to be in complete responsible charge of a work package. As a

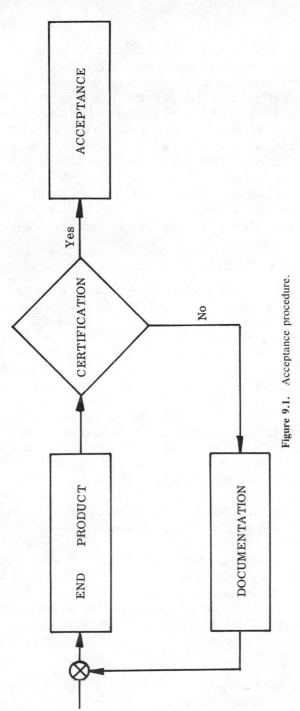

Figure 9.1. Acceptance procedure.

low-level member of the design department he has no control over contractual and business matters that determine his allocation of resources. He does not have complete control over technical matters, because many of the technical people contributing to the work package do not work for him. This is particularly true with respect to the functional units. For example, the thermal group, which has its own limited budget, will concentrate on those thermal problems that it knows to be most urgent and will attempt to get by with a minimum of work on problems that are not so urgent. The cognizant engineer, who may believe that the thermal problems in his work package should receive more attention, has no way of enforcing his opinion. There is no way for him to exercise the complete control over the work package that the customer wants him to.

9.5.5 Bottom-Up Breakdown

The WBS and PERT utilize a top-down breakdown. The complete system is shown on the top level and is broken down into its major subsystems on level 2. Each major subsystem is then broken down into its major elements on level 3. This procedure continues through several levels. The major question involved in the design of this breakdown is how many levels are required to achieve the objectives of the WBS. The author knows of no application on a major program where more than 10 levels have been used.

If the purpose of the WBS is to provide financial reports to government customers in a standardized manner, the number of levels is not a significant point. Below the level where the data enters the WBS, the contractor must have his own system, any system he likes, as long as item identification and similar requirements are met in the reports to the government.

If the purpose of the WBS is to serve as a management tool to control the accomplishment of work, the question of levels is pointless. For control, it is necessary to control the lowest actual elemental level, whatever that may be. The number of levels cannot be set arbitrarily; as soon as this is done, there will always be items that go below that level and are out of control.

This is, for example, the problem of PERT. The NASA, PERT-II, EDP program can collate 2000 items. A major systems development program cannot be managed within these limitations. Theoretically there is nothing wrong; but with the stated limitations, the lower-level items defy control.

Drawing Trees

There is, however, a traditional control system in all engineering departments that inherently provides all the necessary elements required for all types of control, namely, the drawing tree. The drawing tree starts at the bottom, with the smallest of parts: a bolt, a nut, a cotter pin, a rivet, or a transistor. The lowest-level drawing always describes the part in detail sufficient to identify, fabricate, and inspect it. When supplemental information is required, the drawing calls out the necessary specifications or test procedures. This set of documents provides complete technical control.

Lower-level (trailer) drawings are called out on the next higher assembly (header) drawings in the same sequence as the parts are generally assembled. At the lower levels of the drawing tree, manufacturing drawings predominate; at higher levels the drawings call out more assembly and installation operations, but a manufacturing drawing can actually appear on any level except the top one. On every level several related trailer drawings are combined into a header assembly on the next higher level, until the final system is shown completely assembled on the top system drawing at the tip of the tree.

Contractural Reference Units

All specifications and technical reports can be related to drawings. Drawings, specifications, technical reports, and related change documentation (plus technical support to other departments in directly related work) constitute the technical output of an engineering department. Specifications and technical reports also are organized in family trees, but these always can be related to the drawing tree. There is, therefore, no contractually covered work done on a project that cannot be related to an element of a drawing tree.

For example, a stress or thermal analysis may refer to a wing or radar set, both of which are defined by a drawing number. Much correspondence with vendors and customers concerns production or testing of items specified by number on a drawing. Project management activities could be charged against a numbered monthly progress report that, in turn, could be charged against the basic system drawing.

Levels are never a matter of discussion in the organization of a drawing tree, as the number of levels are determined by the number of assembly operations. There is no uniform lowest level. Some drawing chains may go down eight levels, others 20 levels; whatever is functionally necessary.

A piece of equipment procured from a supplier may be defined by either a specification or a source-control drawing. In most complex equipment both documents are required; the purchase order cites one, which then calls out the other. The author prefers to give precedence always to the drawing, that is, to maintain the drawing tree as the controlling tool and to use specification call-outs to supplement the drawings.

Control by Drawing Numbers

The numbers of the drawing tree provide a natural numbering system for accounting and management control purposes in an engineering organization. The use of drawing numbers automatically guarantees completeness, since there are no items of hardware included in the system that are not shown on drawings; and there is no duplication. The drawings control all aspects of the hardware; the hardware system is changed only through changes in the drawings. The drawings are the most up-to-date representation of the hardware, since the drawings control the hardware. All hardware work packages and header assemblies can be identified by drawing numbers.

Drawing numbers are always well known to all who are associated with the hardware. They think in terms of these numbers. The introduction of any other numbering system into an engineering department is necessarily artificial and merely adds confusion and decreases visibility. Control of work in an engineering department based on any numbering system other than drawing numbers, and associated specification and technical report numbers, always can be bypassed or circumvented. Control of an engineering department is an exceedingly difficult task, and the use of an arbitrary numbering system makes a difficult task more difficult than necessary. Unfortunately, some government entities have had to abandon the drawing tree as a basis for control because of an inability to recruit and maintain the necessary expert staff.

9.6   SCHEDULING

9.6.1   Gantt Charts

Given all the tasks identified in a plan, the manpower allocations, and the calendar time required to accomplish each task, the tasks are laid out in proper sequence on standard Gantt charts. The techniques are described in standard management textbooks such as refs. 26 and 27.

The starting point in constructing the charts is the system delivery date. Working backward in time, each task is represented by a line whose scaled length signifies the calendar time required to accomplish the task. Simultaneous tasks are listed on separate lines, but starting and ending dates are shown in proper reference to the true scale.

9.6.2   Manpower Loading

The manpower requirements can be noted on the chart for each task and for each time interval (such as a month). The total manhours required for a time interval can be computed by adding vertically all the requirements falling in that time interval. This can be done for all the individual time intervals along the horizontal axis, yielding the manpower loading curve over the program life. On the first try this manpower loading curve will exhibit sharp peaks and valleys that cannot be staffed from a practical viewpoint. For efficient staffing it is important that this curve be fair (smooth) and as level as possible, with a gradual build-up and let-down. To achieve a curve of this shape, individual tasks must be moved back and forth to obtain the best possible manpower loading. Care must be taken to assure that all tasks remain scheduled in proper sequence. With the many constraints involved in this task, it is not normally possible to achieve a manpower loading curve of ideal shape.

In view of the large amount of routine labor involved in this scheduling task, many organizations have developed computer programs to perform the routine operations.

9.6.3   Equipment Availability

The delivery dates of equipment (both contractor-furnished and customer-furnished) also should be noted on the schedule. Again, on first try, these availability dates will not coincide with the need dates for systems integration and other testing. Equipment delivery dates are controlled primarily by the required lead times starting with pruchase order or subcontract release dates, which in turn depend on when the system design is sufficiently firm to determine equipment specifications. Also on the first try, the availability of engineering drawings will not meet the dates needed for manufacturing. All of these events must be laid out on a Gantt chart in proper sequence. A great many trial-and-error revisions are performed in attempts to find the best solution. Sometimes all of the tasks cannot be fitted into the available time, and shorter calendar time requirements for certain tasks must be bought at a cost of higher manpower allocations. All of these inconsistencies must be renegotiated with all of the individuals involved until a satisfactory schedule is worked out listing all tasks in proper sequence.

9.6.4   Scheduling Breakdown

Scheduling takes place at various levels. At the top is the system master schedule which can only list the major milestones. Individual portions of the master schedule are then shown highly enlarged on separate sheets. In turn, individual portions of these schedules are shown on schedules for components. The process continues on separate sheets until all items in the system are properly scheduled.

On the lowest level good time estimates can be established on many tasks based on past experience and logical thinking. However, there are some tasks where it is very difficult to estimate the calendar time required. Two specific areas that have been mentioned before are design of items involving advances in the state of the art and system integration.

Good scheduling is possible when a specific sequence of operations can be determined, and each operational task is small enough so that all operations can be visualized or associated with such past experience as may be available in a handbook or manual. The processing of a machine part, no matter how complex, usually can be broken down into a number of tasks that must be performed in an established sequence, even though the individual operations may take place in geographically separated plants. There may be more than one sequence possible, but for each sequence an estimate can be prepared. Given the material, cutting rates, the amount of material to be removed, and so on, the calendar time required for each operation can be determined and a schedule established.

In all projects involving advances in the state of the art, design work cannot be broken down into such an orderly sequence of tasks, or will each task necessarily proceed as planned because of unforeseen occurrences, commonly known as "errors." In such cases, scheduling becomes an art requiring great insight into and basic knowledge of very sophisticated processes; yet it is necessary, at the outset of a development program, to have a schedule that, hopefully, can be met.

9.6.5   Systems Size, Complexity, and Errors

Today's advanced systems and the associated calculations have become very complex, providing more opportunity for errors. Complexity in this case may be indicated roughly by the number of people involved in a development effort. Reference 135 states that the preliminary design of the XF-104 was accomplished by less than 20 men. Some 17 years later, the same company required 1200 men for the equivalent preliminary design phase of the C-5A. Although the C-5A is a much larger airplane, this in itself does not explain this manpower increase.

It is interesting to consider what these men do. Figure 6.5 shows analysis as a feedback control loop to the design activities. The increase in manpower has been not so much in the number of designers, but in the number of analysts, stemming in part from the relentless emphasis on attaining superior performance. For example, to design an airframe structural configuration, applied loads are required. However, applied loads for flight-design conditions are based on the results of wind-tunnel tests. To build the wind-tunnel model to run the tests, the structure and aircraft configuration must be known. Thus, the answer must be known to establish the prerequisites for computing the answer. To overcome this difficulty, the various engineering groups assume or guess at the prerequisites that enable them to progress with their own specialty. These outputs are passed on to their successor activities, which in turn can then refine their results on the next cycle. Although each group is kept informed regularly by the project coordinator of what all the other groups are doing, it is an impossible task to coordinate the work of 1200 engineers in this type of operation, so some "errors" are inevitable.

Furthermore, scale effects, the impossibility of simulating certain operating conditions in wind-tunnel tests, and unavoidable limitations on the accuracy of wind-tunnel test results, are additional sources of error affecting the design of aircraft that extend the state of the art. These errors add a significant degree of uncertainty to the scheduling operation.

9.7 PROJECT CONTROL

Section 9.4.1 establishes a set of objectives for a systems planning, scheduling, and control system. *Performance* in this context is defined as meeting these objectives. The following control techniques should refer to this definition of performance.

9.7.1 Status-Control Charts

Control is normally exercised on the basis of status-control charts that plot "performance" versus "plan." Status refers to the amount of resources consumed and to the number of scheduled items completed as of a given date; the plan indicates which resources should have been consumed and how many items should have been completed. There are many statusing techniques available. They are discussed in standard textbooks, so there is no need to describe them further at this point. Let it be noted that none of these techniques gives any indication of the quality of the work, although in some cases they presume to measure "technical performance."

To be meaningful, control charts are prepared in family sets, following the drawing-tree breakdown. Low-level control charts are prepared for the lowest-level items. The information from a number of related low-level charts is combined on a chart representing the next highest level. On the highest level, the control chart covers the entire system. All important or critical charts are normally maintained and displayed in a central location variously designated as the chart room, command center, or action center.

9.7.2   Critical Items

In the development process there are certain events that must be watched more closely than others. The submittal of design and source-control drawings required by the customer for approval must occur in a timely manner. A purchase order to a vendor or subcontractor normally cannot be issued until those customer-required drawings have been approved. A delay in starting the work at a supplier's plant normally cannot be tolerated if schedules are to be met.

During systems integration testing normally there are many interrelated activities occurring simultaneously. Not everything will go according to plan. Each day a status report is prepared to determine the activities that will make best use of available resources for the next day. There are many people involved in this operation, and it is very difficult to maintain management visibility.

9.7.3   Technical Quality

Technical men working either for the customer or the contractor feel that they share a unique appreciation of the fact that money and time are linear variables in the contracting process, whereas technical quality level can be highly nonlinear. That is to say that each working day costs roughly the same and consumes the same amount of schedule, yet does not necessarily produce the same amount of technical value.

Usually, near the end of the allotted time, after tests have been run and it is clear that performance is not altogether satisfactory, it is recognized for the first time how technical performance could be dramatically improved. For example, great improvements in structural performance can often be realized by the judicious use of rivets larger than required by pertinent analyses to fasten parts that are inherently strong. However, such structural improvements usually are introduced as the result of the analysis of failing-load tests. During the testing phase, design engineers typically shift creatively into high gear, and the rate at which value is added to the product is greatly accelerated. Unfortunately, this is the time when the planning, scheduling, and control system is seeking to conclude the activity.

This conflict is so typical and occurs so frequently that strong feelings are generated. Design engineers can become so committed to achieving the results they see as just around the corner, that they interpret management's desire to terminate the effort as a "selling out" of their objectives (see Section 8.5.1). To the extent that engineers earnestly believe in these objectives and have confidence in their immediate prospects for increasing product quality and technical performance, they feel justified in thwarting management's desire to terminate the development. A very effective countermeasure is to convince management that this is not a discretionary matter at all; that management has no alternative. Marginal test results will be portrayed as totally unsatisfactory, indicating financial disaster with respect to the contract. Management typically responds to such threats of disaster by reallocating resources in a more bountiful fashion. The engineer may get some unplanned overtime, which may or may not offset the adverse image he has created for himself in failing to do his job properly at the right time. But he succeeds in obtaining the opportunity to make his improvement to the product, and this seems to be his overriding motive.

9.7.4 The Race for Next-to-Last

The impact of a plan and schedule on a large engineering group must be thoroughly understood. Absolute control in accordance with a preestablished control chart can be maintained only in an isolated group that has no contacts with other groups. As soon as there is cooperation and communication between various groups working together on a single project, the control system will revert to a "relative mode," and there is nothing that the manager can do about it. Groups of engineers working together watch the progress of the groups around them, and coordinate their own progress with that of the other groups. Without enough time to do a job the way he would like to do it, the designer fights for as much time as he can get, and it will never be enough. He, therefore, hangs on to his drawings or technical reports and keeps working on them until he is forced to release them. This point is in fact determined by the "pacing" or last item. Each group takes care not to be last, but they watch carefully the man who is last, and attempt to come in just ahead of him. It is a race for the next-to-last place (25).

In this situation, if the engineering manager desires to increase the speed of the pack, all he has to do is to identify the group that is last in line and speed them up. All of the rest will be quick to observe and to adjust themselves automatically.

9.7.5 Countermeasures by Those Being Controlled

The manager should be aware of several other deficiencies of formal control systems. The introduction of any control measure will always result in the development of countermeasures by those being controlled. This applies in particular to the paperwork systems servicing command centers. Normally such paperwork is carefully edited at lower levels of the organization to reduce the impact of sensitive data long before it reaches the command center.

The normal countermeasure against control-chart monitoring is the inflated estimate by low-level designers. Such individuals have a motivation different from the manager's, and once they know their output will be monitored on a control chart, it becomes impossible to get an honest estimate out of them. When they later underrun their inflated estimate, it makes them look good.

On the other hand, it must be realized that top management will never support a system with the capability of making it look bad. An experienced management, therefore, will be sympathetic to the plight of the designer.

9.8 MONITORING TECHNICAL QUALITY

The discussion in Section 9.7 identifies three basic weaknesses inherent in traditional planning, scheduling, and control systems, namely:

- The planning, scheduling, and control system measures the consumption of resources as related to a preestablished plan, but yields no information on the quality of the work.
- Low-level designers will edit the information input to the system so that it will not represent true project status.
- Designers invariably become most creative at the end of the project when time and resources are running out.

There exists a problem, therefore, not only of measuring technical quality, but also of making designers more productive earlier in the program.

9.9 SUBCONTRACTOR MONITORING

The relationship between a prime contractor and his subcontractors and suppliers is similar to the customer-contractor relationship, with certain exceptions. The prime contractor always retains the full responsibility for the quality of the product. Although contractors have a great deal of latitude in selecting subcontractors and suppliers, large subcontracts and purchase orders above a stated sum (such as $50,000 or $100,000) require customer approval. All source-control drawings are subject to customer control.

9.9.1   Economic, Political, and Strategic Considerations

The requirement for customer approval reflects customer interest in both technical and nontechnical areas. In the nontechnical area, it has been mentioned that the government is often interested in the economic development of certain parts of the country and will encourage the contractor to place his purchase orders there. Sometimes the opposite is true. Certain areas may be particularly vulnerable to attack, and the government may not wish to rely on a supplier located in such an area. For example, during World War II all U.S. aircraft used a small patented item that was manufactured along the U.S. seacoast in a spot considered potentially vulnerable to submarine attack. In actual fact, enemy submarines were sighted less than 100 miles from this location. An attack could have severely handicapped all U.S. wartime aircraft production. Thus, for strategic reasons, the government often is interested in developing additional sources of supply for certain types of products.

9.9.2   Standards and Commonality

In the technical area, the customer is concerned that all of his applicable specifications and other requirements have been properly called out on the source-control drawings. The customer may be trying to achieve maximum interprogram standardization and commonality. He may also wish to perform an independent evaluation of the subcontractor's capability to comply with the requirements of the purchase order.

9.9.3   Purchase Orders

The formal aspects of subcontractor monitoring resemble closely the procedures employed by a customer to monitor a contractor. A purchase order replaces the contract, but the layout and format of all the purchase orders on a single project may follow the configuration of the parent contract very closely, for ease of coordination.

9.9.4   Design Control

Because final responsibility for quality rests with the prime contractor, his relationship with subcontractors and suppliers is much more intimate than the normal customer-contractor relationship. The prime contractor is the systems designer and subsystems integrator; he must have a thorough detailed knowledge of all technical aspects of the supplier's equipment. Contractor and subcontractor/supplier designers work very closely together throughout the development period.

The formal paperwork follows the general pattern of the approval loops shown in Figure 6.5. Design reports and test plans must be reviewed, and approved or disapproved within a specified time limit, usually 30 days or less. Timely review must be maintained; if reports are not disapproved within the specified time limit, they are considered to be approved automatically. The prime contractor is then financially responsible for any necessary corrective action or schedule slippages that might occur if the particular design decision is not compatible with other system parameters or interfacing equipment.

Actually, the 30 days is not too long a period considering the interplant mail distribution in a large company. Each report may have to be reviewed by 10 or more different specialists (such as representatives of weights, materials, structures, thermal) as well as the various performance disciplines. Normally, these groups review copies of the reports concurrently and then submit their answers to a work-package engineer for coordination.

9.9.5   Volume of Paperwork

The amount of paperwork continuously being processed in a prime contractor's facility during the development cycle cannot be described in words; it must be experienced, otherwise it would not be considered possible.

Monitoring of a normal subcontract requires 25 or more different technical reports per week. Several copies of each report are submitted, sometimes as many as 25. With 25 major subcontractors and over 1000 suppliers, the technical reports pile up overnight into a respectable mountain.

The major problem is to assure that each report will be properly read and evaluated, and that approval or other action will be taken in a timely manner. There must be a central library for such documents, and it must exercise rigid control over all of these documents and assure that timely approvals or other actions are issued.

9.9.6   Report Review Procedures

Under the terrific pressure that always exists in an overworked office during development, a contractor's engineer often "reads" a subcontractor report by telephoning the subcontractor's engineer who wrote the report and discussing it with him. Often those two men will then make oral agreements over the telephone as a result of this review of a report. When this is done, it is necessary that it be followed up with proper paperwork so that all others involved in the approval cycle will be aware of the agreement. This practice is often paralleled by government engineers in reviewing and approving reports submitted by prime contractors, government laboratories, and other government facilities. Oral telephone approval has the advantage that it saves valuable time compared to letters, which frequently must be routed through the organization all the way to the top for signature. Approval is a contractual committment, however, and should always be followed up promptly with proper paperwork. Often these approvals are recorded in the minutes of the design reviews conducted at contractually scheduled milestones.

9.9.7 The Vertical Cut

Subcontractors design and supply many of the advanced subsystems for a new system, and it is essential that their operations be properly integrated into the overall system-development program. The prime contractor has many formal tools for monitoring subcontractor performance on a current basis including: design reviews; regular periodic technical and financial status reports; and personal liaison of contractor's personnel with their counterparts in the subcontractor's organization, particularly in the business office, in design and manufacturing engineering, and in quality control. On large hardware-system development programs, these monitoring systems tend to become so large and complex, and there are so many geographically dispersed subcontractors, that at times it is difficult to maintain clear top-management visibility.

The prime contractor's program manager carries the overall responsibility for the success of the system development. There may be times when he will have difficulty in maintaining full confidence in his monitoring system. With critical delivery dates still two or more years in the future, many critical problems still awaiting solution, and with the overruns piling up, the program manager may wonder about the capability of a subcontractor to fulfill his commitments. Neither formal progress reports nor liaison between his experts and their counterparts in the subcontractor's organization necessarily provide sufficient credibility to allay the personal qualms of the program manager.

In this situation *the vertical cut* may provide a simple complementary device to check the credibility of the formal monitoring procedures. It must be emphasized that the vertical cut is not a substitute for any existing formal monitoring device.

Experienced Investigator Required

The vertical cut must be performed by an experienced investigator with a broad background who enjoys the full personal confidence of the program manager. The vertical cut technique enables an experienced investigator to obtain a remarkable amount of information in a short time, such as one or two 8-hour days. The investigator must have free access to all pertinent subcontractor operations. This free access is normally provided for in subcontractual arrangements, but sometimes a subcontractor may be hesitant to cooperate fully.

Selection of Critical Part for Investigation

The technique is based on the investigation of a single small part, preferably a critical part such as a small machined part, that is completely designed, fabricated, tested, and assembled in one facility (that is, not sent out to another facility for some of the processing) and that is identified by serial number. The part should be selected at the contractor's plant in consultation with cognizant contractor personnel before the investigation is started. It is preferable to select a part that has a record of troubles, but this is not essential. The investigator must be thoroughly familiar with the history of all troubles and attempted corrective actions associated with the part selected.

Interview with Designer

Upon arrival at the subcontractor's plant, the investigator asks for the part drawing by number. The drawing is examined and the responsible designer identified from the title block. The investigator requests the privilege of interviewing the designer at his drawing board (*not* in a conference room).

The drawing is spread out on the designer's board, and the designer is requested to explain the function of the part, the function of the next assembly, and of the end product. Further questioning should cover design requirements, environmental requirements, selection of materials, analytical design control procedures, methods of manufacturing, quality control, and qualification. In respect to materials, it should be determined whether the material characteristics are taken from handbooks or whether the subcontractor has run tests of his own. Penetrating questions must be asked with respect to development testing, qualification, maintenance of quality during follow-on fabrication, and, in particular, the demonstration that this particular part satisfies all of its requirements. All previous failures and corrective actions should be discussed in detail.

Internal Relationships

The questions to the designer should be deliberately formulated so that he will not have all the answers. He will have to get out assembly and installation drawings, possibly his original layouts. It should be determined what alternate configurations were considered and why this one was selected. The questioning should require the designer to call on his boss for assistance, and he in turn may call his boss. Further questions should be deliberately formulated so that they cannot be answered by those present to make it necessary to call in representatives of other groups: materials, analysis, test, manufacturing, and quality control. One of the objectives of the procedure is to initiate a discussion, focused on the part in question, between a representative cross section of subcontractor personnel in their natural working environment. Under the impact of his questions the investigator observes how these men talk to and support one another; he particularly watches for conflicts and communications difficulties between organizational entities. Within an hour or two, a skilled investigator can develop a good feel for the technical competence of the various subcontractor groups and their internal relationships.

Inspection of Part

As soon as this aspect of his mission is accomplished, the investigator explains that he would like to see how the part is handled and stored in the shop and requests to be taken to the stockroom and shown a part. He should take the part drawing with him and observe how the stockroom clerk goes about finding the part, how the part is stored and handled. He should inquire about the organization of the stockroom, as well as about pertinent management policies. The investigator accepts the part from the stockroom clerk, examines it and compares it with the drawing, makes a note of the serial number, and returns the part to the clerk, observing the procedure for returning the part to the correct storage location.

Review of Paperwork System

The investigator now requests to see evidence that the part that he has just seen, identified by serial number, is actually made from the material or materials called out on the drawing. The observation of how this evidence is gathered is just as important as the critical examination of the documents themselves. Advance systems contracts require that all procedures be in writing, and that all material be traceable. The investigator should ask to see the written instructions controlling all activities that he has just observed, determine where these written instructions are maintained, and ascertain the extent to which those involved are acquainted with the existence and contents of the written instructions. Material review reports, scrap records, and corrective action reports should be reviewed in detail. This review should give the investigator a good feel for the quality, completeness, and credibility of the subcontractor's internal paperwork system.

Plant Tour

In the process of reviewing the paperwork, the route card (or cards) for the serial number under discussion should be located, reviewed, and retained. The investigator should request a plant tour based on the route card. He should visit each manufacturing and inspection station in the order listed on the card, starting with incoming material inspection. At each station he should talk both to the foreman and the operator of the station. The operator should be shown the drawing and asked to explain what operations he performs, what tools, set-ups, and instruments are available (the set-up will probably be dismantled by this time, but the description should suffice). Where applicable inquiries should cover written manufacturing, process, and inspection instructions or specifications; instrument calibrations; scrap rate; corrective actions; training; and so on. This tour should give the investigator a good feel for the efficiency of the manufacturing process, plant layout, and technical competence and capability of personnel.

No Diversions

All subcontractors working on advance systems will have developments going on in their plants of which they are particularly proud and which they delight in showing off to visitors. An investigator following a route-card tour should not permit himself to be distracted by such sideshows, since he must follow the actual route of the material to get a true feel for the process. Only after he has completed the route outlined on the route card, and if he still has time, should he permit his guides to take him back and show him their "goodies."

Interview with Subcontractor Executives

Only after completion of an investigation outlined here should the investigator talk to company executives. With a better understanding of their problems, their words will have more meaning, and he can talk to them on much more equal terms. Typical subjects of discussion are reasons for cost overruns, probability of maintaining the schedule, and so on. A review of the accounting system might be in order. Typical questions an investigator might ask of top subcontractor officials are: Why do you think your product will work? Will it be ready on time? Will it be reliable, easy to maintain? The answers to these questions should be given by the responsible top subcontractor officials themselves in easy conversation; under no circumstances should they be permitted to call in subordinates to supply technical answers.

Recording of Observations

Notes and observations made during a tour of this type should be dictated into a tape recorder no later than that same evening. Whatever is on tape or in writing does not have to be remembered and relieves the mind of the investigator.

The foregoing merely describes the basic information-gathering process. If there are troubles with the part or in the plant, or if adverse observations have been made, an analysis is required and recommendations for corrective action must still be developed, possibly in close cooperation with the contractor's cognizant engineers.

Recognizing Shortcomings

Large organizations normally do not maintain an equal level of quality in all activities. In some technical areas they may be very strong, while in others they will be weak, but they may not always be aware of this discrepancy. By following a simple critical part through its entire design and manufacturing cycle, it is often possible for an experienced investigator to recognize the strong as well as the weak areas. This is the first step necessary for developing an action plan to reinforce the weak links. The vertical cut method described here provides a skilled investigator with much of the information that he needs to assure the program manager of the credibility of the formal monitoring system.

9.10   COMPUTER APPLICATIONS

The amount of information required to manage a modern system development project is so large that the use of a computer is indicated. The author believes that the computer should be an integral part of the management system. It should serve as the basic tool to collect, store, collate, and retrieve data (bookkeeping), and should also serve as a basic communications and forecasting device. Each key engineer and manager should have a CRT console-telephone combination on his desk, connected to the central data bank. This instrument should be his means of all record keeping and of obtaining immediate status information. Much of the current paperwork should be eliminated. The security system for protecting the information in the central data bank must be foolproof.

Some very successful work has been done to computerize the design control functions (28). There are also numerous "management information systems" in operation, in particular those associated with PERT (98) and trouble-reporting systems (99). A complete system would integrate all aspects of planning, scheduling and control, the work breakdown structure, and subcontractor monitoring. The following remarks pertain only to a few of the specific needs of the planning, scheduling and control system.

9.10.1 Categories of Data

The data to be stored and processed by computer must be carefully defined. Two types of management data are handled in an engineering department: pick-off data and abstract data. Pick-off data are data that a trained reader can pick off a document and classify by a computer code on the basis of immediate recognition, such as a document number, title, and release date. Abstract data are all data not suitable for pick-off. For example, information normally transmitted by a drawing is abstract data. Although today a great deal of graphics is being done by computer, a special effort is required to enter the data into the computer, so this type of data is excluded as pick-off data. The same is true of the technical information in analytical or test reports and test plans. However, the identification of a drawing or report and their respective filing locations are pick-off data. The bulk of the information needed to manage a systems development project consists of pick-off data recorded on individual documents.

9.10.2 Stored Data

The bank should consist of two types of pick-off data: stored data and current-input data. Stored data would be, for example, the contract requirements and due dates. A processing group is required to analyze documents and input all of the required stored data into the bank. This processing group is responsible for the integrity of the stored data.

9.10.3 Current-Input Data

Current-input refers to daily operations and comes from various sources. Whenever possible the originator of the information should be made responsible for the correct input into the bank. The best method is to make the computer part of the information generating process.

For example, if an engineer requires the assignment of a drawing number, he types the title into a typewriter console, together with next assembly, his identification number, and other pertinent data; he receives from the computer on the typewriter the drawing number and due date. The data entered by the engineer is instantaneously collated with all other interfacing data stored in the computer and becomes available to anyone in the company with a need-to-know who questions the computer for status information. The bank should be organized so that only individuals responsible for the generation of certain data can input that data. On the other hand, supervisors can have keys to permit interrogation of portions of the bank. Each key should open that portion of the bank corresponding to the supervisor's responsibilities.

9.10.4 Status Reports

The computer also could be programmed to issue periodic, management summary status reports, with details provided on request, such as:

- Document status versus requirement and due date
- Percent of manpower budget consumed
- Manpower consumption versus labor category, per drawing, report, contract paragraph, or other combination of items (work packages)
- Overdue notices

Detail data on manpower expended against a document (drawing or report) by labor category, which can be totaled in any desired combinations such as contract paragraphs, work packages, internal accounting codes, or other groupings established in the future, will yield an information base that can be easily correlated with other jobs and extrapolated for estimating new jobs. This makes it necessary to establish in advance the types of engineering labor on which records should be kept. This includes such categories as original design, redesign to correct defects, redesign for cost reduction, weight reduction or other design improvements, analysis, design control, test, administration and secretarial, and so on. This list should be as extensive as necessary and can be changed from time to time, but changes cannot be made retroactive.

Outputs

The outputs of the central bank would consist of various types of periodic runoffs and special printouts prepared on request. Lower-level individuals would receive periodic runoffs in a fairly well established format showing items for which they are responsible, the established budget and due date, anticipated completion date, manpower expended, percent completed, overdue items, and the like.

Reports for top management consist of periodic status reports and special reports indicating where action must be taken, as well as special reports made up on request. Reporting for top management must be quite flexible, since it is difficult to anticipate what initial questions will be asked, and any answers usually lead to further questions.

An important consideration is, for example, response time to customer requests which may or may not be associated with incentive payments. The capability to coordinate instantaneously an updated set of contractual requirements, completion dates, due dates, and available manpower, eliminates almost all time loss for administration and permits use of almost all available time for actual engineering work. This is an example of how the data bank could serve as a management aid; the bank itself cannot reduce response time.

9.10.5   Program Flexibility

In a developing situation there will always be many computer program changes. It is impossible to design such a complex man-machine system in its final configuration and expect no changes in the initial design. Trial-and-error procedures are essential to arrive at a good workable program. The approach is to design a program and put it into use. As deficiencies are uncovered and new requirements established by the many individuals learning to use the new system, the program must be revised and improved. Responsiveness to changes is a first requirement, and computer-program flexibility should be traded-off against program efficiency; that is, an "efficient" program is one that consumes less machine time than a flexible one. During the development of a computer-based management system, computer-program flexibility is much more important than minimum machine time. Once a program is stabilized, the program can be redesigned for minimum machine time usage.

9.10.6   Retrieval System

A key consideration in the design of any automated management information system is the structure of the retrieval system. It is suggested that the most universal top breakdown is provided by the requirements of a contract, listed by numbered paragraphs. For any customer (Army, Navy, Air Force, NASA), contracts are written in a more or less standard format that changes only slowly with time.

Implementation of this recommended procedure involves analysis of each contract and applicable top specifications for identification of all engineering data requirements by paragraph number, title of item, and due dates. These titles represent a complete and unique description of all of the funded engineering work required on the contract. If any engineering work is attempted that does not fit into this breakdown, it should come to the immediate attention of management, as it may be an indication that the contract is deficient and funding arrangements are inadequate.

9.10.7   Contractual Changes

Provisions must be made for the timely incorporation of contract changes; and dollar or manpower budgets should be listed for each requirement, as negotiated with the customer.

These contractual requirement titles can be grouped as desired into work packages and milestones, and assigned to various engineering groups. There also can be a parallel assignment into PERT events. These groupings are flexible and tend to change from project to project, and may even change during the life of a project.

9.10.8   Work Breakdown

The contention being made here is that the structure of our contracts is the most basic, permanent, and all-inclusive top framework we have for the organization of engineering work. At the other end of the scale, drawings and reports are the most elemental engineering work unit. These concepts of funded contract requirements and of drawings and reports do not change significantly with time.

Except for certain overhead functions, there is practically no engineering work performed under contract that cannot be referred to a drawing or a technical report, specification, or similar basic document. All correspondence, engineering orders (EOs), memos, and other documentation written under a contract always can be referred to a drawing, report, or a provision of a contract or related specification.

Furthermore, drawing and report numbers are already organized in drawing trees or other unique header-trailer structures culminating in numbered contractual paragraphs. These unique numbering systems appear to provide a ready-made workable breakdown structure that satisfies all of the objectives.

Between contract requirement and corresponding drawing or report there are many possible combinations of work groupings and assignments. This intermediate structure should be held flexible, as it is subject to continuous change. In this intermediate area, the computer program must have the capability of making up any combination of groupings or assignments as desired.

9.10.9   Real-Time On-Line Rolling-Wave

The program suggested here is a real-time, on-line system. However, the entire bank need not be accessible in this manner. For example, data on events over say 1 month old and more than 8 months in the future could be stored on tape, available within 24 hours on request. The working on-line program would consist of a 9-month rolling wave encompassing the previous 30 days and the forthcoming 8 months with respect to the current date.

This proposed central computerized data bank appears to have the capability of providing management with the means for meeting the objectives listed in Section 9.4.1 and represents a major improvement over available current procedures. It is important to keep in mind that today's records will be past experience tomorrow, to be used at that time to make estimates for the day after tomorrow.

9.11   QUALITY OF PAPERWORK

There are some paperwork systems absolutely necessary to carry on a systems development project. These systems must include documents such as contracts, purchase orders, personnel records, specifications, drawings, engineering orders, technical reports (analyses, test plans, test reports), work orders, route cards, material tags, instructions, procedures, handbooks, policy manuals, letters, and memos.

There is a multitude of literature and courses available on the philosophy and correct preparation of each of these document types. Each document type has a defined function to play in system development and cannot readily be replaced by any combination of the others. They are, however, all interrelated and supplement each other. To obtain a desired result, it is necessary to utilize the one single document type that is designated (and designed) for that purpose. For example, drawings and specifications supplement each other, but one does not perform the function of the other.

9.11.1   Format and Function

For each document there is an established format and associated procedure that allows it to perform its function in the most efficient manner. To achieve efficiency, the manager must thoroughly understand the function of each document and enforce its correct format and use.

9.11.2   Optimum Amount of Paperwork

The optimum paperwork system exists on a project when the minimum basic types of documents are used to control the system development processes described in Chapters 2 and 6, such as, for example, the functions illustrated in Figure 6.5. On a large project, this absolutely minimum necessary paper already reaches mountainous proportions as mentioned in Section 9.9.5. Quality of paperwork is a most critical subject that rates top-management attention and requires constant monitoring. The area is so sensitive that with just a little neglect, it is easily possible to strangulate a project.

9.11.3   Means of communication

In studying paperwork systems, it is important to realize that the paper is only the carrier of information. If a paperwork system appears to be in trouble, the defect may lie with the information and not the paperwork. In a system development project, one phase progresses on the base of preceding phases much like each story of a building is built on a base provided by the lower floors. If certain jobs in earlier development phases are defective, subsequent corrective actions can become extremely costly.

For example, it has been mentioned that initial specifications may be vague in certain respects; in the area of quality and reliability the initial vagueness may be deliberate. This is often a trap for the project manager; a thinly veiled invitation for him to go hang himself. If he does not find some surreptitious way to compensate for the essential features of those deficiencies, at least in a manner compatible with his conscience, the project is doomed. It is like building a house on an inadequate foundation, and later trying to install the roof and reinforce the sagging foundation at the same time. This kind of situation will be reflected first in the paperwork system, and it is important to realize that the fault is with the health of the project and not with the paperwork. On the other hand, a well-organized, intelligible, and orderly paperwork system is one (but only one) of the characteristics of a well-managed, healthy project.

Several major companies are so saturated with paperwork that the paperwork-distribution system limits the amount of work which can be done. In one instance, an incoming piece of paper required three weeks for delivery to the intended recipient via the interplant mail system. When the company increased the capacity of the distribution system, paper generation also expanded so that the delivery time remained the same. As a test, the company contracted the delivery facilities, and paper generation also contracted to maintain the three-week delivery time. During the life of the project, the company was never capable of providing sufficient delivery capability to reduce the delivery time.

9.11.4 Duplication of Data Collection Systems

It must be realized that a project organization is a temporary establishment; when a need for a new piece of information arises, the tendency is to design a new form, together with a new data collection, analysis, distribution, and retrieval system. Except for the new item of information, all other items on the form will be duplicates of items on other existing forms. Many project personnel may not be very well acquainted with established corporate systems, and may not realize that they could obtain the same information at much less cost by a minor redesign of an existing form, or by use of an existing form being used on a companion project.

Such special purpose data collection systems are responsible for considerable economic loss. Although an individual office may have a legitimate need for certain data items, once a system is established to collect the information, it often is used exclusively to satisfy the requirements of the one office. Other offices may have a requirement for the same data but do not know that it is already being collected and may establish their own system to collect the same data. There is much unjustified duplication in the area of data collection.

9.11.5 Distribution of Paperwork

The distribution of paper also is frequently inefficient. Names are added to distribution lists because sometimes individuals without a need-to-know are interested in the type of information, or the orginator of the document thinks they should be. Many individuals spend unproductive hours each day scanning their incoming mail to see if it contains information of interest. Usually there are a few such items, but finding them is like finding a needle in a haystack. In a well-run office, most of the documents are identified and discarded immediately, but sometimes they are carefully filed and preserved for no purpose whatsoever.

9.11.6 Need for Paperwork Control

Paperwork systems always tend to grow and if not carefully and intelligently controlled, may constitute an invisible and uncontrolled source of economic loss.

9.12 QUALITY OF ENGINEERING

Section 9.4.1 lists in "Collateral Requirements for the Accounting System," as an objective of the planning, scheduling, and control system, the provision of indices for the establishment of the requirements for quality of engineering services based on the following:

1. Minimum overall end product cost.
2. Life cycle costs (if applicable).
3. Anticipated production quantities.

This is a controversial subject, since professionals tend to think in terms of only one level of quality of service—always the highest. However, Figure 8.1 depicts the ratio of income versus development costs as a function of production quantities. Production quantity is selected here for the sake of simplicity as a parameter to represent "total value," but there are other factors involved. It is a singular fact of life that there is a relationship between the funds available for engineering and the value of the work. In planning the work the wise planner will be aware of this situation and will assure that available funds will be expended in the most useful way. Consistent with value engineering concepts, the planning should include only those tasks absolutely necessary to assure that the product will perform its mission and be compatible with reliability, safety and other essential requirements.

9.12.1 Depth of Documentation

When large production quantities are anticipated, there will be more funds available for engineering, but manufacturing and operational requirements will also be more demanding. Large production quantities usually are associated with less skilled manufacturing, operating, and maintenance personnel. If only a few systems are being built, it is possible to maintain effective personal communications between design engineers and the manufacturing shops, and the documentation need not be complete in any great depth. Even here, though, care must be taken to counteract such shop foibles as the tendency of some shop personnel to make an item in accordance with what is actually drawn, ignoring critical information that is specified by notes rather than by lines on the drawing.

A project-oriented fabrication shop is geared to producing a small number of complete systems and can work out many integration problems on the shop floor; with adequate engineering liaison, there is no need for elaborate records. They do not need the same depth of documentation as do the production shops that make only one kind of item in large quantities and may not know anything about the intended use of the item. These latter shops normally have no information available to them other than as provided by the documentation, and they therefore have no means of making allowances for any deficiencies in the documentation.

9.12.2 Safety Factors

If there are only a few advanced systems in operation, such as manned spacecraft, the number of operators will be small. Probably they will be well trained and intimately acquainted with all of the characteristics and limitations of their equipment and will be careful not to exceed the allowable operating limitations. As soon as there are many systems available, operators and maintenance personnel will not be as carefully trained, and there is a higher probability that the system will be subject to abuse. In this case, more consideration must be given to safety factors.

State of the Art

A critical factor in determining the depth of engineering needed on a project is the anticipated advance in the state of the art. Many engineering projects that do not involve advances in the state of the art do not need the depth of systems engineering described in Chapters 2, 6, and 7. Even very large and complex structures such as bridges need not be and are not subject to qualification testing to demonstrate compliance with design requirements, although qualification and rigid acceptance testing may sometimes be required for certain subassemblies, materials, and joining practices. In such cases, the objective of much of the analytical work is to provide assurance that the structure complies with applicable building codes, the requirements of insurance companies, or other regulating authorities (see "Risk Management," Section 9.15).

In any engineering project involving advances in the state of the art, a decision must be made establishing the depth and quality of engineering to be maintained on the project. Unfortunately there exists as yet no formula or model that indicates, for desired output variables, what the input requirements in terms of quality of engineering should be. This question is similar to the "recipe" for "quality" of engineering discussed in Section 9.5, "How Much Planning?" By default, the quality of engineering on a project is normally determined by the availability of funds, which are generally provided on the basis of contractual negotiations and parametric estimates (section 9.5.2).

9.12.3   Need for Formal Drawing System

Within this framework of establishing the necessary depth of engineering records, there is still room for many trade-offs. Most companies involved with the development of complex advanced systems have learned that even in cases of low production, a complete set of formal drawings is absolutely essential. If unexpected failures occur, as they invariably do on advanced systems, one of the first actions is to compare the failed part with the "as-designed" records to determine if the failed part was constructed correctly. However, drawings may be of the layout type, showing many interacting parts on a single sheet. The higher the production quantities and more specialized the shops, the more the drawings must be broken down into individual details. The same rigor is not always maintained with respect to specifications.

9.12.4   Design Engineers—Best Judges of Required Depth of Engineering

Design engineers and analysts are in a good position to determine the quality of engineering needed in specific instances. The reliability engineer, for example, must be capable of identifying those components that are subject to random failures so that he can include these in his analysis. In a broader sense, the design engineer can identify those components that are affected by advances in the state of the art, and he can determine the depth of engineering needed by the circumstances. The procedures should be adequate only to provide assurance that systems effectiveness, safety, reliability, and life cycle cost requirements will be met.

In areas where they have control, engineers generally will not go beyond necessary quality levels of engineering service. If there are any extra resources available, such as time or funds, designers will use them to improve the product itself and not the quality of engineering services such as documentation. An exception to this generalization may occur under the temptation to "gold-plate" that is encouraged by cost-plus contracting.

9.13   CONFIGURATION CONTROL

Some of the conditions in the weapons industry that led to the establishment of requirements for configuration control are described in a quote from a speech, made by a high official of a U.S. military procuring agency (188), and which states:

Expensive and large systems were being delivered with no information as to how they were built and how to keep them running. Deficiencies and failures in tests started long sleuthing operations to determine what had failed and what design alternatives were available to solve the problem. Parts lists appeared months, sometimes even years after the equipment, and were usually inaccurate and incomplete. We qualified certain components and tests and then found usually after they failed in service, that the production design had been altered, and we found that only the original manufacturer could supply the system or its components because we couldn't tell anyone else how we made it.

To understand how this situation came about, consider Section 6.2.3 where it is stated that pre-World War II aircraft production contracts consisted of not much more than 12 units. At that time the contractor was required to furnish a complete set of drawings only for the last article of the series, since this aircraft would be the starting point for any potential repeat orders. Then came the explosive expansion of World War II with its cost-plus procedures and emphasis on financial record keeping to assure that no funds were spent for unauthorized purposes.

With this background, the aircraft industry entered the post-World War II systems development era involving large teams of sophisticated subcontractors with all the needed stimulating and exciting technical skills and facilities, but without an adequate appreciation for the rigorous and dull configuration and interface control procedures required for cost-effective management. These procedures have, therefore, all been developed as a result of an urgent need to provide the customer with the documentation he requires to conduct his operational testing, Boxes 22 and 27 of Figure 3.1, and to facilitate the transition from the prototype to the production phase.

9.13.1   Contributing Factors to Lack of Control

The difficulties in maintaining hardware-configuration control are caused by a large number of factors including:

1. Size and complexity of new systems. The development costs of new systems run between hundreds of millions and billions of dollars. The magnitude of these projects is so great that individual managers have a

difficult time in maintaining the overview necessary to coordinate all critical details properly and effectively.

2. Number of organizations involved. A large proportion, possibly more than half of the design and development work, is subcontracted to a relatively large number of geographically dispersed specialty houses. Corporate organizational boundaries and geographical distances restrict quality, volume, and timeliness of communications between lower-level working individuals who should be closely coordinating their activities.

3. The large number of creative people involved, generally on rotating assignments. The technical difficulties associated with systems development require large numbers of highly creative technical individuals who move from one difficult technical problem to another. As soon as a top creative engineer finds an acceptable solution to one problem he is reassigned to another problem, and the execution of the task is left in the hands of less creative individuals.

4. Requirements for extensive software and complex support.

5. Compressed schedules—see Figure 2.2.

6. High incidence of engineering changes. In any system involving growth potential and advances in the state of the art, hardware changes will continue during the entire production cycle. The number of changes will be largest during the operational test phase and initial deployment.

7. Time lag of 6 months or more between drawing release and implementation of a change.

These factors describe an operating environment that is fundamentally different from a small operation where the boss knows everything that is going on, or from a large commercial, mass-production facility that has been making the same product for years and where every operation has been thoroughly analyzed, all procedures have been worked out in detail, and all significant information is documented. For example, to replace a gas cap on a given 1976 model automobile at any repair shop, the mechanic simply looks in his manual and determines the gas cap number which fits all cars, say from 1973 to 1976. His manuals define the configuration of every model car and identify all interchangeable parts between specified models.

9.13.2   Special Effort Required for Advanced Systems

During the development of an advanced system a special effort is required to generate configuration control information. In the acquisition of advanced systems, the customer purchases not only the physical end product, but its design, development, and the operating and support software. The operator of sophisticated advanced systems requires precise knowledge of the configuration of his equipment to obtain the required performance, operational efficiency, logistic support, systems readiness, and safety of operations. The concept of configuration control is closely associated with quality control of the paperwork necessary to describe the system.

9.13.3   Updating Delivered Equipment

For example, electronic equipment is checked out on test benches driven by taped programs that control the test inputs to the equipment and check acceptability of the resulting outputs. Any change in the equipment configuration normally requires a corresponding change in the test tapes. It would be intolerable for an operating unit to have several configurations of the same piece of equipment, each configuration with its own peculiar test tape. It would be impossible to always coordinate equipment configuration with the proper test tape. To cope with the situation, an operational unit will stock only equipment of the same configuration, which can all be tested by the same tape.

In these circumstances equipment improvements must be introduced in blocks so that when an operational unit is supplied with a new improved configuration all pieces of the equipment are exchanged at the same time. The process of updating existing equipment is facilitated by the use of modular concepts. Improvements are then introduced by replacement of obsolescent modules.

Configuration management procedures provide a basis for control in conditions such as these by assuring the precise definition of the product configuration during its life cycle. The actual configuration of each hardware system or blocks of identical systems must at all times be properly recorded in pertinent documentation. There must be no discrepancies between contract, internal paperwork, specifications, hardware drawings, and inspection documentation. The procedures must assure customer and manufacturer that the delivered product corresponds with every aspect of the contractual documentation.

9.13.4   Scope of Configuration Control

Configuration control procedures apply mainly to paper, there is little contact with actual hardware. Furthermore, the review applies mainly to paper originated by others rather than by the configuration control unit. The major concern is with critical interfaces such as:

User—	Procurement agencies
Procurement agencies—	Prime contractor
Prime contractor—	Subcontractors/vendors/suppliers
Subcontractor—	Subcontractors/suppliers
Prime equipment—	Support equipment/software
Contract—	Delivered hardware

9.13.5 Baseline

The basic procedure is to identify the product at an interface as described by a mutually convenient *baseline*. Once this baseline is established, all subsequent changes from the baseline are monitored and tracked to assure that paperwork on both sides of the interface are in agreement and reflect the change correctly. The baseline is a significant point since it is normally a decision milestone for the customer and provides him with the required visibility and control for making such critical decisions as continuation, cancellation, or modification of a project (Box 19, Figure 3.1). The baseline assures the customer that he is getting the product contracted for and provides him with a means for assuring that the product will be compatible with other interfacing elements in his complement of available or planned equipment.

9.13.6 Configuration Identification

The configuration of an item is defined as the physical and functional characteristics of the hardware as documented in the *configuration identification* (pertinent drawings, operating support documentation, and specifications). It is derived during development, determined during design, usually approved by the customer via appropriate documentation, established during production, and maintained during operational support. As a product proceeds through research, concept design, development, detail design, first-article production, and follow-on production, the identification of its configuration becomes progressively more definite and precise, but it may be continually modified throughout the service life of the hardware, necessitating periodic updating of operational support documentation.

9.13.7 Project Phases

The major project phases included in most contracts are: conceptual, definition, and acquisition. The corresponding baselines are often referred to as "functional," "allocated," and "production."

Functional Baseline. Configuration identification applies to specifications defining project requirements.

Allocated Baseline. The project requirements are allocated to various lower-tier component elements of the product. Component design requirements are defined by means of lower-tier, "design-to" specifications. To accomplish this allocation, the weapons system is broken down into subsidiary categories to establish appropriate hardware identification points, and a baseline is established for each such point (see Section 9.5.4). These points define the levels at which identification will take place; they are not the same as quality-control inspection points, since the purpose is to identify the configuration and not to control the manufacturing process as discussed in Figure 10.12. A replaceable assembly, a vendor-supplied component, and a work package are obvious lower-tier identification points. However, certain higher-tier assemblies supplied as spares are also logical identification points.

Production Baseline. Configuration identification starts here at the lowest hardware identification point with the review of "build-to" documentation: specifications, engineering drawings, test plans and reports, and other detailed technical paperwork. All engineering change proposals (ECPs) and design change reviews are monitored, and the implementation of all changes is tracked. The "build-to" documentation is compared with quality-control records for verification. The identification procedure progresses from lower-tier identification points in an incremental progression to the next higher-tier points, until the entire system is identified and the documentation is completely validated.

At each identification point an audit and review procedure takes place to assure that the physical and functional characteristics achieved in the hardware item match those specified in the configuration identification, and that the requirements of the contact are satisfied.

A baseline is not a design freeze; the system baseline evolves toward the final configuration in an incremental manner over a considerable period of time. Each configuration identification point is audited separately at the optimum point in the development cycle of the hardware element involved. The system baseline then consists of a number of discrete steps leading to a final audit that encompasses all lower-tier baselines and changes.

9.13.8   Status Accounting

The configuration control function on most large projects is so complex and specialized that it is performed by separate specialty groups. One of the responsibilities of these groups is status accounting. The establishment of each baseline is recorded, and all proposed and approved changes from the baseline are continuously tracked and coordinated with other pertinent data. In administering a configuration control program, it is important to keep in mind at all times the purpose of the program, otherwise the job can be overdone and result in a waste of resources without achieving any useful purpose. It is not easy to determine the depth to go to be most cost-effective. The configuration control program should be carefully tailored to be consistent with quantity, size, scope, nature, and complexity of the hardware items involved.

9.14   ACCEPTANCE PROCEDURES

Advanced systems are expensive items often involving safety of life. In military systems, the security of the nation can be at stake. Formal procedures have been developed for the final acceptance of such items by the customer.

9.14.1   Acceptance Data Package

Adherence to the procedures described here results in the accumulation of a considerable volume of paperwork over time when applied to a complex advanced system. To assure completeness and integrity of this paperwork, current contracts call for a final review of an *acceptance data package* as part of the acceptance procedure for a system or subsystem. The acceptance data package may, but need not, coincide with the production baseline of Section 9.13.7. Selection of the baseline depends on the program and is scheduled as a matter of convenience.

9.14.2   Scope of Acceptance Procedures

As shown in Figure 9.1, three controlling items are involved in the acceptance procedure as follows: ·

End product.
Documentation.
Certification.

The end product consists of the completed physical system in deliverable condition.

Documentation consists of a package describing the system (drawings, specifications, engineering orders, pertinent analyses, test reports, etc.).

The certification consists of a package of papers verifying that the system conforms to its description, noting all exceptions (inspection records, material certifications, acceptance-test reports, waivers, etc.).

During the acceptance procedure this material is examined and checked in detail, and any deficiencies must be corrected before acceptance. To assure acceptance and on-schedule delivery, many suppliers perform a dry run of the acceptance procedure before the formal run. In addition to this procedure, in the space program, NASA required its system contractors to make a formal statement that they knew of no reason why the product is unacceptable.

9.14.3   Periodic Requalification

There may be concern about the ability of a supplier to maintain over time the capability that he demonstrated in the qualification test. This is often particularly true when value-engineering or other types of design-improvement changes are proposed without substantiating data at least as good as those that substantiated the original design. In such cases the supplier may be required to repeat the qualification test procedure periodically, as on every twenty-fifth or hundredth unit. It may not be desirable for the supplier to know in advance which units may be subjected to test; instead of identifying a specific unit, a specimen may be selected by lot at stated periods by a customer representative.

9.14.4   Introduction of Product Improvements

If improvements are proposed for a component of a subsystem subject to periodic qualification, the acceptability of this improvement can be demonstrated in a periodic qualification and accepted in associated production blocks.

9.15   RISK MANAGEMENT

One phase of systems development management that deserves special attention is that of risk management. The foregoing discussion of systems development management has been concerned with the customer-contractor relationship. In the case of systems that involve the public interest, in particular where safety or environmental hazards or impact are involved, as discussed in Section 5.12, a third public participant enters into the development process.

Apollo Fire

A number of references have been made in this book to the outstanding advanced systems development management procedures that made possible the unique success of the Apollo. It should be noted that these outstanding procedures were not all in force at the beginning of the Apollo project.

A fire occurred in the Apollo Spacecraft 012 on January 27, 1967, during a space vehicle integration test on Launch Complex 34 at Kennedy Space Center. Three astronauts were killed. The test was conducted with a pure oxygen atmosphere within the spacecraft.

The NASA Administrator convened a Review Board to investigate the accident (100), and subsequently there were extensive hearings before the U.S. Senate Committee on Aeronautical and Space Sciences (101). It was determined that Project Management had not maintained adequate visibility over lower-level hazardous operations, and that the safety procedures discussed in Section 5.12 had not been rigorously applied. There was real concern that Congress would cancel the program.

The fire had a profound effect on every professional associated with the Apollo program. From that day on there was a new dedication and determination to make the program a success. Lower-level individuals fought like tigers to get those things done that only they knew had to be done to make the program work, and they would not take "no" for an answer. Section 1.2.2 points out that the dedication of individuals reached new heights of personal courage and involvement (119).

Origins of Risk Management

After the fire in 1967, Lederer, then NASA Director of the Office of Manned Space Safety, introduced the concept of *Risk Management* to replace the term "safety" whenever feasible (204).

The term "risk management" implies that hazards are present in advanced systems, that they must be identified, analyzed, evaluated and controlled, or rationally accepted. It calls for the need to explore *all* foreseeable options to correct a hazardous situation and to put them in a suitable form for management decision. It recognizes a mandatory requirement for a system of *formal documentation* to assure that management knows and understands the risks it is assuming or rejecting. The techniques required to support a decision to accept or reject risks include broad considerations of available resources in terms of funds, manpower, and equipment; cost benefits and cost effectiveness; schedule interruptions; cost and consequences of failure; performance (weight, space, drag, reliability, maintenance); human factors; employee/management relations; public safety; employee health and welfare; public acceptance (noise, pollution, cost, hazard); prestige; state of the art; competition; waivers; risk levels; uncertainties; probability of success and its effect on the rest of the system; design trade-offs; and political factors.

It should be noted that today the term "risk management" is also widely associated with the determination and evaluation of the risks and uncertainties involved in attaining specified advanced system performance. In the following discussion, the term "risk management" is applied exclusively in its original NASA safety-associated definition.

Effect of Failure on Human Performance

Men rise to their highest level of performance when motivated by some serious disaster or threat of disaster. Almost all successful advanced systems whose details are known to the author have been designed under the pressure of imminent disaster, of failure of preceding programs, or of failure of similar programs in competing organizations. It appears that a good program manager is one who has survived and learned both from his own past mistakes and those of others.

Building Codes

A common example of risk management is provided by municipal and state building codes. A contractor building a structure for a customer must not only comply with the customers requirements, but also with those of various regulatory agencies.

In many jurisdictions a building permit will not be issued until the drawings and specifications have been reviewed by a regulatory agency to determine that they comply with the applicable safety codes. During construction there may be a number of step-by-step inspections; and the next phase of construction may not proceed unless previous inspections are satisfactorily completed. For example, a foundation may not be poured before the inspector certifies that the concrete mix complies with the specification.

Building codes, however, are not only concerned with safety, sometimes they also involve environmental and artistic factors. When factors other than safety are involved, compliance with code requirements may add a substantial amount to the cost of construction.

An example of such increased costs is provided by the S.S. United States, then the pride of the U.S. Maritime Fleet. There were over 100 regulatory agencies involved in the construction of this ship, and only a

few of these agencies were concerned with safety. The various requirements included the ability to convert the ship quickly in case of a national
emergency to a troop transport with the ability to pass through the
Panama Canal. The SS United States was the largest ship ever built that
had the capability of passing through the Canal. All of these regulatory
requirements increased the construction cost of the ship by a factor of
almost 5.

However in most cases of intelligent risk management, the safety codes
require no more than sound engineering design practices, tests, inspections, operator training, and documentation. Many studies, primarily in
the aerospace field, have shown that the application of these procedures
produce superior systems with reduced life cycle costs which, over time,
will much more than compensate for the higher initial costs.

Scope of Risk Management

The purpose of this section is to discuss the basic operational principles
involved when public or operational safety is a factor and requires a
regulatory body to participate in the advanced systems development
process. For the sake of simplicity it is assumed that only a single
regulatory body is involved, providing a three-way relationship between
customer, contractor, and regulator.

The basic elements of risk management include:

Regulatory codes.
Regulation.
Assurance.
Enforcement.

Each of these elements is discussed in the following material.

9.15.1 Regulatory Codes

The role of codes and standards has already been mentioned with reference to design requirements, standardization, and commonality in Sections 3.2.2 and 5.12. The following discussion concentrates on safety.

Central Analysis Agency

It takes considerable time and perseverance to produce a good regulatory
code. There must be a technically competent permanent central organization that collects and analyzes over time all pertinent experience, in
particular accident data, and maintains the code in an up-to-date condi-

tion. Since safety requirements often increase initial costs, requirements are not included in the codes unless there exists overwhelming evidence that they are necessary. For example, most building codes grow with time, each new provision being added as a result of some spectacular disaster in which a number of lives have been lost.

ASME Boiler and Pressure Vessel Code

An excellent example of an accepted code is the Boiler and Pressure Vessel Code of the American Society of Mechanical Engineers (ASME), mentioned in Section 3.2.2 (196). The code is written and maintained by a committee that consists of over 650 volunteer technical experts. The committee investigates all failures of equipment within the scope of the code, and the code is continuously revised and expanded to eliminate deficiencies, improve requirements, and to take advantage of new technology and materials. The code was first published in 1914; at that time some 2000 people were being killed each year in the United States because of boiler explosions. Today the number is practically zero. The ASME Code is used and accepted in most of the industrialized nations of the world. It has been tested in the courts and survived all challenges.

MIL-HANDBOOK 5

Another successful code is MIL-HANDBOOK 5 (91). This handbook contains the material properties used in aircraft design. It is revised semiannually by a dedicated committee that has been in continuous operation for over 35 years. Every major structural test performed on an aircraft in the United States, every new material or technological innovation, every structural failure and aircraft accident involving material failure is analyzed, and any new conclusions are included in the next revision of the handbook.

Unfortunately, the need for a safety code to guide the designer is not universally recognized as the first requirement for an effective regulatory safety program. For example, this need is completely overlooked by Offner (189).

9.15.2 Regulation

Given a code, there must be a statutory regulator to require its enforcement. For example, all states of the United States and most industrialized countries have laws requiring the use of the ASME Boiler Code. Building and operating permits and insurance policies will not be issued unless it is demonstrated that all boilers and pressure vessels covered by the code are designed, constructed, inspected, and tested in accordance with the code. Authorized inspectors provide step-by-step inspection as an item is being constructed. Aircraft will not be licensed by civil and military authorities unless MIL-HANDBOOK 5 has been used in the design of the structure.

The Proceedings of the 1976 Annual Reliability and Maintainability Symposium contain two papers on risk management (190 and 191). Reference 190 describes a system designed by NASA for the risk management of the very large liquefied natural gas installations planned for New York City. A code for the design and operation of liquefied natural gas facilities and equipment (103) was prepared by NASA-Kennedy Space Center (KSC). One of the background documents used in this project is a survey by Allan (104). Reference 103 was prepared at the request and with the active participation of New York City because of NASA's experience with hazardous materials used in the space program. At the time the paper was published the plan had not yet been implemented by New York.

The plan represents what NASA would do if it had the responsibility for the problem and New York's resources to work with. For a report of the first cautious unloadings of liquefied natural gas in New York City see ref. 192.

Reference 191 describes similar techniques applied to the proposed Baltimore and Atlanta subway systems. Another example of the safety management of a complex NASA ground operating system is provided in ref. 102.

9.15.3 Assurance

It must be admitted that sophisticated advanced systems are always associated with residual risks; these must be identified and proper provisions made. For example, insurance companies recognize and evaluate the risks involved in commercial power plants built in compliance with the ASME Boiler and Pressure Vessel Code and provide for them in their premium charges. One of the reasons for the status of the ASME Code is the large amount of pertinent data that insurance companies have collected; there exists no other comparable data base for the evaluation of risks within the scope of the ASME Code.

The NASA assurance procedure consists of six major elements: hazard identification, risk analysis, contingency planning, training of operating personnel, hazard accountability, and project reviews.

For a similar system of assurance published by the American Gas Association for the gas industry see Ref. 29.

Hazard Identification

This process is discussed in Section 5.12.1 and involves the systematic comparison of every feature covered in a safety code with the pertinent requirements of the code. This process may require the preparation of extensive checklists. Noncompliance with a code requirement constitutes a potential hazard. Each potential hazard is followed up by a risk and accountability analysis.

Risk Analysis

Risk analysis is the procedure for determining that a potential hazard has been eliminated or reduced to an acceptable level. The most efficient way to resolve a hazard problem is to design it out of the system. Risk analysis is not a straightforward mechanical procedure; its implementation requires competence and judgment. The elements of risk analysis involve the following:

- Determining the need for an analysis.
- Choosing the applicable technique.
- Recognizing the technical level of competence of the personnel conducting the analysis.
- Understanding the results of the analysis in terms of acceptance of the risk associated with the hazard.

The techniques used are similar to those used by reliability engineers (see Section 7.6 and also ref. 29).

Contingency Planning

This procedure applies to residual potential hazards that cannot be reduced to an acceptable level without additional protection and safety systems. The purpose of the exercise is to identify and provide "before-the-fact" the additional resources needed to cope with a residual hazard, rather than to provide such resources as a costly reaction to a major disaster.

Training of Operating Personnel. In conjunction with planning, provisions must be made for training of operating personnel and assuring that in an emergency each man knows what to do. In cases requiring special skills, licensing of personnel may be required.

Section 5.12.1 states that major accidents do not occur without some prior warning signals. A special list of these warning signals should be prepared for all residual hazards, and operators should be trained to watch for these signals and to recognize them. This list should also provide for routine safety checks which should be performed and recorded on a scheduled basis to assure safe operation. Strategically located warning posters and instructive labels should provide a constant reminder to operators to remain alert. As stated in Section 5.12.1, operators should be drilled on exactly what to do when they detect a warning signal: whom to call, what alarm to activate, what valves and switches to turn off, where to go. These procedures should be as simple as possible; a man who has recognized a danger signal should not be required to perform any further analysis or do any thinking; he must be trained to act automatically, decisively, and quickly.

Unfortunately, the value of this training and drilling is insufficiently recognized in many hazardous occupations. It has been demonstrated that properly trained people in an emergency will generally act exactly as trained. An outstanding example has been the performance of cabin attendants in airline crashes. They are trained to evacuate the aircraft in an emergency in less than 90 seconds. Disciplined females shouting standard preestablished orders at the top of their voices command a degree of attention and cooperation that prevents panic and assures an orderly evacuation in a minimum of time. The outstanding performance of these heroic women has been responsible for the saving of many lives.

Hazard Accountability

A hazard accountability system should track the status of all hazards during the resolution process. The system should also record the closure of each hazard or the acceptance of a residual hazard—and the names of the engineers responsible for eliminating the hazard. Such a system should be computerized and have an on-line capability of issuing reports at any time on demand, as well as periodic reports to provide visibility to cognizant management personnel.

Project Reviews

Project reviews should be scheduled as strategic milestones during system development and include representatives of all interested parties: customer, contractor, and regulator. The status of all outstanding hazards should be reviewed, the accountability determined, and an appropriate management plan developed to cope with the residual hazards.

9.15.4   Enforcement

The effective enforcement of a risk management system depends to a great extent on the integrity of the personnel involved. Much of the significant risk management effort in the Apollo Program was conducted by a small group of competent and dedicated personnel. A risk management system is in effect a system of resource controls which can be maintained and improved by a permanent team of a few good engineers who can exercise good judgment and make timely decisions in a closed-loop iterative process. Continuity and quality of the team is essential. A large organization or bureaucracy and expensive "red tape" will poison the effectiveness of the risk management system.

Unfortunately, many inspection laboratories take shortcuts; and payoffs to inspectors for accepting unsatisfactory work is a practice in many industries (189, 195). In the case of the ASME Boiler Code, insurance companies generally will not insure an item within the scope of the code unless it conforms with the code requirements. This financial responsibility is quite effective in assuring that inspections are properly carried out and has been one of the features contributing to the success and acceptance of the Boiler Code.

9.16   MANAGEMENT EFFECTIVENESS

Given an unexpected event, such as a premature failure, the amount of time required by a team to analyze the problem, and to devise and take effective corrective action, is believed to be a good measure of the effectiveness of the team. A more skilled engineering and management team will properly analyze the causes of premature failure and institute effective corrective action in less time than will a less skilled team. In particular, it is assumed that the total time between the initial observation of an unsatisfactory condition and the completion of an effective corrective action is an index of program management capability to cope with its problems.

9.17   CHAPTER SUMMARY

This chapter provides an overview of the techniques required to manage a large dynamic systems development program. Some historical observations trace the growth of development programs from pre-World War II fixed-price competitive practices and small production contracts to the major cost-plus expansion associated with World War II, calling for new management and control procedures. The exciting technical breakthroughs following the war added a new dimension to the size and integration challenges involved in development, giving birth to new systems engineering disciplines. This technical growth was achieved at astronomical cost; one of the basic challenges now is how to maintain and continue this technical progress while simultaneously substantially reducing cost.

In view of the large numbers of creative individuals involved in a modern systems development, management attitudes towards its personnel is of critical importance. Individuals must be made to feel responsible and challenged to fulfill their potential. Because of the large size of development organizations and the constant turnover of personnel, emphasis is placed on good and complete policy manuals.

The functions and mechanics of planning, scheduling, and control are discussed; they provide the basic tools for allocating resources, establishing the quality levels to be attained, and providing management visibility. However, all control systems have their limitations, since those controlled invariably devise countermeasures to further their own interests. It is therefore suggested that management must find additional means for monitoring the technical quality of the development.

A large portion of any systems development will be subcontracted to specialty houses that complement a contractor's capabilities. The technical coordination, monitoring, and approval cycles are associated with great volumes of paperwork that must be processed in a timely manner. A computer program is essential to provide management visibility over the large flow of data.

Other somewhat intangible but fundamental considerations are quality of paperwork and quality of engineering.

A systems development project cannot be run without mountains of paperwork. The system manager must assure that the correct document is used for the proper purpose, and that document types are kept to a minimum. Paperwork systems all have an inherent tendency to grow, and, if not monitored carefully, they can strangulate a project and become the source of uncontrollable overruns. Development engineers are extremely difficult to manage; they are always striving to improve the design or process, and they always come up with their best ideas after funds have run out. They cannot be expected to be profit oriented. The systems manager must establish the standards for technical adequacy and must enforce them firmly under conditions where every effort will be made to frustrate his profit oriented objectives.

Formal configuration control has become a necessity because of the size and complexity of modern systems and associated factors. It demands a new degree of discipline in the paperwork system. It is a distasteful activity to technical development personnel since it imposes an additional burden on them without providing any assistance in system development. The problem is that the system may be useless to the customer if configuration control is not maintained.

The development concludes when the system is accepted by the customer. The elements involved in the acceptance procedures are briefly listed.

The above discussion concentrates on the customer/contractor relationship. Whenever public safety or an environmental impact is involved, a regulatory agency becomes a participant in the development, providing for a three-way relationship. This relationship is described in the section on risk management. The discussion is broken down into four elements: regulatory codes, regulation, assurance, and enforcement.

The chapter closes with the suggestion that a measure of management effectiveness is the amount of time it takes a management team to identify a problem and implement a solution. Management procrastination in making decisions is considered one of the most common sources of high costs.

10

Principles of Quality Assurance

Quality is discussed by Edwards (205) as a concept which changes with time and circumstances; it varies with "wants" and "needs" from buyer to buyer, and a single buyer will change his concept of quality with time, even between the time of contract award and the delivery of the end product. In any specific case, the concept of quality will have many dimensions, some of them intangible. In a technical sense in systems procurement, the term quality has meaning only when *wants* or *needs* are intelligibly defined by *contract requirements*.

There exists a wide divergence between what progressive quality engineers believe should be their responsibilities and the actual scope of their activities. If the desired full scope visualized by quality engineers can be represented by a line several yards long, then their actual activities can be indicated by a very small portion of this line, not more than an inch or so in length. In the past quality control has concentrated on assuring manufacturing conformance with engineering requirements. More recently, there has been a movement for "total quality control" to involve the quality engineer in policy matters affecting product quality. Along this line there have been serious efforts made to establish programs to improve the "quality of design." Although this is a top-priority national need, most of the procedures and tools needed to implement the concept still must be developed (see Section 7.6.13).

Historical Development

Part of the quality engineer's problem is historical. As the original one-man operation in the initial stages of industrialization began to grow, the owner was forced to delegate some of his functions: the bookkeeping and financial matters to his wife, the sales function to a salesman, supervision of the shop to a foreman. As long as he could, however, he himself would inspect all outgoing articles. This was the last function to be delegated.

However, the inspection function itself does not establish the quality and workmanship levels in a plant. The establishment and maintenance of quality levels is a complex function involving many business trade-offs, which the owner himself executed. When he delegated the inspection function, he did not delegate these other functions. This is the crux of the quality control concept dilemma today.

Quality Policy

Our manufacturing plants have become very large corporations, run by teams of managers. Owners (stockholders) and top executives often no longer identify themselves with the product. Control of the design and production process has been largely delegated to specialized teams of professional managers who have full responsibility within a limited scope of total corporate activities. The quality engineer sees his best chances of contributing to the establishment of quality policy by gaining recognition as a member of this management team. Today a number of progressive corporations have a management official with a title such as: Vice President for Quality Assurance. Reference 129 develops the following definition of the quality manager and engineer of the future:

In view of the increasing awareness in industry of its broader responsibilities to serve mankind, the quality manager will emerge as the corporate executive responsible for developing and assuring the implementation of corporate policy for quality. The quality engineer will be responsible for implementing corporate policy to achieve quality of products and services to meet consumer needs.

Quality Assurance Codes

Individual industries have developed quality assurance codes to fit their peculiar needs; for example ref. 138 serves the nuclear power plant industry. One of the most progressive concepts of the quality assurance function in the United States today appears in MIL-Q-9858A (73), and this document is used as the basis for the following discussion. This specification requires a defense systems contractor to prepare and implement a quality assurance program that will assure his compliance with the requirements of his contract. This procedure makes the contractor himself responsible for the quality of his own product; this constitutes a historical reversal of the age-old principle of "caveat emptor." One of the basic reasons for this development is that modern advanced systems have become so large and technically so sophisticated that the customer no longer can staff his organization so as to be able to judge quality realistically by analysis and inspection.

Legalistic Approach

Within the framework of advanced systems procurement described in Section 2.2, it is possible to visualize two attitudes that a contractor may take toward his contract. One extreme viewpoint, based on the mistaken belief that the contract specification is a true and complete description of the customer's requirements for a new system, is to examine each clause of the contract and then so lay out the work that every requirement of the contract or specification is satisfied with a minimum of work and a maximum of profit. Since specifications may be, and too frequently are, out of date, obsolescent, or defective, blind adherence to this viewpoint will result in a less than satisfactory product.

Adaptive Approach

The opposite viewpoint acknowledges the fact that a new system cannot be fully described before it is developed, and a contract is largely an agreement to work with the procuring agency on the development of a new system. Throughout the development a continuing analysis is made of what needs to be done at any given point to accomplish the desired end objective, which may change as the design progresses. The work is planned and scheduled in what is considered, at the moment, to be the best manner to achieve the final goal, regardless of any technical shortcomings of the contract. If the funding is inadequate, the deficiency is pointed out to the customer as soon as it is discovered. It is assumed that the procuring agency does not want something for nothing but is obligated to maintain effective controls to assure that it gets its money's worth. The written contract is considered a dynamic document that in good business practice records the status of agreements at any given time.

Most advanced systems contracts contain provisions whereby both parties, that is, customer and manufacturer, are invited to propose amendments to contract specifications and designs in consonance with the foregoing discussion. These proposed amendments are reduced to writing. However, to protect the interests of both of the parties, the contracts and designs proceed as though no amendment had been proposed until both parties have accepted the contract amendment.

The Role of Specifications

In actual practice, neither of these two extreme viewpoints is applied in its pure form. Although most companies lean one way or the other in their basic outlook, the approach for any one contract will be a mixture of both viewpoints.

Approximately 70,000 to 100,000 pages of specifications are required to control the procurement of a modern advanced system. With such a large amount of documentation, and considering the time schedules and the frailties of human nature involved, there will always be deficiencies in the specification structure in most technical areas. The purpose of specifications and related procurement contracts is to procure a satisfactory system; the establishment of a perfect and consistent specifications structure is not in itself a major objective. Specifications will, therefore, always be secondary and subject to some interpretation by reasonable and intelligent men. Nevertheless, every effort should be made to keep specifications as up-to-date as possible, and once a contract is let, to keep specification changes to a minimum.

Assuring Customer Satisfaction

These dynamic contractual relationships provide the framework for the quality assurance program. In almost every practical case, a good quality assurance program is expected to do more than just assure compliance with the formal system requirements. It must anticipate the customer's needs. The advanced system should provide customer satisfaction in actual use, and this invariably requires the advanced system to have characteristics not explicitly stated in the specifications. Nevertheless, it behooves the customer to stipulate, so far as is feasible, any and all critical design requirements that the customer considers must be complied with to attain an effective, satisfactory advanced system

The following sections contain a simplified description of the principles of quality assurance. These principles must be applied with skill and imagination. They will provide a competent and aggressive quality assurance engineer with the tools to get done those things that possibly only he knows need to be done to achieve eventual customer satisfaction.

10.1 DEFINITION OF A QUALITY ASSURANCE PROGRAM

In the early days of quality assurance, the major emphasis was placed on inspection and on organization (197). Although the inspector's job was to pass and accept good work, he achieved his end by performing a negative function, by rejecting bad work. To enforce this negative function, two requirements had to be met: first, he had to be an unbiased judge of the facts, that is, be independent of engineering and manufacturing; and second, he had to have the authority to make his decision stick, at times against formidable opposition created by the need to meet tight schedules. To command this authority, the head of the quality assurance organization has to be at least on the same organizational level as the heads of engineering and manufacturing, so that he cannot be overruled by them without recourse to higher authority.

In many companies today, the major emphasis of working quality assurance engineers is still on the screening function and the associated statistical considerations. However, MIL-Q-9858A (73) now goes far beyond this concept. Although MIL-Q-9858A is today called out in most major military and space contracts in the United States, its concepts are so all-encompassing that it is doubtful that there is any program in existence at the present time that complies fully with all the requirements of this forward-looking specification.

10.1.1 Quality Assurance Program Requirements

MIL-Q-9858A lists tasks to be accomplished but does not specify any organizational structure for the implementation. A contractor is free to accomplish these tasks in a quality assurance program of his own design to best fit his own organization. Many of the tasks listed are not accomplished by quality assurance personnel, but they inherently affect quality and must, therefore, be recognized as part of the total quality assurance program.

Today there are quality engineers who believe that quality assurance departments should divest themselves of the routine inspection function and should associate themselves with systems engineering.

A quality assurance program describing the contractor's quality objectives, procedures, processes, and management plan normally is submitted as part of a proposal. However, the procuring agency may not agree fully with all of the provisions of the proposal, and these items will be negotiated. The final plan is revised accordingly and submitted to the customer in technical report format for approval shortly after contract award. On approval, it becomes part of the contract.

The quality assurance program of MIL-Q-9858A applies to all areas of contract performance. Specifically mentioned as applicable areas are: design, development, fabrication, processing, assembly, inspection, test, maintenance, packaging, shipping, storage, and site installation. Figure 10.1 presents the concept in its most simplified form as applicable to any one of these areas or "functions."

10.1.2 Monitoring a "Function"

Given the necessary funding and specifications, the quality assurance program monitors the performance of the "function" to assure compliance with the specification requirements. This quality assurance activity is shown as a feedback-control loop, as it is a continuous process that controls the performance of the function on a real-time basis.

Although the monitoring tools appear to be negative, the purpose of the quality assurance function is a positive one, and a great deal of thought is being expended on developing better motivational techniques, as discussed in Section 10.6.

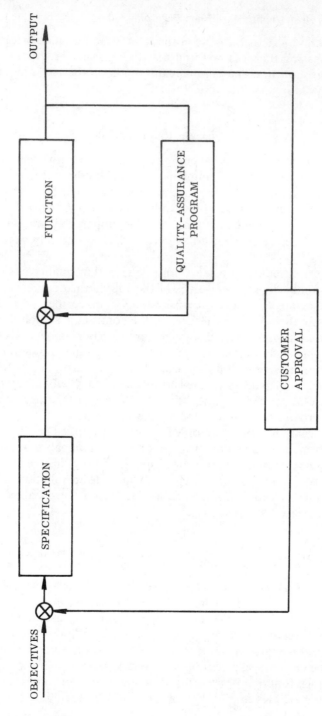

Figure 10.1. Definition of quality-assurance program.

The only way to deliver 100% good parts is to fabricate 100% good parts in the first place. If the fabrication process produces a mixture of good and defective parts, and reliance is placed on the inspection system to identify and eliminate the defective parts, some defective parts, with certainty, will occasionally escape detection.

Quality assurance must, therefore, be concerned with the primary factors that assure that all parts produced will be good parts. These factors include worker and operator training, motivation and morale: encouragement of craftsmanship, development of a sense of personal responsibility in all concerned, and motivational programs. Increasing attention is being given to the detail identification and elimination of the causes of scrap in all phases of fabrication (193).

10.1.3 Incoming Materials

When procuring critical material from vendors (raw material, piece parts, or components), it is essential for a contractor to maintain a technical staff that thoroughly understands the design and fabrication processes involved in producing the items being procured. Without such technical expertise there is no way for the contractor to assure himself that the supplier is complying with the technical requirements of his purchase order.

The first formal inspection point in any operation is concerned with incoming material. The most effective method of detail inspection of incoming material is at the source. In critical areas the contractor will maintain a permanent or transient inspector at the supplier's plant to assure that the supplier's quality assurance program is operating satisfactorily, and to inspect all material before shipment.

Incoming raw material is normally certified to be in accordance with the procurement specification. Depending on the criticality of the application, the contractor may then perform a chemical analysis and other tests on a spot-check or batch basis.

During the manufacturing cycle the satisfactory completion of all pertinent operations is checked, verified, and recorded as the work moves from one department to another.

10.1.4 Acceptance Criteria

Although in advanced systems procurement the theoretical standard of comparison to determine acceptability is the specimen that passed the qualification test, in practice it is not physically feasible to use the successful qualification test specimen as the standard for comparison, and acceptance criteria are established by comparisons with drawings; material, workmanship, and other derived standards—all of which must indicate the acceptable tolerances.

Acceptance criteria must be established in terms of a go/no-go test and should leave no latitude for judgment in marginal cases. If an expensive part is marginally defective but still serviceable, it should be examined by a material review board consisting of customer and contractor experts authorized to grant waivers to specification requirements in individual instances. Such waivers may be coupled with price adjustments.

In the case of devices involving precise flows or electrical characteristics, measurable geometric similarity of interacting parts may not be sufficient to establish the required characteristics, and additional performance tests may be required. Such tests are performed on special-purpose test rigs which must be as simple as possible and which must be based on an analysis of the qualification test results. Normally, they will verify conformance with one or two critical points on a curve established during the qualification test.

10.1.5 Licensing of Mechanics

There are processes, such as welding and soldering, of which the integrity cannot be completely determined by any available nondestructive inspection method. When these processes are used in the fabrication of critical parts, such as engine mounts or aircraft arresting hooks, or critical electronic parts, it is common practice to require that the work be done by licensed mechanics. Furthermore, such parts usually are subjected to a proof-load test before acceptance. These proof-load tests are screening tests that provide evidence that parts are unacceptable when they do not pass the test. The tests are not sufficiently comprehensive to provide evidence that parts that do pass the tests are completely acceptable. In particular, these tests do not confirm design integrity completely in those cases where deficiencies are not highlighted by the test results.

10.1.6 Traceability

Another requirement for critical parts specifies the maintenance of traceability records. Should a defective element be discovered in a given assembly, it must be possible on the basis of the part and serial numbers of the defective part to identify the lot or batch from which the defective element was taken, and to trace the disposition of all the other elements manufactured in the same lot or batch, that is, to identify the assemblies into which they have been installed and their location in those assemblies, so that all the elements from the same lot can be inspected and replaced with like parts from another verified lot, should the circumstances warrant.

10.1.7 Customer Monitoring of Quality System

Figure 10.1 shows the customer monitoring the operation of the quality assurance system, specifically by constantly comparing the output (product) with the quality program objectives. Should the customer be dissatisfied with the output, the contractor's quality assurance program should be improved as necessary to attain corrective action. If this fails, it may be necessary for the customer to modify the contractual specification requirements to make the contractor's quality program more effective.

This type of feedback loop appears in almost all of the figures in this section. As with similar feedback loops discussed in previous sections, the loops leave much to be desired. They hardly ever perform with the desired effectiveness. The more people in a loop and the more organizational lines that must be crossed, the more complex the system or function, and the more lead time involved, the less effective the feedback loop will be.

Some of the reasons for the lack of effectiveness of the feedback loops are:

1. Problems are often not well enough defined, making it easy for those responsible for taking corrective action to pass the buck.
2. There are communication barriers between groups, especially across organizational boundaries; groups do not understand each other's needs and language.
3. There is too much reliance on oral communication. To be realistically effective, formal written documents are necessary.
4. There is a general lack of a formal, written follow-up system.
5. There is too much of a tendency to substitute lip service for adequate corrective action.

10.2 INITIAL QUALITY PLANNING

Given the task of establishing a quality assurance program for a specific function as shown on Figure 10.1, a master plan first must be prepared involving the elements shown in the boxes of Figure 10.2.

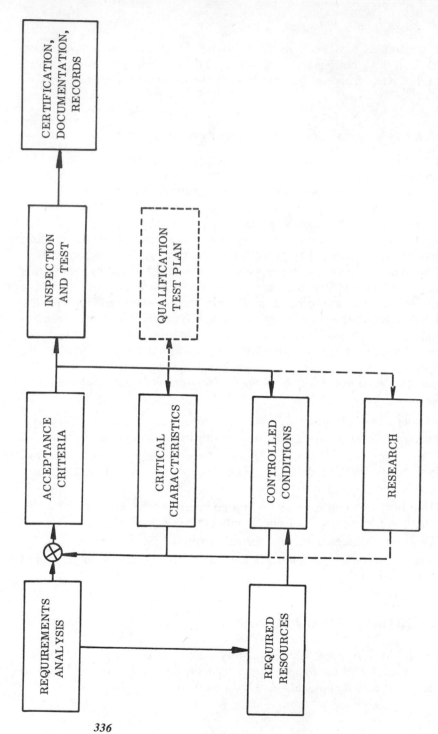

Figure 10.2. Development and use of acceptance criteria.

10.2.1 Required Resources

The requirements of the contract must be reviewed and analyzed to identify the *required resources,* that is, the personnel, facilities, processes, materials, test equipment, tooling, skills, special controls, and other elements that will be required for assuring quality. The master program plan must make proper and timely provisions for these resources.

In some cases, the contract requirements may exceed the technical state of the art or know-how of the organization; it is important to recognize at this stage that research may be needed in these areas.

10.2.2 Critical Characteristics

A collateral result of the review and analysis of the contract requirements is the determination of the *critical characteristics* that the end product must exhibit to perform its mission. This analysis normally culminates in the qualification test plan, Section 6.10; this is straightforward systems engineering. However, it is a key operation that affects quality, and it is listed in Figure 10.2 as one of the building blocks of a quality program.

In the initial stage of a program, the qualification test plan will not yet be formulated; nevertheless, a preliminary list of the critical characteristics can be established, and this must be done, even though changes will be made as the design progresses. An analysis of the critical characteristics provides the basis for the establishment of the *acceptance criteria* (Section 10.1.4).

10.2.3 Controlled Conditions

Given the list of required resources, the critical characteristics, and the acceptance criteria, the next step is to define the *controlled conditions* in which the function is to be performed. Controlled conditions are established by documented work instructions; adequate tools, equipment, and facilities; and the special working environment as needed. In particular, no physical examination, measurement, or test may be conducted except under controlled conditions.

The work instructions must cover all of the work that affects the quality of the product. The procedures must be prescribed in clear and complete documented instructions appropriate to the circumstances; they must provide the criteria for performing the work functions and must be compatible with the acceptance criteria. In particular, work instructions must include the criteria for acceptable or unacceptable "workmanship."

The critical characteristics, the controlled conditions, and the research program are shown in Figure 10.2 as the major factors acting in a real-time feedback loop for controlling the acceptance criteria. During the course of the program these four elements will continue to change; the feedback loop is intended to indicate their mutual interdependence and the fact that they must all be compatible.

10.2.4   Inspection and Test

The *inspection and test* activities are derived from the acceptance criteria in a straightforward manner.

In a production function, complete or direct inspection is sometimes impossible, such as with structural welded joints. In such cases indirect methods must be used: the monitoring of process methods, equipment, and personnel, and the destructive testing of samples. If an operation is critical, personnel may be required to be specially trained and licensed. In many cases both physical inspection and process monitoring are used. Adherence to the selected prescribed methods must be complete and continuous.

If corrective measures are necessary, those methods must be used that yield adequate results in the most timely manner—by machine operators, automated inspection gauges, moving-line or lot sampling, set-up or first-piece approval, production-line inspection station, inspection or test department, roving inspectors, or any other type of inspection. Acceptance criteria must provide means for identifying approved and rejected parts.

10.2.5   Quality Records

The final block of Figure 10.2 indicates that the quality assurance program must provide for the preparation, maintenance, and use of records and data essential to the economical and effective operation of the quality assurance program. Records are considered one of the principal forms of objective evidence of quality assurance so they must be complete and reliable. There must be provisions for the analysis and use of appropriate records to serve as the basis for management action, and in a like manner, the records also must be available to the customer for his review and analysis on request.

Records for monitoring work performance must indicate the acceptability of work or products and actions taken in connection with deficiencies. Inspection and test records must indicate the nature of the observations together with the number of observations made and the number and type of deficiencies found, if any.

10.3   QUALITY ASSURANCE OF A FUNCTION

Figure 10.3 shows the relationship of the various activities involved in assuring the quality of a specific function. This figure is generalized; the procedure applies to all functions subject to quality assurance control, both in the contractor's plant and in the facilities of all subcontractors and vendors.

Given a set of requirements, Figure 10.3 shows the function being performed under controlled conditions.

The performance of the function is monitored to assure that all work is being performed in accordance with the work instructions, and that the output will meet the requirements. The monitor must detect promptly any condition adverse to quality and initiate appropriate corrective action.

The output of the function is subject to inspection and test to assure that it complies with the acceptance criteria. Rejected output is subject to review upon request of the producer. Data on rejected material are subject to analysis and must result in corrective action to prevent further occurrence of the defective condition. All records produced in the acceptance procedure are subject to analysis and review to serve as a basis for management action.

Corrective action, as a minimum, includes the following activities:

1. Analysis of records and data and examination of the rejected output to determine the extent and causes of the defective conditions.
2. Analysis of trends in processes or performance of work to prevent nonconforming output.
3. Introduction of required improvements and corrections, initial review of the adequacy of such measures, and monitoring of the effectiveness of the corrective action taken.

10.4   CONTROL OF ENGINEERING DOCUMENTS

Figure 10.4 shows in highly simplified form the basic procedure involved in maintaining control of engineering drawings and specifications. The purpose of these documents is to define the product to be built, and they provide the basis for the operation of all downstream company departments such as manufacturing, planning, purchasing, manufacturing-support, and shipping.

In addition to meeting the contract requirements, drawings and specifications must comply with applicable customer standards and handbooks as well as in-house standards and manuals. *These standards, handbooks,*

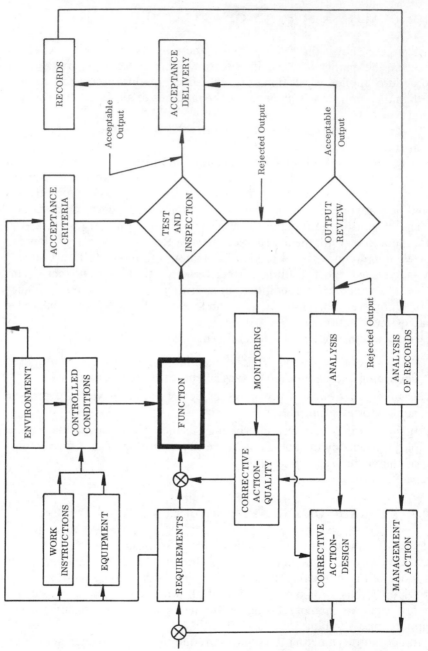

Figure 10.3. Quality assurance of a function.

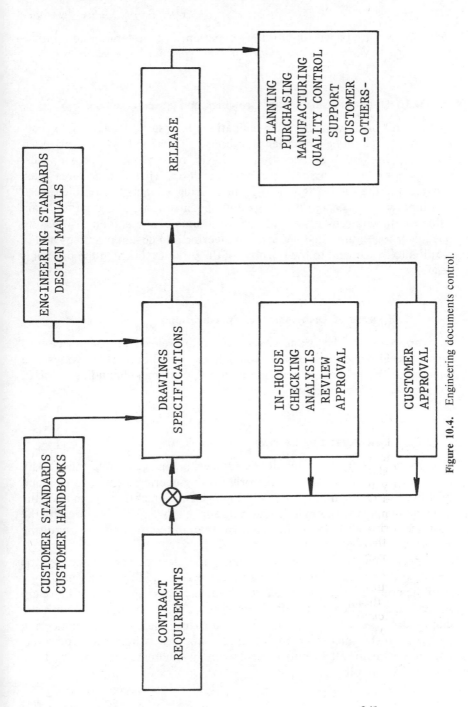

Figure 10.4. Engineering documents control.

and manuals are based on previous experience of customer and contractor; *they are, therefore, the heart of the experience retention system.*

10.4.1 Quality Assurance of Drawings and Specifications

The quality of drawings and specifications is assured by real-time feedback loops containing a number of in-house clerical and systems engineering procedures such as checking, analysis, review, and approval. Customer approval is provided in a separate loop. These procedures must provide for the evaluation of the engineering adequacy, completeness, and timeliness of design drawings and specifications, and for the establishment of requirements for effectivity points. They should also assure adequacy in relation to standards engineering and design practices and in relation to design and to the purpose of the product to which the drawings relate.

10.4.2 Release of Drawings and Specifications

After approval, drawings and specifications are released through a central point, which is also responsible for delivery of all correct drawings and equipment and process specifications to the customer, including those of subcontractors and vendors.

10.4.3 Control of Engineering Changes

The procedures for the control of engineering changes are identical with those for the initial control of the original engineering drawing and specification, as shown on Figure 10.5. A changed design must be of at least the same quality as the original design; this applies to changes required by the customer as well as those initiated by the contractor and not requiring customer approval.

In preparing an engineering change, the designer is guided by the analysis of the defect he is trying to correct. In addition, he must cope with the constraints of the existing design and possibly with parts already manufactured. The control procedures must assure compliance with the contract requirements for proposing, approving, and effecting engineering changes. In the release of engineering changes it is important for reference purposes to retrieve obsolete documents and parts or to assure their proper disposition.

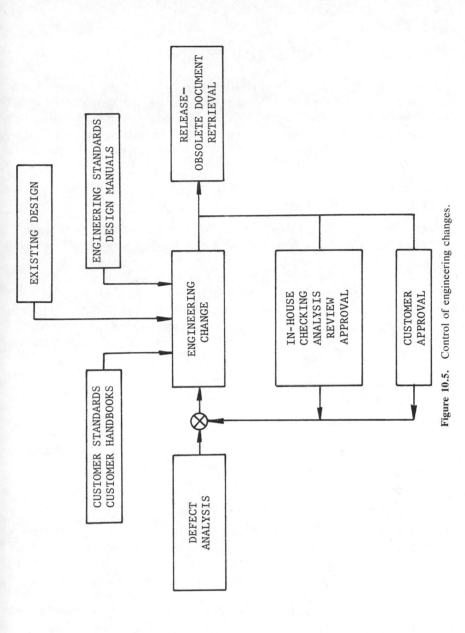

Figure 10.5. Control of engineering changes.

343

10.5 CONTROL OF MEASURING DEVICES

A quality assurance program must have provisions for the control of the accuracy or calibration of measuring and testing devices as shown in simplified form in Figure 10.6. Test and manufacturing plans must be analyzed to determine the requirements for such equipment and the accuracy needed to assure that the output will comply with technical contract requirements.

10.5.1 Calibration Plan

Specifically, the quality assurance program must include a calibration plan that provides for the periodic calibrations of all gauges, measuring, inspection, and testing devices against certified measurement standards that have known valid relationships to national standards at established periods. The calibration schedule must assure that inspection and test equipment is adjusted, replaced, or repaired before it becomes inaccurate. Adequate calibration facilities must be provided, including protection for the equipment. MIL-C-45662A (105) is a specification for a recognized calibration procedure.

10.5.2 Subcontractor Calibration System

The quality assurance program must contain provisions to assure that only those subcontractors and vendors are selected who have calibration systems that adequately control the accuracy of their test and inspection equipment.

10.5.3 Review of Test and Manufacturing Plans

The analysis of test and manufacturing plans, in particular, must examine these plans to determine whether or not any precision measurements are needed that might exceed the known state of the art. Such conditions must be identified and reported in a timely manner to both company management and to the customer. The contractor's special measuring devices and personnel must be made available to customer personnel, on request, to permit the customer to make occasional periodic checks.

10.5.4 Production Tooling

The control of the accuracy of measuring devices also must include production tooling whenever such tooling is used as an inspection device.

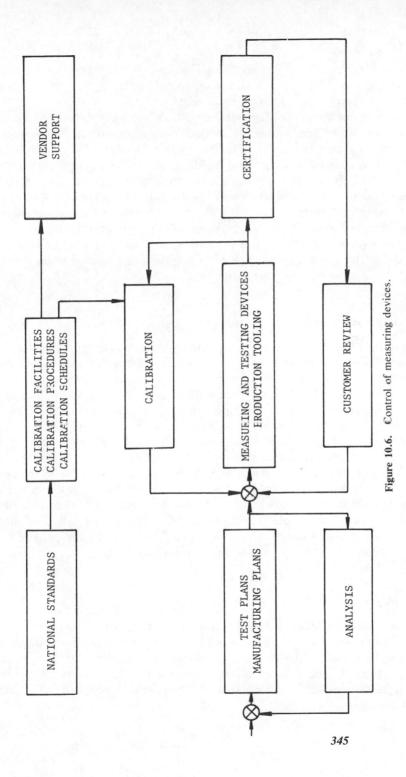

Figure 10.6. Control of measuring devices.

For example, many jigs, fixtures, tooling masters, templates, patterns, and the like, are often used as measuring devices, and they must be treated as any other measuring device.

The prime function of master templates, for example, is the accurate transfer of, or verification of, dimensional information. To protect the integrity of these masters over the project life time, they must be subjected to controlled usage and handling procedures. They may be used only in verifying the manufacture of production templates; masters may never be used in the production of parts. Subcontractors must have a set of duplicate masters, and all differences between the subcontractor's masters and the contractor's masters must be a matter of record. If large dimensions are involved, the subcontractor's masters must be stored and used at the same temperature and environmental conditions as the contractor's masters are stored and used.

10.5.5 Complexity of Fixtures

Good control of measuring devices must include activities to assure their most effective utilization. For example, the most accurate master template is a flat plate with holes normal to the flat surface. The size and location of such holes can be held with considerable accuracy. The jigging of out-of-plane points, or planes angular to the base plate, increases tooling complexities by orders of magnitude. Hence, designers should be encouraged to base their designs on the use of flat plate masters, at least with respect to key dimensions affecting mating or easy interchangeability. This type of control is executed by becoming involved with the design in its early formative stages and possibly in assisting the designer to find a configuration that meets the design requirements but also can be tooled with a series of simple flat plate masters. The development of such a design may require considerable ingenuity, but if well done, will be worth the effort by reducing tooling and production costs.

10.5.6 Protection of Measuring Equipment

The protection and handling of measuring equipment also must be given consideration. Sometimes, to emphasize the necessity for careful handling, it is helpful to disclose the value of sensitive test or measuring equipment to the individuals who use it. If a test stand or master is unique, users might be made aware of the cost that might occur as a result of damage or loss.

10.6   CONTROL OF PERSONNEL PERFORMANCE

The achievement of quality is in large part dependent on the acts of individuals. Hence, the control of these acts, that is, the control of personnel performance, is a vital factor in quality assurance. As in all functions involving the human element, the exercise of such control is not a simple matter. The problem of control can no longer be approached on a purely technical basis; use also must be made of the social sciences, psychology in particular. One of the important tasks in this area is the development and application of effective motivational techniques.

Figure 10.3 shows the prescribed method of a function being performed under controlled conditions, that is, in accordance with written work instructions on adequate equipment in a favorable environment (Section 10.2.3).

10.6.1   Work Discipline

Discipline is discussed in some detail in Sections 5.12.3 and 5.12.4. The theories in these sections are applicable to quality assurance in all cases where the actions of individuals affect the quality of the output. The occurrence of a major quality defect can be equated with the occurrence of a major accident. Defect prevention can be equated to accident prevention. If major accidents can be prevented by the suppression of the causes of minor accidents, major quality defects can be prevented by the elimination of the causes of minor quality defects. It follows that the way to assure quality is to establish proper procedures and enforce strict compliance with those procedures. The normal practice of excusing minor infractions, discussed in Section 5.12.3, breeds the environment in which it becomes only a matter of time before major quality defects are certain to occur.

Dimensional standards, for example, can be used to maintain the work discipline essential for the production of quality systems. Maximum acceptable tolerances are established, and these must be held regardless of whether or not they are required from a functional or assembly viewpoint. In the aircraft industry, except for special applications, such overall tolerances are customarily ±0.010 inch for machine parts and ±1/64 inch for sheet metal parts. These overall tolerances are not so tight that they increase the cost, yet, by requiring employees to hold these tolerances whenever closer tolerances are not required, workers are forced to maintain their skills, which are then available when needed. Every part is treated as a high quality item, creating an environment that assures that sensitive parts also will be handled with the required care in a routine manner.

10.7 CONTROL OF PURCHASES

In advanced systems many purchased parts and equipment are special, that is, unique to the particular system. Because of weight, size, performance, and other constraints, there is not much standardization among systems, except for small parts and materials.

Special parts and equipment usually are purchased from single source suppliers, especially in the case of complex or technically advanced equipment. However, the responsibility for the quality of purchased parts and equipment rests exclusively with the prime contractor. When equipment fails in service use, it is of no interest to the operator where the faulty part was made. Responsibility for quality assurance is something the prime contractor cannot delegate. A supplier's design and manufacturing facility is in effect an extension of his own.

A major advanced systems program may include some 25 subcontractors for specialized major subsystems, and several thousand suppliers. The purchasing climate is established by the delivery schedule and competitive bidding procedures. Such schedules, planned prior to and established at the time of contract award, usually are quite optimistic; certainly, they include little slack. Competitive bidding procedures normally require that bids from at least three qualified bidders be evaluated.

10.7.1 Selection of Subcontractors/Suppliers

In normal contract work the main control a contractor can exercise over a supplier is derived from the contractor's power to terminate the contract and transfer the work to another supplier. In the tight schedule environment in which advanced systems are developed, it is practically impossible to change suppliers once the development has started. The initial selection of suppliers is, therefore, a critical operation, the importance of which cannot be overemphasized. The selection of an incompetent supplier is an error difficult to correct and may endanger the success of the entire program.

As a result of these circumstances, advanced systems contractors have well-developed procedures for the selection of qualified suppliers (Figure 10.7). These selection procedures, and the nature and extent of control that prime contractors exercise over suppliers, depend on the material being purchased. Suppliers must satisfy both customer standards and contractor standards. Although price is a consideration in the competitive bidding process, it is by no means the deciding factor.

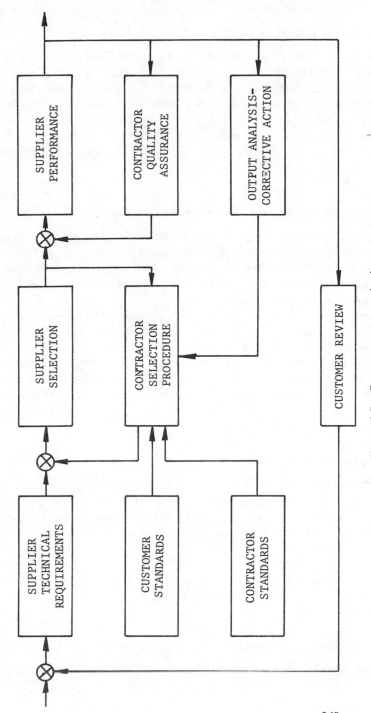

Figure 10.7. Contractor source selection.

10.7.2   Geographical Distribution

Government customer requirements often include more than technical and financial considerations. Government funding of the magnitude involved in major advanced contracts has an important economic impact on the contractor's geographical area. For many social, economic, and political reasons, the government customer may be interested in a broader geographical distribution of funds—particularly in support of small business operations, or in the development of alternate sources of suppliers for specialized equipment.

10.7.3   Complementary Skills and Ability to Cooperate

The prime contractor selects his subcontractors and suppliers from the viewpoint of best complementing his own skills and specialized facilities. He wants to select suppliers whose brains he can pick to enable him to develop a superior system. He often prefers a supplier with whom he has worked well in the past.

Source selection, therefore, is based on demonstrated supplier capability to perform and cooperate well with the prime contractor. In all important cases, selection is on the basis of an output analysis of past history, including supplier records and performance. Review and assessment of supplier effectiveness and integrity is a continuous contractor activity, consistent with the complexity and quality of the purchased items. As a minimum this evaluation is based on delivery and inspection records, test reports, source inspection, and other certificates.

10.7.4   Qualified Products List

Inclusion of a supplier's name on a "qualified products list" published by a procuring agency has little real meaning to a contractor, since the inclusion of a name on such a list does not relieve a contractor of any responsibility. Inclusion of a supplier's name on such a list normally indicates that at one or more times in the past the supplier successfully demonstrated his capability to produce a product that conformed to requirements. The list offers no assurance that supplies currently being offered for delivery in any way conform to a contractor's current quality standards.

10.7.5   Monitoring of Subcontractors and Suppliers

Depending on the complexity of the purchased product, the prime contractor will establish a quality assurance program to monitor the supplier's output. If warranted by the complexity or technical difficulties, design and/or quality assurance engineers may be stationed at the supplier's facility on a permanent, temporary, or transient basis. This program will be supported by procedures for the transmission of design and quality assurance requirements, the evaluation of the adequacy of the procured items, a system for the early feedback of quality and production information, and for the efficient correction of the results of nonconformances.

10.7.6   Developing Alternate Sources of Supply

Since source selection for important equipment is based primarily on the evaluation of demonstrated past capability, the number of suppliers available to one contractor in any one commodity area tends to be limited. Major contractors, therefore, are always on the lookout for new sources of supply, and they maintain teams of engineers that survey potential suppliers. Without a specific contract, these survey teams visit a potential supplier and assess his capability. On an initial visit the team looks for basics: good housekeeping, record keeping, gauge control, lot control, and control by commodity. They look for workmanship specifications; if there are none, they try to determine what governs workmanship. One of the best ways to evaluate a small plant is to issue a series of orders, starting with a small, simple one, and based on the vendor's performance, gradually increase the size of the orders or the complexity of the product, or both.

10.8   SUBCONTRACTOR QUALITY PROGRAMS

10.8.1   Paperwork

The procurement of special products in the advanced systems industry is associated with considerable paperwork. The skill and intelligence with which this paperwork is administered has an important bearing on the quality and cost of the product. Quality of paperwork has already been discussed in Section 9.11.

If uncontrolled paperwork growth is permitted to occur in a procurement system that is already so heavy on paper, the movement of paper rapidly becomes the critical factor in the procurement operation. Critical action will be delayed until the necessary movement of paper has been accomplished. This becomes most critical when action is required to cross organizational lines, as between contractor and supplier. This places the operation in danger of becoming saturated with paper and then of stalling completely.

The paperwork of a subcontractor's quality assurance program must be designed to provide a base for the establishment of a close technical relationship between prime contractor and supplier, which is absolutely necessary to integrate the supplier into the contractor's development team. This program objective runs counter to many existing factors in the normal business environment. Large multidivisional subcontractors sometimes compete with the primes for contracts and often resent their status as a subcontractor. They may be working simultaneously with a contractor's competitor on a competing program. Such a subcontractor may be involved with a number of conflicting interests. He may make a higher percentage of profit on his commercial operations than on his government controlled advanced systems contracts. Each system contractor must be alert to assure that his own program does not suffer.

Although the importance of complete and correct paperwork cannot be overemphasized, it is not an end in itself. It merely provides the framework for the creative technical work that is the basis for achieving quality assurance. However, without a sufficient and albeit extensive paperwork base, it is impossible to coordinate all of the various activities necessary to develop a modern advanced system; technical contributions cannot be exploited in a systematic and effective manner, and necessary quality levels cannot be maintained.

10.8.2 Competitive Subcontractor Selection

The products under consideration here are primarily those specially designed and developed for integration into a particular advanced system; they do not exist at the time of contract award. The process of deriving component requirements from system objectives is described in Section 6.4.2. Since supplier selection is normally based on a minimum of three bids, the invitations to bid or requests for proposals must describe the item to be built and the required delivery schedule in a complete, clear, and detailed manner which cannot be misunderstood. Each competitor must understand what he is bidding on and must base his proposal on the same baseline so that a fair comparison can be made.

This work is normally done under tremendous pressure. Approximately 50% of modern defense systems consist of items purchased from suppliers or vendors. Not only are there many vendors and suppliers involved, but also they must all be selected and placed under contract quickly, since any delay in this process reduces the design time available under the already established system delivery schedule. (Note that design time is the only item that suffers; manufacturing, assembly, and testing time cannot be reduced. The reduction of design time necessarily adversely affects quality.)

10.8.3   Impact of Design Changes

During the component design and development phase, there will be many design changes. These changes create a complex paperwork chain; a change in one piece of equipment in the system normally requires some changes of a more or less extensive nature in other contiguous or associated system components. Many of these changes affect the original scope of the subcontract and will, therefore, require adjustments in funding and schedules. To administer such constant changes it is necessary to have a well-defined, established base to build on and a responsive administrative procedure to keep track of changing commitments and associated adjustments in funding and schedules, in a timely and reliable manner.

10.8.4   Quality Requirements in the Purchase Order

The purchase order to a supplier, which is the result of negotiations based on the invitation to bid and the supplier's responding proposal, therefore, must include a complete description of the equipment being offered for procurement and a set of requirements. Among the requirements will be those for a quality assurance program, Figure 10.8. In general, a contractor tries to model his relationship with his suppliers on his relationship with his own customer, and he formulates similar requirements. However, *the relationship can never be quite the same;* the contractor is responsible for the quality of the system, including all supplier parts; *this is a responsibility he cannot delegate downstream.* Hence, the contractor must exercise greater quality control over his own suppliers than the customer exercises over him. Basically, the supplier is required to submit a quality assurance program for approval in a procedure similar to the plan that the contractor submits to the procuring agency. However, contractor participation in a subcontractor quality assurance program will be substantially dependent on the type of product.

Furthermore, there must be provisions for customer review and, in particular, for customer input based on analysis of his service experience with the supplier's product. The purchase order should provide for the supplier's responsiveness to data based on service experience.

Briefly, the purchase-order technical requirements should include all applicable customer and contractor standards and specifications and all

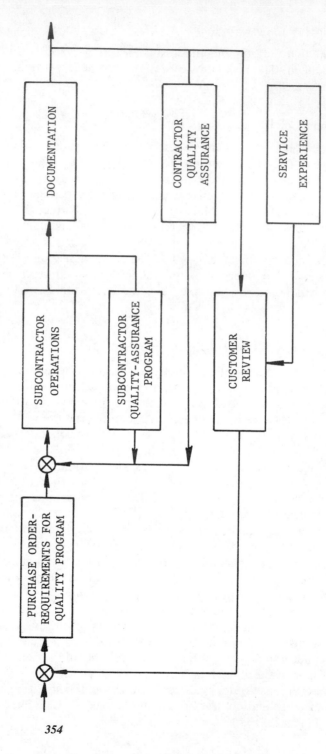

Figure 10.8. Subcontractor quality assurance program.

special requirements for manufacturing, inspecting, testing, and packaging, including customer and contractor inspections, qualifications, and approvals. The technical requirements cover items such as performance, drawings, engineering change orders, specifications, quality assurance program, reliability, safety, weight, maintainability, demonstration, logistic support, and other special requirements, together with procedures that apply throughout the system for identification of models, special revisions, traceability, and serialization. Provisions must be made for contractor inspection at source as necessary. Inspection requirements for raw materials may include lot identification and specify the desired chemical and physical testing and certifications. The design changes that require contractor approval should be specified. If any material is to be shipped from the supplier to a location other than the contractor's main plant, special instructions must be given.

10.8.5   Subcontractor Coordination

Subcontractors and suppliers of similar equipment on the same program may be sharp and bitter competitors. The equipment from such competitors must be designed to work together in an integrated manner to yield effective performance at minimum weight. The designers of mating pieces of equipment should be sitting at design boards located physically side by side to best integrate their concepts. In actual fact, all communication between the two geographically separated firms must go exclusively through the prime contractor, who must exercise due care to protect the proprietary rights and concepts of each company.

10.8.6   Supplementing Subcontractor Capability

Although the supplier brings some unique and/or necessary skills or facilities into the program to complement those of the contractor, by no means is the contribution of the supplier in a neat and complete package. In advanced systems equipment, independent entities hardly ever exist. The prime contractor himself, therefore, must maintain staffs and some facilities with technical competence in the supplier's specialized fields. These staffs must have a broader outlook and better knowledge of requirements than the suppliers and must know at least as much as the suppliers about technical feasibility, state of the art, performance characteristics, and development tests and demonstrations. In many cases, a supplier will have some but not all of the technical capabilities required to satisfy the new and advanced requirements of his purchase order. The contractor must be capable of stepping in and providing the necessary technical and management assistance to keep the supplier on schedule.

Introduction of Reliability Technology

An example of this contractor support to subcontractors on an industrywide basis is provided by the introduction of reliability technology. Formal, statistical reliability requirements were first conceived in the Office of the Secretary of Defense (83) as a result of the operational deficiencies of electronic equipment observed in the field. From this top level, reliability requirements moved down through the procuring agencies to the prime contractors. It took a tremendous effort over a 15-year period on the part of the prime contractors to develop the necessary skills in their suppliers. As a result of this prime contractor effort, defense suppliers generally have developed a good understanding of reliability requirements and a new capability to comply with them.

Aircraft Structural Reliability

Another even earlier example is provided by certain structural requirements established by the Navy, as discussed in Section 6.2. Although not referred to as formal reliability requirements, structural design and proof-of-design criteria for several decades have required various analysis and tests whose primary objectives are to assure reliability. The evolution and monitoring of one category of these structural reliability requirements is worth the following short review.

About 1939 it became reasonably obvious to Navy airplane design personnel that deliberate actions during design were necessary to avoid classical flutter in future Navy airplanes. Accordingly, Navy airplane procurement contract specifications were modified to include requirements for flutter calculations, and in some cases flutter model testing to confirm flutter design integrity. Most Navy prime contractors, at that time, lacked specialists in airplane flutter prevention technology and, therefore, had to acquire competent personnel in this highly specialized field. This government action, in advance of actually experiencing classical flutter in field operations, was deemed to be essential because of the dire consequences that could be foreseen without that action. So far as is known, these flutter prevention procedures have been completely effective in preventing losses of Navy airplanes attributable to classical flutter.

10.9 GOVERNMENT SOURCE INSPECTION

U.S. Government contracts normally reserve the government's right to source inspect materials at the supplier's plant, and the government maintains extensive inspection facilities throughout U.S. industry. Government source inspection, Figure 10.9, does not constitute acceptance

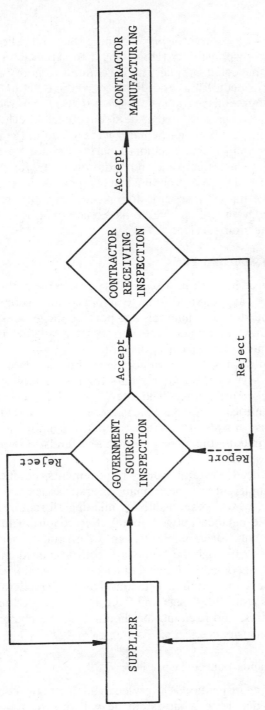

Figure 10.9. Government source inspection.

and does not replace contractor inspection. It does not relieve the prime contractor of any responsibility. However, if the contractor rejects government source inspected supplies, he is required to notify the responsible government inspector and coordinate corrective action with him.

Government source inspection is a service that the contractor may use at his descretion to whatever extent he deems feasible. The use of government source inspection is particularly widespread in the procurement of raw materials and of small standard parts common to a number of systems. Some standard electronic and hydraulic parts, for example, may be used in over 100 different programs. In such cases, it is more efficient for a single government inspector to provide the uniform inspection service for all programs rather than to have each contractor maintain his own source inspection service.

10.9.1   Use of Standard Parts

Government source inspection is part of the government standardization effort. Contractors are required to have standarization programs and are encouraged to use standard materials and parts to the greatest practicable extent. Standardization reduces the variety of available materials and parts; this permits more effective use of production and distribution facilities and contributes greatly to economy, both in initial cost and in maintenance. In service the requirements for the variety of spare parts that must be stocked are reduced and interchangeability is improved. The use of larger quantities of standard parts increases the size of the market for standard items; this results in more competition among manufacturers who are now all making the same product, and tends to improve quality and reliability.

In most advanced systems some requirements conflict with one another, for example, aircraft performance versus space and weight. The prime contractor, who is responsible for meeting all requirements, must find an acceptable balance between conflicting requirements. To attain this objective he will, for example, design for maximum performance at minimum weight. Although he will start with standard materials and elementary parts, at some higher level of assembly, he must start utilizing special components and relinquish standardization. The determination of this level follows from the design and from the associated technical and schedule difficulties, and it cannot be dictated by the customer without compromising performance.

10.9.2   Contractor Source Inspection

Government source inspection is a service available to the contractor, but he himself must decide to what extent he will rely on it. Normally he makes full use of the service with respect to raw materials and small standard parts. In technically advanced systems, contractor monitoring of subsystem development must not be compromised, and the contractor establishes his own inspection service. Between these two extremes, there is some point at which the contractor relies less and less on government source inspection and starts to depend more on his own facilities.

10.10   RECEIVING INSPECTION

Contractor's quality assurance programs are required to include receiving inspection (Figure 10.10). Contractors maintain receiving inspection procedures to assure that the materials and products being introduced into their fabrication and assembly processes conform to technical requirements. It is a costly, time-consuming, and frustrating process to invest time and labor in processing work or assembly only to discover later that the material is defective and the product must now be reworked or scrapped.

10.10.1   Raw Materials and Purchased Parts

Receiving inspection must assure that raw materials and purchased parts conform to applicable physical, chemical, technical, and other performance requirements as listed under acceptance criteria in the purchase order. Most raw materials and standard parts will have been government-inspected at source, and in such cases a part of the incoming inspection procedure consists of a paper review of test and inspection records and of certifications. However, contractors maintain complete physical, chemical, electrical, and mechanical laboratories for incoming inspection purposes, and samples are taken on a selected basis from various lots to confirm the certifications. If any doubts arise, or if there is any suspicion of irregularity, a full laboratory test is conducted.

Suppliers are required to use equivalent procedures for control of their raw materials and purchased parts. Prime contractors will often supply the vendor with the necessary raw materials and individual piece parts for critical parts and components.

10.10.2   Separation and Protection of Inspected Material

Rigid procedures must be used to separate raw materials and parts awaiting testing from material that has been tested and approved, and also to prevent the inadvertent use of material and parts that have failed to pass acceptance tests. Disposition procedures for rejected material must be rigorous; rework and scrap procedures must be closely controlled. Some-

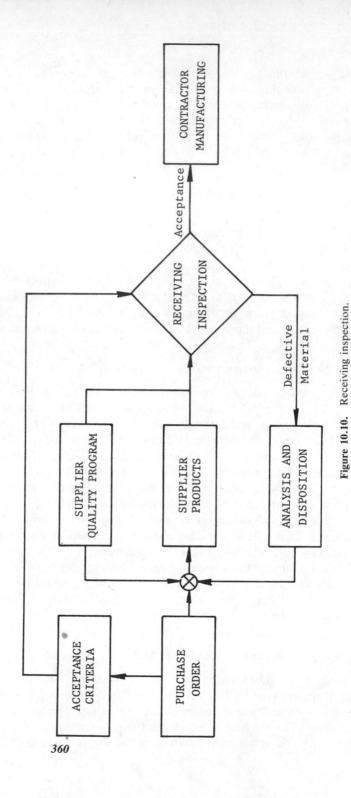

Figure 10.10. Receiving inspection.

times elaborate precautions must be taken to assure that rejected material returned to a supplier is not shipped back as good material and again offered for acceptance, or offered to the customer as spares. All individual items of accepted material must be positively identified and controlled until the identification is lost by processing.

10.10.3   Inspection Status

There must be a positive system throughout the organization for identifying the inspection status of all parts and assemblies. Identification may be accomplished by means of stamps, tags, route cards, move tickets, totebox cards, and other commonly used control devices. The contractor's stamps and devices must be different from the customer's control devices.

10.10.4   Tailoring of Inspection Procedures

Inspection is a nonproductive labor activity, and inspection executives are always short of manpower to do those things that they believe should be done to assure quality. As a result, a great deal of effort is normally expended on running an efficient operation. Receiving inspection procedures are adjusted according to commodity type and criticality and also are based on knowledge available of the supplier's processes and quality program, and of the quality and reliability of his records. Emphasis is placed on not duplicating tests and inspections already performed, provided that there is acceptable evidence that they were done correctly.

10.10.5   Statistical Techniques

Additional savings are achieved by the use of statistical sampling techniques. Acceptance sampling is a method of evaluating the quality of a large lot of material by taking random samples from the lot and basing the decision to accept or reject the lot on the quality of the samples.

Structural Applications

In structural applications involving unconventional methods of construction, materials, or combinations of materials, there may be reasons for suspecting that the structural characteristics of manufactured components are variable to an unknown degree because of problems involved in maintaining ideal control over the manufacturing processes, such as in glass-fiber reinforced plastic or metalite construction and composites. In such cases the following may be necessary:

1. The structural design be predicated on an ultimate factor of safety greater than the customary value of 1.5.
2. The statistical sampling technique be amplified to include static failing-load and fatigue-life tests of randomly selected components throughout the production run, as well as some logical testing to determine time-dependent effects, such as weathering and erosion.

It follows that the manufacture of the necessary additional components for failing-load static and fatigue-life tests must be planned at the request for proposal phase.

Sampling Plans

Acceptance sampling must be applied when the determination of the critical characteristics of a material involves destructive testing, as in the case with the tensile strength of steel. However, sampling techniques also have broader application in the inspection of many mass-produced parts such as nuts and bolts. These methods have a great advantage in that they reduce the expense that would be incurred if each and every item in the lot had to be inspected.

Sampling plans may be used when records, inherent characteristics of the product, or noncritical application indicate that a reduction in inspection or testing can be achieved without jeopardizing quality. It is necessary to know and understand the supplier's manufacturing processes, and to know that his process controls are effective. Sampling plans are subject to review by the customer.

The mathematics of sampling have been well worked out. In their efforts to reduce inspection costs, quality assurance engineers have done a great deal of work in this area, and there are many textbooks on the subject. Calabro (30) has a chapter on sampling methods and lists a number of other books on sampling in its bibliography.

In designing a sampling plan, the problem is to determine the number of test specimens that should be randomly selected from a given lot. If too many specimens are taken, the process of testing may be prohibitive in cost and time. Yet, if too few specimens are selected there may be doubt that reliance can be placed on the results obtained from so small a number of specimens. The ideal is to select randomly a number of specimens so that the tests will assure—with maximum confidence at minimum cost—that the quality of the lot has been properly assessed. This is the objective of a good sampling plan. It will specify the number of test specimens that should be selected, and the maximum number of defects that can be tolerated in the sample to assure with a certain confidence that the lot is as good as, or better than, the specified quality. Detailed instructions for the design of sampling plans are included in refs. 106–110.

10.11   GOVERNMENT FURNISHED MATERIAL

Some subsystems in a defense system are government furnished equipment (GFE) and are procured directly by the customer rather than under the jurisdiction of the contractor. This equipment is supplied to the contractor for integration and installation into his system. Generally, these are subsystems common to a number of different systems, such as power plants, guidance and navigation systems, and the like.

10.11.1   Contractor Responsibilities

As mentioned before, when a system fails in service, the user is indifferent to the identity of the manufacturer of the failed part. This attitude applies as well to GFE. When such equipment fails in service, the user looks to the prime contractor for redress; in an emergency situation it is ineffective for a contractor to take the position that the failed equipment is customer furnished and he, the contractor, is not responsible. As a result of this situation, contractors have established quality assurance programs to assist as much as possible in controlling the quality of GFE. They maintain close contact with the suppliers, and as a minimum, a contractor's quality assurance program, Figure 10.11, will include:

1. Examination on receipt for damage in transit.
2. Inspection for completeness and proper type.
3. Periodic inspection and precaution to assure adequate storage conditions and guard against damage from handling and deterioration.
4. Functional testing, before and after installation, to determine satisfactory operation.
5. Identification and protection from improper use or disposition.
6. Verification of quantity.
7. Procedures for reporting to the customer any damage, malfunction, or other unsuitable condition detected.
8. Diagnostic procedures for determining and recording probable cause of failures occurring during and after installation.

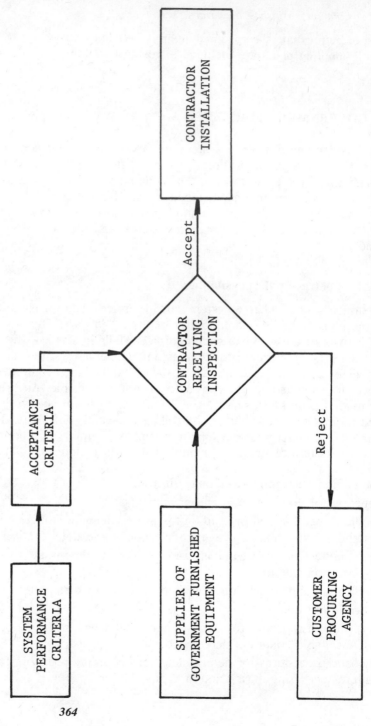

Figure 10.11. Contractor receiving inspection for government furnished material.

10.12 PROCESS CONTROLS

Process controls include those controls, tests, and inspections that provide assurance that manufacturing functions are performed in accordance with prescribed procedures and that the characteristics of the output of the manufacturing function are held within the prescribed limits. Process controls usually must be specially designed for the functions that they monitor.

Manufacturing methods have a significant impact on the cost and quality of products. In a competitive market, there is a constant search for improved value, and this results in a great variety of available manufacturing methods and processes.

10.12.1 Classification of Manufacturing Operations

Four basic methods of organizing a manufacturing operation are listed here. Any one or more of these methods, or any combination of them, may enter into the manufacturing of most products during the various phases of fabrication.

1. Discrete Serial Operations. Material is assigned in sequence to a series of workers; the first one performs a series of discrete operations, and the partially manufactured article is then transferred to the next worker who performs the next series of operations. The process continues until all manufacturing operations have been completed. Small parts usually are manufactured in job-lot quantities; very large units such as aircraft are assembled one unit at a time. A machine part is manufactured in a series of discrete operations. The greatest number of parts of an advanced system is produced by discrete serial operations.

2. Moving Assembly Line. The work moves on a moving assembly line past various work stations. A worker at each station performs a limited number of operations. Automobiles are mass-produced on moving assembly lines.

3. Continuous Flow Operations. Raw materials enter one end of a processing plant and flow through various automatic equipment stations in a continuous operation. The finished products flow in a continuous stream out of the other end of the plant. This process is applicable to liquid and gaseous products. An oil refinery is a continuous flow operation. So are paper, wire, and some rod and plate manufacturing operations.

4. Automatic Processing and Assembly. This process is similar to con-

tinuous flow, except that it consists of a series of discrete operations performed by automatic machinery. A cigarette manufacturing machine is an automatic processing and assembly device. Today, more and more discrete operations machinery is being controlled by computer programs (numerical control), increasing the degree of automation. However, most so-called automatic assembly lines today still include some manual operations.

10.12.2 Scope of Discussion—Discrete Serial Operations

Because of the complexity of the subject, the following discussion is limited to a simplified illustration of the use of process controls in a discrete serial operation as might be found in the advanced systems industry (Figure 10.12).

10.12.3 Work Orders

The manufacturing procedure is initiated by the issuance of a numbered work order by manufacturing planning. In Figure 10.12 the first work order is issued to Foreman 1. A work order is issued to one department only; work orders do not cross department lines. A work order is known by many names in the industry, such as "route card," "traveller," or "operations sheet."

The work order refers to drawings and specifications by applicable revision identification (change letters) and outstanding engineering orders, lists the material requirements, and indicates in proper sequence all operations, tests, and inspections to be performed. It may include also the required completion date. For convenience, operations usually are described by applicable process specification numbers, since all critical operations are described in detail in individual process specifications. Requirements for part identification and serialization also are included.

10.12.4 Stockroom Material Control

Upon presentation of the work order at Stockroom 1, the necessary material is issued to Workstation 1, and Stockroom 2 is notified of the transaction. The material "belongs" to the stockrooms and is "loaned" to the worker. After all operations under Foreman 1 have been completed and inspected both the work order and the partially finished work are surrendered to Stockroom 2.

If an assembly is too large physically to be conveniently transferred to a stockroom, it may remain on the factory floor while just the work order is

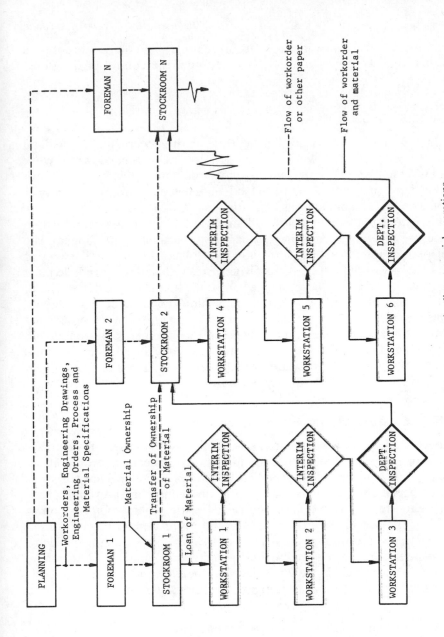

Figure 10.12. Process control—discrete serial operations.

surrendered. It is, nevertheless, useful for material control purposes to assume that the physical article is now under the control of the stockroom.

10.12.5 Inspections

Figure 10.12 shows the work flowing through three workstations, with an interim inspection following each workstation. This breakdown of work stations and inspections depends on the nature and criticality of the work to be done. Inspection, to be effective, must be performed in proper sequence with manufacturing operations.

For example, the fit of a close-tolerance bolt is a critical structural item since the fit affects load-carrying capacity, particularly in fatigue. The acceptance criteria, that is, the allowable tolerances on the diameter of the hole, are called out on the drawing. Every close-tolerance hole must be inspected with a go/no-go gauge before the bolt is inserted in the hole. The successful completion of this inspection (compliance with the acceptance criteria), is recorded by the inspector on the work order.

After all the work operations scheduled for Foreman 1 are completed, the product undergoes another department-outgoing inspection before the product and the work order are turned over to Stockroom 2. The procedure now is repeated in subsequent downstream operations.

10.12.6 Work and Inspection Instructions

The work and inspection instructions must cover adequately all manufacturing operations as well as all handling, storage, cleaning, preservation, packaging, and shipping provisions needed to protect the quality of products. Special attention must be given to procedures for preventing handling damage, deterioration, or corrosion during fabrication or interim storage, as well as to procedures for the protection of quality during shipping. The procedures should assure that products being shipped are accompanied by the required technical documents, and when necessary, packaging must include means for accommodating and maintaining critical environments within the packages and be so labeled.

10.12.7 Control of Nonconforming Materials

The inspection procedures must include a positive system for the control of nonconforming materials, including the identification, segregation, material review, and disposition of such materials. All nonconforming materials must be positively identified, and holding areas must be provided for the temporary storage of these materials. Rework must take place in accordance with documented procedures acceptable to the customer. Acceptance of nonconforming supplies is a prerogative of the customer, and may involve a price adjustment.

Uncontrolled Materials

A recurring problem is uncontrolled material in the hands of shop personnel. Stockrooms issue materials in standard sizes sometimes larger than needed to satisfy requirements. Manufacturing personnel save the excess trim scrap and accumulate a private stock of material. When, on occasion, a part is spoiled during manufacturing, and time is of the essence, the quickest and easiest way to get a new piece of material is to withdraw it from a private stock; there is no red tape. However, without proper material identification and control, the wrong type of material may be used for the new part. In one accident investigation, it was determined that a failed structural control rod was made from soft fuel-line tubing. It

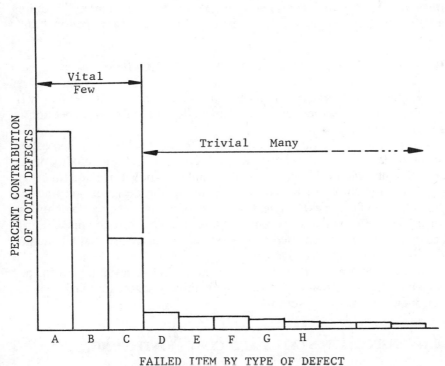

Figure 10.13.   Distribution of defects.

is, therefore, essential that all trim scrap be returned to the stockroom so that it can be placed under proper material control procedures, and that the shops be policed to prevent the accumulation of private material stocks.

10.12.8   Statistical Analysis of Process Controls

The area of process control has always been one of intensive study in continuing attempts to reduce costs in a competitive society. These studies lend themselves particularly well to the application of statistical techniques. Reference 193 is the analysis of the output of a manufacturer of advanced electronic subsystems. Defects are classified by lot size, type of product, department, operation, and type of defect. These defects are ranked in order of frequency of occurrence, A, B, C, . . . , and plotted in the type of diagram shown in Figure 10.13. In this type of analysis, it is invariably determined that a few operations in a few departments are the cause of a majority of the defects. By identifying these "vital few" and taking effective corrective action, appreciable savings can be achieved immediately. Effective action with respect to the "trivial many" requires a different approach such as a motivational program (see Section 10.6).

10.13   COMPLETED-ITEM AND FINAL INSPECTION

Acceptance testing and acceptance data packages are discussed briefly in Section 6.11. The procedure for completed-item and final inspection is shown in Figure 10.14.

This procedure should provide a system for the final test and inspection of completed items and end products to measure the overall quality of the completed products. The tests should simulate product use and functioning. The procedure must provide feedback to the designers on unusual difficulties, deficiencies, or questionable conditions. When modifications, repairs, or replacements are required after the final tests or inspections, products must be reinspected and retested so as to assure that all characteristics affected are satisfactory.

Final inspection must certify identification and serial numbers for acceptance. Rejected material should be positively segregated and disposition indicated (rework, scrap).

10.14   INSPECTION STATUS AND COST MONITORING

Inspection records, as they are generated from day to day in a large organization, provide an authentic base for estimating current status and

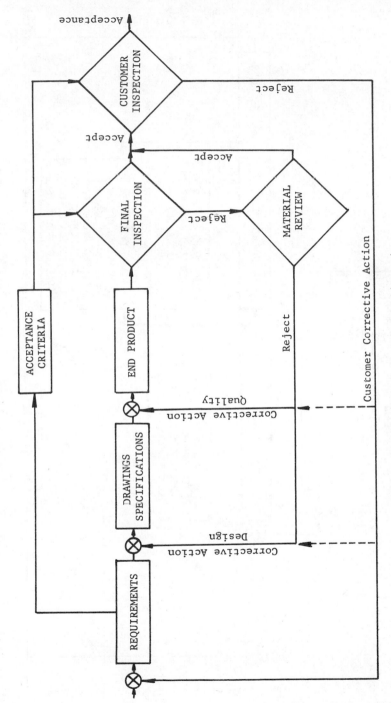

Figure 10.14. Completed-item and final inspection.

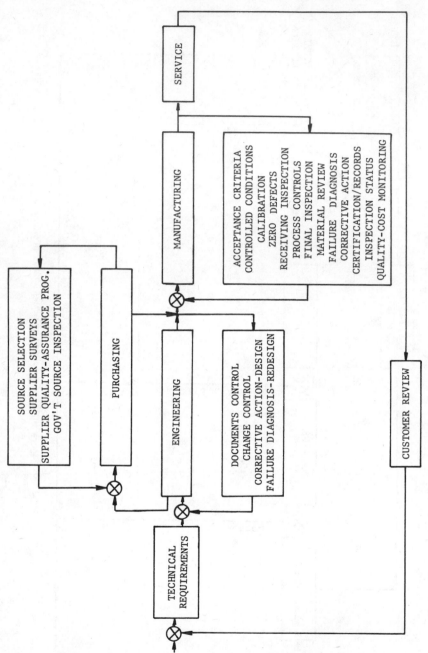

Figure 10.15. Main elements of a quality assurance program.

quality of work accomplished. MIL-Q-9858A(73) requires contractors to maintain positive systems for identifying the inspection status of products.

MIL-Q-9858A also requires contractors to maintain and use quality-cost data as a management element of the quality assurance program. These data must serve the purpose of identifying the cost of both prevention and correction of nonconforming supplies. The records must cover the costs and losses in connection with scrap, with rework, and with scrap data on rework necessary to reprocess originally nonconforming material. Both inspection status and cost records are subject to inspection by customer representatives.

This area of status and cost records is a fruitful field for analysis to serve as a basis for improving management systems. These records generally are not fully exploited in many organizations.

10.15   CHAPTER SUMMARY

The main elements of a quality assurance program are summarized in a simplified form in Figure 10.15, as they apply to the functions of engineering, purchasing, and manufacturing. In each case the quality assurance program is shown as a feedback loop controlling the quality of the function.

The program is subject to customer review in a loop shown leading back to the requirements. Customer control over the program is through control of design and proof-of-design requirements. Naturally, the customer can and will require corrective action anywhere in the process if he finds that deficiencies exist, that is, if there is nonconformance to requirements. But that is not the main function of the customer's quality assurance program. The contractor is responsible for the quality of the end product, that is, the contractor's quality assurance program must assure conformance with requirements. The purpose of the customer's review and surveillance is to assure that the contractor has an adequate program and that it is operating satisfactorily.

The basic concepts presented in this chapter are derived from MIL-Q-9858A (73). It is noted that effective quality control is provided for the basic manufacturing and purchasing processes. These controls are effective when applied to the factors that directly establish quality.

Operational Testing
and Product Improvement

In Figure 3.1 operational testing is broken down into two basic elements: customer engineering tests and customer field (or service) tests, as mentioned in Sections 3.4.1 and 3.4.2. Various customer agencies use different names and breakdowns for this test cycle; and even within a single customer agency the names, breakdowns, and organizational responsibilities change from time to time. However, the basic functions generally remain the same as discussed here. The current Army procedures, for example, are outlined in AR 70-10 (111).

In the ideal simplified diagram of Figure 3.1, the main purpose of these tests is to assure the customer management that the system is ready for production release, Box 30. However, operational tests usually extend beyond production release to provide assurance that the conversion from the prototype to the production phase has been successfully accomplished and that the production systems are acceptable.

The operational tests should provide information and objective technical and operational assessments so that the responsible decision makers can make sound judgments.

Orientation of Test Agencies

Engineering and field tests should be conducted by separate organizations because of their different orientations; but there should be a coordinated test plan to assure that available resources are used efficiently and wisely, in particular that the various agencies do not duplicate each other's tests, including those run by the contractor. Engineering and field tests may, however, be conducted simultaneously.

Correlation of Operational and Qualification Testing

Box 17 of Figure 3.1 shows the qualification test as a discrete single event. In an actual system development this is never the case. On a major system there may be many hundreds of qualification tests on various levels of components, subsystems, and on the system level. Some component qualification tests are conducted early in the development program, whereas others, for a variety of reasons, are temporarily waived and may not be conducted until production is well under way. Customer operational tests are broken down into phases that parallel the contractor's test schedules. Low-level customer operational tests actually may start early in the development as soon as qualified components and subassemblies can be made available. Part of the procedure normally includes customer observation, examination, and evaluation of contractor tests and test data.

Operational Testing Requirements

A significant feature of customer operational testing is that the test plans are completely user-oriented. In Figure 3.1 the operational test plans refer to the customer's original systems analysis (Box 2) and operational requirements documents (Box 3), which at this time were prepared many years ago and may have been incomplete. It is noted in Section 3.2 that in the initial concept definition cycle, compromises are made in interpreting the original operational requirements of Box 3, Figure 3.1, in the RFP, Box 7. Further compromises may have been necessary in negotiating the contract, Box 8, Section 4.1. In order to meet limited funding and tight development schedules, waivers may have been granted during development and qualification, so that the prototype system may not even fully conform with the original already diluted contract requirements in a number of important respects.

Furthermore, in addition to the reference to Boxes 2 and 3, the customer's test agencies formulate their test plans on the basis of their expert knowledge of their current and future needs. The operational test agencies have had no responsibility for the contractual requirements, and their test plans ordinarily make no reference to the contract. The term "requirement" in an operational test plan applies to a different baseline than the requirements of the contract, with respect to product characteristics, operational techniques, and environments. Sometimes these discrepancies may be substantial. Unless this difference in viewpoint and in the meaning of the term "requirements" is understood by all concerned, serious misunderstandings and conflicts arise.

Reliability Growth Cycle

Under these circumstances it is to be anticipated that the operational tests will uncover a number of "deficiencies." These are reported to the customer, Boxes 23 and 28 of Figure 3.1, through the program manager's office. Many deficiencies will refer to nonconformance with contract requirements, but many also will refer to product characteristics that are not defined in the contract but in the new test plans of test agencies. Nonconformance with contractual requirements should be corrected as provided for in the contract, but deficiencies with respect to operational requirements established by the test agencies and not provided for in the original contract may require design and/or manufacturing process changes. In this case, the normal procedure is for the contractor to respond with an engineering change proposal (ECP) and an associated cost estimate. If one or both types of deficiencies reported by the operational test agencies are corrected on the preproduction test specimens, the operational test period is a classical "reliability growth" (test-fix-test) cycle.

Customer Funding of Engineering Change Proposals (ECPs)

On modern complex advanced systems it is unusual for a contractor to make a profit in the development phase. Consider, for example, the always unanticipated difficulties that arise during the subsystem integration phase (Section 6.8). Hence, the contractor sometimes attempts to recover his lost profits by overpricing his ECPs. On the other hand, the customer, at this time, has expended all of his development funds. Additional customer funding at this time usually can be made available only by juggling priorities between the customer's various current programs. This procedure involves painful trade-offs determining how far the customer will back off from his operational requirements in a process similar to the one outlined in Figure 7.1. These are complex management trade-offs balancing available resources against immediate operational needs and operational deficiencies that must be corrected before production release against those that can be corrected later, should the system turn out to be successful. These very difficult trade-off decisions involve funding and scheduling, and customer management may find it necessary to make substantial compromises. Hence, test agencies should evaluate systems with respect to well-defined stated requirements, traceable to accountable sources, and should avoid making statements that the equipment is or is not suitable for service use.

11.1 CUSTOMER ENGINEERING TESTS

The objective of the customer engineering tests, Box 22 of Figure 3.1, as defined briefly in Section 3.4.1, is to measure how well the system meets the user's technical requirements and, in particular, to determine the performance limits and environmental operational envelope of the product. In Box 24 the term *technical integrity* is used to indicate that sound engineering and test procedures have been used in the system development.

Customer engineering tests should start as early as possible in the development cycles, as soon as qualified components are available for test, and should then progress to subsystems and finally to prototypes, preproduction and early production systems. Customer engineering testing should assist in progressively eliminating acquisition risks and in assessing system worth. The procedure should identify critical issues to be examined by testing prior to each decision point and should focus on high-risk areas (such as those extending the state of the art), on the accomplishment of engineering design goals, and on the conversion from research and development prototypes to production models.

11.1.1   Engineering Test Objectives

The major objectives of customer engineering tests are as follows. Some of these objectives are taken from ref. 111.

1. Assess the technical risks involved in the systems development program.
2. Determine whether the design risks have been eliminated and/or are manageable.
3. Measure the technical performance of the system to determine its performance limits and the degree to which the performance meets the applicable operational and environmental requirements established in the operational agency's test plan. Evaluate the adequacy of the environmental, logistics, and support requirements.
4. Demonstrate that the engineering design and production engineering planning are complete, acceptable, and adequate to support the scheduled early production program within the anticipated funding limits.
5. Using representative user personnel, determine compliance with human engineering and safety characteristics wherever the techncial characteristics of the design and associated support, test, diagnostic, and training devices have human factors implications, in particular with respect to accident prevention and error proofing of operating and maintenance actions (countermeasures to Murphy's Law).
6. Demonstrate the reliability, availability, and maintainability of preproduction and early production systems achieved with the skills,

procedures, and resources contained or described in the maintenance test package.

7. Determine that the production engineering planning, manufacutring methods, and facilities are adequate to support the proposed production schedules.

8. Assess the adequacy of the contractor's organization, policies, procedures, capabilities, facilities, and determination to comply with production requirements.

9. Assess the adequacy of the training and maintenance packages; logistics planning, training procedures and devices; test, diagnostic, and support equipment; common and special tools; calibration procedures and equipment; available shop facilities; and the technical adequacy, accuracy, and completeness of manuals, instructions, and other associated software.

10. Determine how well the characteristics of the early production models meet all applicable requirements.

11.1.2 Technical Integrity

In providing the decision makers with an objective technical assessment, the concept of technical integrity, Box 24 of Figure 3.1, is of critical importance. The concept is based on the traditional engineering-analysis approach. It is valid over such a wide range of practical applications and is so ingrained in engineering education and practice that little thought is ever given to defining it in an explicit manner, especially since the approach is difficult to understand. However, it is generally accepted that the application of the engineering-analysis approach requires considerable professional skill and judgment.

In applying generally accepted engineering test procedures, the Customer Engineering Test Agency, either knowingly or unwittingly, recognizes the initial validity of upstream engineering design analyses. These procedures have worked well when applied to hardware systems such as aircraft, ships, and power plants.

Envelope of Operating Conditions and Environments

Given an operational system, the various parameters representing the operating conditions can be plotted on one or more related diagrams. There are many innovative and sophisticated methods for presenting operating conditions in a graphical display. The extreme or limiting values can be identified for each anticipated operating condition. These limiting values can also be interconnected to form an envelope, tent, or multi-dimensional enclosure that contains all operating conditions. The construction of an envelope of operating conditions and the selection of test points are discussed in Section 7.7.1.

11.1.3 Engineering Test Phasing

Customer engineering tests will be conducted in a number of different phases in which different engineering and service use aspects are examined at different locations by different types of testing personnel. Engineering tests always take place under the controlled conditions described in the technical requirements, and measurements are taken as prescribed in an approved test plan to determine feasibility and performance characteristics of the specimen under the specified environments. At times the requirement may be questioned. Use is made of controlled laboratory equipment, environmental chambers, and instrumented field trials. Deterministic models and statistical methodologies are used, as applicable. Human engineering characteristics are measured in a controlled environment as stipulated in the specification and may not necessarily be truly representative of the operational environment.

The three major test phases include:

Tests of Qualified Components and Subsystems during Development. If competitive components or subsystems are being considered, the tests should provide a comparison between the competitive units.

Tests of System Prototypes or Preproduction Models. These determine if the system is ready for production release. The tests evaluate feasibility, technical performance and safety, test and maintenance equipment, training package, support and facilities plan, and reliability, availability, and maintainability.

Tests of Initial Production Systems. These follow the decision to proceed with full production to verify that the system is acceptable when manufactured in accordance with production specifications and quantity production processes, that faults identified in the preproduction testing have been corrected, and that the operating characteristics have not been degraded. These tests provide data to customer management to support their decisions on proceeding with further procurement actions.

11.2 CUSTOMER FIELD TESTING

Customer field or service testing and evaluation is accomplished by user and support personnel, preferably with organizational units and personnel of the type and qualifications of those expected to use and maintain the system when in service, to demonstrate operational integrity, Box 29 of

Figure 3.1. In commercial systems it is always advisable to get the public involved. For example, the emergency evacuation procedures of airliners are demonstrated with representatives of the passenger population. The testing is conducted in as realistic an operational environment as possible. As with the customer engineering tests, the field tests are conducted in phases, of which the following three are typical.

11.2.1 Development Phase

This phase should consist of a user-oriented assessment of the first available qualified components and hardware subsystems to provide the following:

1. An early indication of system worth; potential of the new system in relation to existing capabilities.
2. Refinement of critical issues; confirmation that the list of critical issues is complete.
3. A measure of the relative merits of competing components or designs, if any.
4. Plans for developing employment, logistics, maintenance support, operational organization, and training requirements.

11.2.2 Preproduction System Prototypes

These tests consist of an examination of preproduction system prototypes to assess the utility, operational effectiveness, and safety in a realistic operational environment, and to resolve all critical operational issues. The testing should be conducted with representative operating and maintenance personnel associated with operational organizational units and should determine the following:

1. The system's potential utility, worth, operational effectiveness, and operational suitability, in the proposed operational, administrative, and logistics environment. Although engineering, reliability, and maintainability data are collected during operational tests, the collection of these data is not the prime objective of the tests. Operational testing should concentrate on the mission consequences associated with using the system and its suitability under mission conditions. Required corrections (to, say, improve reliability) should not be restricted to improvements in the system but may include changing requirements, operational procedures, the logistics and support system, or the training requirements.
2. The effectiveness of the safety engineering features. A formal safety release should be granted.
3. The effectiveness of the logistics support planning and maintenance

test packages to assure timely availability of training devices and field manuals and associated software.

4. Whether from the user viewpoint the new system is desirable considering equipment already available and the benefits and/or burdens associated with the new system.

5. The need for any modifications.

6. Adequacy of the customer's organization, policies, procedures, and capabilities to benefit from employment of the new system.

11.2.3   Initial Production Systems

These major field tests are very similar to those of Section 11.2.2, except that they will be broader in scope, may involve more or larger organizational units, and may be conducted on a number of initial production systems. The purpose of these tests is to assure that the system is ready for employment in the field. All outstanding critical issues should be emphasized and resolved, and it should be determined that there are no new ones. The tests should verify that the operational organization, the application policies and procedures, and the provisions for logistics, maintenance support, and training are all adequate—and should identify the benefits and/or burdens associated with the introduction and operation of the new system.

11.3   RELIABILITY GROWTH MONITORING

To achieve a required performance characteristic in a system, the development must be monitored to assure that proper consideration to achieving that characteristic is given in each design decision. Section 6.6 describes the monitoring process for weight control. Similar specialized techniques are used for structural integrity, payload, speed, and all other design characteristics. The most effective control technique is a test performed as soon as a concept is formulated. Section 6.6 describes how weight control engineers in the aircraft industry prepare weight estimates as soon as a designer completes the layout of a proposed design configuration and how the first piece of hardware is immediately weighed as soon as it is finished, to confirm the weight prediction. The monitoring techniques vary with the design characteristic being controlled, but the basic principles remain the same.

The techniques for establishing system reliability requirements, apportioning them by means of a systems model through various subsystem and assembly levels to the lowest level components, and the procedures for predicting reliability during the development cycle are all described in

Section 7.6. To support these analytical procedures reliability testing of components and integrated subsystems should take place as soon as hardware becomes available.

Sometimes it is possible to select components that have been subjected to statistical reliability testing, but, because of tight development schedules and funding constraints, statistical reliability testing is normally impossible on most newly designed components. Only critical parts such as dynamic helicopter components are an exception to this rule. Hence, other indirect methods must be used to assure systems reliability.

11.3.1   The Test-Fix-Test Cycle

Failures occur in any development program on many components and subassemblies during routine development, prequalification, qualification, and especially systems integration testing. Most such testing is conducted under representative operational and environmental conditions; occasionally some tests are run at limit conditions. However, sometimes the reliability engineer can arrange for overstress testing which, when skillfully performed, can be a very effective technique for increasing reliability.

As failures occur, are fixed, do not recur, and do not cause other failures to occur, the reliability of the system increases, that is, reliability growth takes place (test-fix-test cycle). There remains a management question whether the process of reliability growth is taking place at a rate sufficient to attain the system reliability requirement on schedule. One answer to this question is provided by reliability growth monitoring techniques.

11.3.2   Budgeting Reliability Growth

Duane (139) discovered that reliability growth during the development cycle takes place at a characteristic rate. In Figure 11.1 he plots on log-log paper the cumulative failure rate versus cumulative operating hours for five different types of advanced equipment, designated in the figure by the following encircled numbers:

1. Complex hydromechanical devices
2. Complex hydromechanical devices (different function)
3. Complex aircraft generator (after equipment was placed in operational service)
4. Complex aircraft generator (early stages of in-house test program on a new and different type of generator)

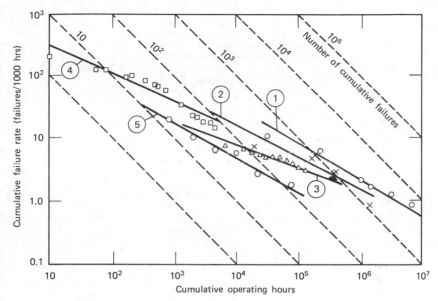

Figure 11.1.   Reliability growth monitoring during development.

5. A complete aircraft jet engine during early stages of introduction to service

It is noted that in every case the reliability growth over time is described by a straight line on log-log paper. Duane's general findings have been confirmed by much subsequent research (156, 157). It has been determined that for certain equipment in a continuing management environment, the slope of the reliability growth curve within certain bounds is an indication of the degree of management support and of the vigor and competence of the reliability personnel.

Given a Duane diagram for a specific system type, it is possible to construct a "budgeted reliability growth curve," Figure 11.2, against which actual progress during development can be measured. This figure is taken from ref. 112, which states that generally one dollar spent on achieving reliability during early development will save $100 in retrofits—if a defective part is allowed to go into production and then fails in service. References 113 and 114 contain detailed instructions for constructing budgeted reliability growth curves.

The Duane reliability growth curve is used by the customer to determine the rate at which the contractor is correcting defects. As time goes on and the reliability of the equipment improves, there are fewer and

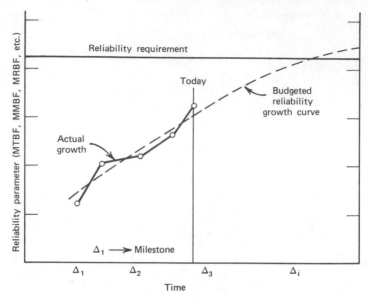

Figure 11.2. Budgeted reliability growth.

fewer defects to correct, and the logarithmic scales provide a realistic tool.

11.3.3 Reliability Growth Monitoring as a Management Tool

The reliability growth monitoring technique, if properly used, can provide management with the following valuable information regarding:

- Allocation of resources to achieve goals on schedule and within cost constraint.
- The focus of management attention on failure to meet goals in order that corrective action may be properly taken.
- Perspective into the relationship of requirements to the stage of materiel acquisition.
- The focus of attention on the need for quantitatively tracking reliability throughout the materiel acquisition process.
- The establishment of a logical basis for projecting reliability growth so that this consideration may be included in the decision-making process.

As with every technical monitoring technique in advanced systems engineering development, reliability growth monitoring techniques must be applied with understanding and professional skill, otherwise management will be provided with a distorted view of the process. It is pointed out in Section 6.6. that even though weight control procedures are technically very straightforward, almost all modern advanced aircraft turn out to be overweight. Dollars are also easy to count, and cost prediction and accounting systems are based on technically sound techniques; yet cost overruns persist. The fact that reliability growth monitoring techniques are based on acceptable empirical concepts does automatically guarantee that their mechanical application to a complex advanced systems development program will yield valid results. The results of any such technique are never any better than the professional skill of the analyst.

11.3.4 Difficulties in Applying Reliability Growth Monitoring Techniques

Among the sources of difficulty in applying reliability growth monitoring techniques are:

- Defining "failures" and deciding what failures to include in the count. This becomes a touchy issue when the designer feels that he has corrected the deficiency and the failure will never recur, that is, that the test-fix-test procedure was part of the normal development process. Many failures are also the result of deficient manufacturing and quality control practices which are readily correctable when the production practices are finalized.

- Failures that occur in tests under overstress conditions or during systems integration, which may not be representative of failures that occur under operational conditions.

- When, to meet schedules or available funding constraints, a test cannot be stopped and the test specimen withdrawn for modification, and a known weak component continues to fail and is repeatedly replaced so that the test may continue. The designer's plan may be to take corrective action after the test is completed, or on the follow-on unit, or on some predetermined downstream follow-on unit.

- Other types of failures that are difficult to classify include:

  o GFE failures—generally not counted against system reliability.

  o Items replaced in troubleshooting and subsequently found to be satisfactory.

  o Components that operate satisfactorily in one system but not in another; that is, the components are not interchangeable, and selective fits are indicated.

○ Failures due to replacement parts that passed acceptance tests but will not operate when installed in the system.

○ Failures detected by diagnostic equipment but that did not affect mission success.

○ Failures of diagnostic or support equipment—generally counted as support equipment failures and not as system failures.

○ Acceptance test failures—usually not counted when the defect is corrected before acceptance.

○ Items replaced per maintenance schedules—chargeable only to maintenance.

○ Directed removals—such as precautionary safety considerations—are generally not counted.

11.4 PRODUCT IMPROVEMENT

On many advanced products, product improvement and customer operational testing continues after qualification and acceptance of the initial units. Sometimes product improvement extends throughout the entire production cycle.

The nature of the product improvement program depends to some extent on the characteristics of the product. To illustrate this point, three types of products are included in the following discussion: aircraft systems, gas turbines, and space systems. Product improvements may be achieved by changes in design, manufacturing, quality control, and/or procedures (software).

11.4.1 Aircraft Systems—Major Phases of Program Life

Modern military aircraft systems are incredibly complex. The development of a new aircraft system or product is an iterative process and proceeds through several definitive stages. In most systems at least three major stages can be identified: (*a*) system development, (*b*) reliability growth, and (*c*) system growth; see Figure 11.3.

System Development Phase

In view of the amount of new advanced technology and the high costs involved, in the system development stage the major effort concentrates on demonstrating that it is possible to build a system that will meet the performance specification. If the essential features of performance expectations are not met, the project is terminated. In this stage trade-offs are

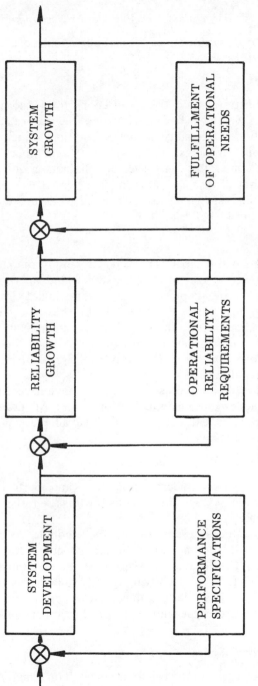

Figure 11.3. Major phases of program life.

387

generally biased in favor of performance. Reliability must be adequate only to demonstrate performance.

Reliability Growth Phase

Once it has been demonstrated that the required performance can be attained and the system's worth has been determined, the major task is to provide whatever additional reliability is needed to meet operational requirements in performing missions. In many new systems, because of funding and scheduling constraints, the level of reliability required for operational service cannot always be attained in the initial system development, and a reliability growth phase is needed to assure that the system attains its inherent design reliability.

System Growth Phase

Systems that prove to be worthwhile in service usually enjoy an extended system growth phase. Once a system is in existence and operating, it becomes obvious where many improvements can be made. Furthermore, as operators learn how to use the new system, they also develop new requirements for its use beyond the originally planned capability of the existing equipment. In the system growth phase, both the basic system and its components may be extensively redesigned to increase both performance and reliability.

To perform this phase effectively, it is necessary to have access to the technical reports and data mentioned in Section 6.4.3. All drawings, design analyses, test plans, and test reports should be available to provide a baseline for design improvements.

Availability of Resources

All successful aricraft systems undergo reliability and performance growth during their useful life cycle. The reasons for this are rather basic and should be understood.

Under normal circumstances, there are never adequate resources available to do a technically optimum job during the contractor's development of prototypes. It is always necessary to make an almost endless series of compromises to take best advantage of available resources. This is why it is normally impossible to attain the full performance and reliability potential in the prototype and early production models (see Figure 11.4).

Example: Complex Aircraft System X

Figure 11.5 provides an example of the reliability growth on a major U.S. military aircraft program. In its day, this aircraft system included one of

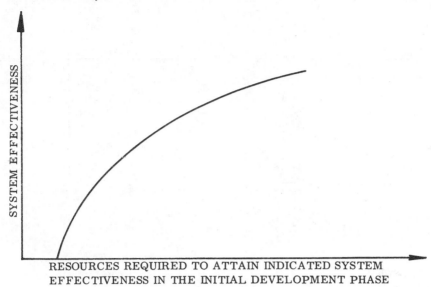

RESOURCES REQUIRED TO ATTAIN INDICATED SYSTEM
EFFECTIVENESS IN THE INITIAL DEVELOPMENT PHASE

Figure 11.4. System effectiveness versus resources.

the most advanced, complex, and sophisticated airborne electronic systems designed in the United States. The reliability values in Figure 11.5 are given in relative terms. During the initial design stage of this project, the following values were determined:

1. Estimate of the reliability needed for satisfactory mission accomplishment: P.
2. Inherent reliability estimate of the proposed design: Q.
3. Reliability estimate for the proposed design: R.

After some 17 years and the deployment of many of these aircraft, these estimates have proven to be remarkably accurate. Over this time, the aircraft system has been produced in three versions: A, B, and C.

The A version may be considered the system development phase, wherein the aircraft system demonstrated that it possessed the capability to meet all performance requirements, although the reliability of the system was low, as predicted, and just barely sufficient to demonstrate system capability under controlled test rather than routine fleet operation.

Once the feasibility of this very advanced design was successfully demonstrated and the design proved to be exceptionally worthwhile, a massive effort was initiated to improve system reliability. So many complex and extensive design changes were necessary that it was decided to

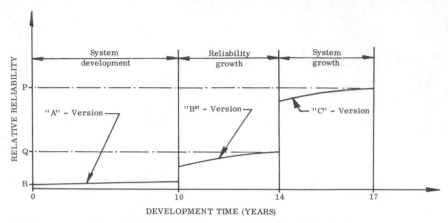

Figure 11.5. Reliability growth of aircraft system X.

designate the changed model as a B system. However, these were changes in details and not the basic system so that the inherent reliability predicted for the A version could not be exceeded. Nevertheless, the B phase was successful, and the inherent reliability was substantially achieved.

In view of the success of this operation, the design of a C version was initiated. This is now a new design start on major critical subsystems, taking advantage of all advancements made in the state of the art during the design and development of the A and B versions, resulting in a new system. In fact, this particular aircraft program forced and funded the electronics industry to develop integrated circuitry, a major breakthrough in the state of the art, so that the required system reliability could be achieved. There is evidence that the C version now meets the initially established reliability requirement of P. This development was a gamble which has paid off magnificiently, producing a new technology of integrated circuits from which the whole world has benefited.

11.4.2 Gas Turbine Engines

One of the most difficult and remarkable reliability programs in U.S. industry has been conducted by the manufacturers of gas turbine engines. This program has been the most significant single factor in the economic success of the world airline industry.

When gas turbines first were introduced to airline service, engine overhaul schedules were still a major operational problem. The time between overhaul (TBO) for airframes was longer than for the engines; airplanes had to be laid up specifically for engine overhaul, with resultant economic losses.

As the reliability (and TBO) of gas turbines increased, the industry dream was to achieve an engine TBO equal to that of the airframe. Today, gas turbines have exceeded this goal by an appreciable amount; on some turbines the TBO is so long that from a practical standpoint their overhaul requirements have been eliminated.

Empirical Procedures

The reliability of gas turbines is all the more remarkable in that it has been achieved on an empirical basis, without a reliability model (33).

Gas turbines are complicated machines, each consisting of some 25,000 parts making up 300 different assemblies, with an almost infinite number of possible modes of failure. The integration of those parts so that they all work together in a highly loaded, dynamic environment requires solutions to problems that generally cannot be solved analytically. Gas turbines are therefore developed in the laboratory rather than in the design office. Because of severe weight and space limitation, all engine parts are subject to design compromise. Among the specific problems are fatigue and wear, two areas for which satisfactory analytical models do not exist (140).

Laboratory Life Testing

In developing a new engine all individual parts are life tested in the laboratory at load levels higher than the maximum conditions anticipated in service by some selected factor. As parts fail in the test, they are redesigned and retested to assure that the failure will not recur. Before the engine is released for flight, it must pass a 150-hour qualification test. The developmental flight-test program will disclose additional defects that call for corrective actions.

Field Monitoring

All engines placed in service continue to be monitored by the manufacturer. Engine manufacturers maintain excellent field-trouble reporting systems; they make it their business to know the operating procedures of the various airline and military users, and they keep detailed track of all their engines. For example, they know that if a particular airline with a vigorous service run encounters a certain problem, other airlines will begin to report the same problem in 6 to 8 months. During this time the manufacturer has been forewarned and will have studied the problem and developed an appropriate solution to provide corrective action. The cost of this continuous product improvement program is at least of the same order of magnitude as that of the original engine development program.

Time Required for Corrective Action

Once an engine is in service and a defect is discovered, it may take five years to complete the corrective action. Figure 11.6 shows the schedule for the correction of a cracked bearing support on a modern jet engine (206). The problem was first revealed in March of 1962. It took almost a year to issue the engineering change and another year before the change could be incorporated in production engines. Soon thereafter the retrofit program was initiated. During all this time failures were occurring in service, and the premature removal rate was climbing. However, as soon as the retrofit program was initiated these engines did not fail again, and the removal rate curve began to fall. It took almost three years to retrofit all engines and eliminate the problem.

The overall results of this program are shown in Figure 11.7. This plot shows the mean time between premature removals (MTBR) for five groups of production engines released in sequential lots of 200 engines each. The plot shows that the first group had an initial MTBR of about 1000 hours, but as retrofits were incorporated, the MTBR was improved by a factor of 5 as the engines accumulated operating time.

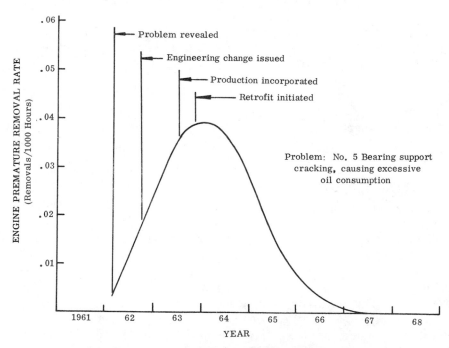

Figure 11.6.   Resolution of typical engine service problem.

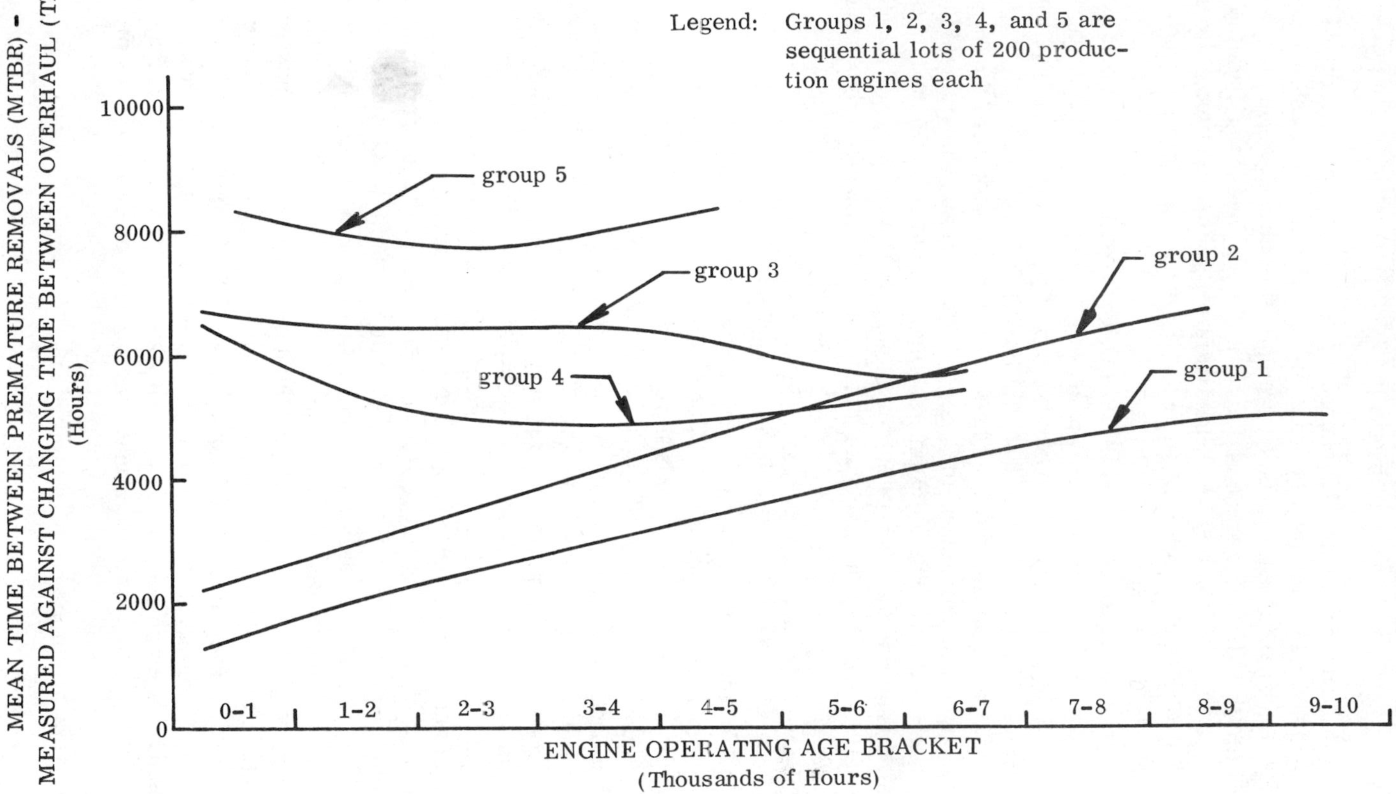
Legend: Groups 1, 2, 3, 4, and 5 are sequential lots of 200 production engines each
MEAN TIME BETWEEN PREMATURE REMOVALS (MTBR) - MEASURED AGAINST CHANGING TIME BETWEEN OVERHAUL (TBO)
(Hours)
10000
8000
6000
4000
2000
0
group 5
group 3
group 4
group 2
group 1
0–1
1–2
2–3
3–4
4–5
5–6
6–7
7–8
8–9
9–10
ENGINE OPERATING AGE BRACKET
(Thousands of Hours)

Figure 11.7. Impact of engineering changes on mean time between premature removal of engines.

Engines from groups 2 and 3 started out at successively higher MTBR values. It should be noted that during this time the time between overhaul (TBO) was also increasing, and "premature removal" is defined with respect to the TBO. The value of TBO is not given, because it is different for each airline, depending on airline operating policy and type of operation. So MTBR is rising even though it is measured against a moving target that is also rising.

The airlines like to increase TBO as much as possible, since it results in more economical operation. However, on this engine the TBO was increased so fast between groups 3 and 4, that the MTBR of group 4 fell below that of group 3, when it should have been higher.

This was a signal that the TBO had been increased too fast; the rate of increase was reduced and group 5 shows a healthy improvement in MTBR to over 8000 hours. The TBO on this engine eventually reached 16,000 hours; this number is so large that for all practical purposes it signifies the elimination of mandatory overhaul requirements for the life of the engine.

Monitoring Dynamic Mechanical Components

This basic empirical approach is used on a number of other complex flight safety mechanical systems such as helicopters and power operated flight control systems. These systems have problems very similar to gas turbines and are also developed and improved in "families." In the absence of analytical models, each improvement step is a careful trial-and-error extrapolation of previous experience.

Helicopter manufacturers, in particular, faced severe structural fatigue problems many years before gas turbine engines became available. At that time, civil and military government agencies refrained from setting forth quantitative fatigue design requirements because means for enforcement were lacking. Under these circumstances the helicopter industry, out of sheer desperation, established an excellent field trouble reporting system. Field representatives of the helicopter contractor monitored all vehicles in operational service and established a proud tradition of making themselves immediately available on the spot wherever trouble occurred. There were many structural failures of dynamic components. With great diligence, the contractors' representatives would obtain the failed parts and expeditiously transfer them to the home plant where an immediate attempt would be made to develop effective corrective actions. At the then current state of the art, the primary objective of this procedure was to prevent the rate of structual failure, which was already too high, from becoming even greater.

11.4.3   Space Systems

Space systems present a different problem because they must obtain a degree of reliability on their first manned space flight that aircraft need attain only after an extensive developmental flight-test and product improvement program. In spacecraft development there is no opportunity for the equivalent of an aircraft developmental flight-test program at increasingly severe flight conditions, and a gradual approach to the extreme flight requirements. The first space flight already takes place at the design load conditions. In spacecraft development, the *product improvement* phase must be telescoped into the period of *prelaunch operations*, which on major space systems may require a time period of a year or more. The unique procedures associated with prelaunch operations are described in ref. (155). The effectiveness of these procedures can be judged by a review of refs. 136, 137, and 187.

11.5   CHAPTER SUMMARY

This chapter completes the discussion of the development cycle shown in Figure 3.1 with a description of the operational test and product improvement cycles.

A significant characteristic of a customer operational test plan is that it usually refers to the original customer operational requirements as modified by the customer's test agency's updated expert knowledge of the user's developing needs. No reference is normally made to the contract. Hence, the requirements of the customer's operational test plans will be different from the contractual requirements.

Under these circumstances the operational tests will uncover two distinct classes of deficiencies. A contractor is usually only required to correct those features that do not conform with his contractual requirements. To correct other deficiencies, he usually first prepares an engineering change proposal (ECP) with the associated cost estimate. Since at this time the customer's funds are usually depeted, it will be necessary to make a number of compromises on the deficiencies to be corrected at the current time and those corrections that can be postponed to a later date.

The operational tests are normally conducted by two customer agencies with different orientations: an engineering and a field/service test agency. The engineering agency assesses technical risks and measures system characteristics under a spectrum of controlled operational and environmental conditions, and establishes the system's performance limits. The field service test agency assesses the utility and worth of the system when it is operated under anticipated service conditions by regular operating and maintenance personnel. The advantages of the new system are determined by comparison with equipment already available. The test results are provided to top management to support their decision on production release.

As deficiencies are discovered and fixed during the operational test cycle, the system reliability will grow. Normally, the reliability of the first prototypes is low, and increases as testing and fixing progresses. For the system to be acceptable, a specified reliability level must be achieved. A budgeted reliability growth curve, based on available empirical reliability growth models can be prepared. This curve will show the rate of reliability growth required for the system to meet its requirements on schedule. Reliability growth should therefore be monitored with reference to this curve. Deviations of the actual attained reliability from the curve provide management with an indication that resources must be reallocated or other action taken to assure that the reliability requirement will be met on schedule.

On some systems the product improvement cycle continues as long as the system is in production, as with aircraft and aircraft engines. On spacecraft the reliability requirement must be achieved on the first space flight, since flight conditions cannot be approached gradually in incremental steps as on aircraft and other earthbound systems. All reliability growth must take place during prelaunch operations. Examples of these three systems are discussed.

12

Professional Responsibilities

In Section 8.1 reference is made to the social objectives of the aerospace industry. Prime advanced systems contracts may be so large that they affect the economic welfare of large areas of the country. Sometimes contracts are used by the government to encourage the development of certain geographical areas. The requirements of MIL-Q-9858A and the associated motivational programs have social implications. Section 9.15 discusses the responsibility of regulatory agencies for safety and environmental impact.

These are only a few of the indications that there is a growing requirement for the consideration of social factors in the application of modern technology. Social factors provide an important element in establishing the requirements for quality during systems development. The purpose of this section is to place the subject in some perspective and to provide a better understanding of the issues for those who are attempting to achieve the required performance and quality in advanced systems.

12.1 PROFESSIONAL ORIENTATION OF ENGINEERS

Before the turn of the century the engineering profession in the United States was exclusively oriented toward technical achievement. Thus, engineers were not particularly efficiency-minded. The basic problems of the nation were associated with taming a vast new continent. The country desperately needed railroads and canals; the west coast had to be joined with the east coast, and cost was not the major factor. Bridges had to be built; factories were going up; and people were populating the cities. The need was for quick housing; America became the land of "instant cities."

The main representatives of the engineering profession at that time were the civil engineers, and they responded with enormous and massive productions, clearly visible to the public. In an environment of urgent need, where cost was secondary, safety was achieved by massive over-building, or "overdesign (141).

12.1.1 Economic Constraints

The United States emerged from World War I as a nation unified from coast to coast. The continent was truly conquered; all the physical elements needed for the "good life" had been created and were becoming available. As our factories started to pour out goods, a surplus of productive capability built up, and competition became a dynamic and effective new force. If a man went to buy a can of paint, he found a choice of brands for the first time. Manufacturers then looked to the engineers to reduce the price while improving the quality of the paint to better compete in the marketplace. In addition to technical adequacy, the engineer now had to satisfy economic constraints. The engineer built the plant on a river where he had a good water supply on the upstream side and a cheap waste disposal facility on the dowstream side, in addition to cheap transportation. Everybody was happy for a while, until our rivers and our air became polluted and the natural beauty of our countryside damaged almost beyond repair. Now the engineer is being called upon to consider social, environmental, and esthetical factors in addition to the economics.

12.1.2 Quality of Life

Today, in an age of affluence, there is no need for any man to be hungry or miserable. The legitimate question arises, "Why should conditions of hunger and misery exist?" Suddenly the engineer is involved in very complex questions involving quality of life and the effect of the application of technology on man's whole environment.

The emphasis has shifted from massive straightforward civil engineering works to mass-produced consumer products to advanced systems involving the quality of life. Everything from mass transportation, urban renewal, health delivery services, energy generation and distribution, food production and distribution, communications, and even mass entertainment is trying to adapt itself to take advantage of revolutionary new advances in technology, in the state of the art. Unknowingly, all are in need of the systems approach.

Our society has become so very much more complex and developed many new sophisticated needs. We do not have the resources to satisfy them all. Yet, we do not know how to mobilize all of the human resources that we do have; a significant portion of our population is unproductive. Or do we appear to have the mechanism for establishing a generally accepted and intelligent list of national priorities. As a result, there is a smaller than desirable proportion of our national wealth available for the application of systems engineering to incorporate the needed advances in the state of the art across a wide range of applications affecting all of society in an effective manner. Engineers must learn to be more productive with the resources allotted to them; there is no room for waste.

One contributing factor in assuring that an advanced systems development project achieves minimum life-cycle costs is the intelligent application of systems assurance and risk management techniques.

12.2   THE CHALLENGE OF OUR TIMES

In discussing the impact of technology on society, the following definition of the responsibilities of the engineer is of interest. This quote is from a lecture by J. Douglas Brown, Dean of the Faculty of Princeton University, and himself not an engineer (142).

The professional engineer must act as the moderator and the interpreter between the two worlds of science and the humanities. He is even more responsible than the medical doctor, who, in a sense, is a biological engineer. The doctor still has to work with a machine designed by God . . . The professional task of the engineer is to design . . . The engineer is designing another physical world around us based on new scientific knowledge.

The problems which engineers must help solve in the future are in part those which engineers helped to create; the congestion of our cities, the dangerous crowding of our highways and airways, the fouling of our atmosphere . . . I would claim that the present conditions are the result in large part of inadequate engineering statesmanship.

The profession of engineering must seize the initiative in acting as a bridge between science and human needs. The scientist is not asked to apply his findings through design. The business man or the politician does not know what science offers or how to apply it. The engineering profession is the channel by which science can greatly improve our way of life, *provided it assumes the initiative of leadership rather than the passive role of the hired consultant.*

These definitions present a formidable challenge to the engineering profession to design systems with proper consideration for a broad spectrum of social values. In an attempt to come to grips with some of the problems involved, a group at Princeton is trying to develop the concept of the engineering critic, analogous to the art critic. The engineering critic would analyze typical works of engineering from the overall viewpoint of society and would establish on a logical basis the desirable and the undesirable features (143). Eventually, such analyses would provide the basis for engineering quality standards, indicating what is "good" and "bad" engineering, and would define the social values that would have to be considered in the design and assurance of a "good" system.

12.3 TECHNICAL ADVANCEMENTS BY THE AEROSPACE INDUSTRY

Lacking an effective recognized mechanism for establishing what constitutes a socially "good" or "bad" system, one development that has had the greatest impact on modern civilization has been the capability to convert technical advances into functional systems. Within this context the aerospace industry, and over the last half century *only* the aerospace industry, has demonstrated throughout its history a remarkable, continuing capability for incorporating new technical advances in its products.

The significance of this accomplishment can only be appreciated when it is compared to the record of genuine technical progress in all other areas. Although there are today more cars, more ships, more railroads, more television sets, more touch-tone telephones, more houses, and more plumbing, once the basic design of any of these products or systems is created, debugged, frozen, and the means of production and distribution established, further significant technical advances in the product itself come very slowly. The research and advances apply primarily to manufacturing processes and means of distribution, seldom to the product itself. From year to year, there is usually little that is really technically new under a shiny new stylish exterior; it is the same old wine in a new bottle.

Not so in aerospace. Consider, for example, the progress made. Every new system, be it the DC3, DC6, 707, 747, Mercury, or Apollo, represents a new spectacular milestone in technical advancement, a new plateau of technological accomplishment. Cancellation of a large nonaerospace program, such as a dam or a new model automobile, may have severe local or regional economic impact, but will be of no technical significance. In contrast, had the Apollo Program been cancelled, technical progress on a broad significant front would have been affected.

It is, for example, hard to believe today that the engineers who designed the 707 were taught in college that the jet engine was an "impracticable" concept and would never fly in a competitive environment. Not too many years before the flight of Apollo, none of the responsible designers would have raised an eyebrow as knowledgeable professionals spoke of the "thermal barrier." None of the responsible designers of our supersonic aircraft, in their college years, were permitted to dream that man would ever penetrate the sound barrier.

12.3.1 The Military Aerospace Industry "Formula

Why is this technical progress so characteristic only of the aerospace industry; in particular the military segment? The answer lies in the fact that the military aerospace industry has developed a unique "formula" and personnel attitude for assimilating technical advances in its products. This formula, one of the main themes of this book, encompasses the following features:

1. A well-defined separation of responsibilities between customer and contractor, and regulator, when there is one (Figure 2.1).
2. A unique, integrated system development cycle (Figure 3.1).
3. Rigorous control of the systems engineering process (Figures 6.3, 6.4, and 6.5), the main elements of which are:

   a. Establishment of technical requirements (with specific accept/reject criteria).
   b. Control of the design phase by analytical procedures to assure that the proposed design, when built, will meet the technical requirements.
   c. Demonstration that the product meets all technical requirements.
   d. Continuous feedback—the process takes place under conditions where research, development of technical requirements, design control, demonstration techniques, and service monitoring all interact with and supplement each other in an endless cycle (Figure 6.2).

The formula may be less than perfect, but it is one that works in the bureaucratic environment of huge organizations working with large sums of money on programs involving high risks. It is the *only one of its kind*. It results in a workable, "least lousy" solution. This formula is complex, and the concept is fundamentally different from the commercial practices with which most people are familiar. This makes it very difficult for the uninitiated to understand. Yet, better general understanding is required to assure the continued support of advanced developments on which our continuing progress depends.

12.3.2 Design Discipline

In addition to the military aerospace formula on procedures mentioned above, the personnel attitudes and rigid discipline with which the procedures are applied are significant factors. This discipline is the result of the intense and unique personal concern of design engineers in the military aerospace industry with safety of life while developing products that are extending the state of the art.

As a responsible designer of military airframes, the author has repeatedly been in a position where he, and only he, understood all of the design trade-offs and risks involved in a proposed flight test required to demonstrate a new device or a performance parameter in an untried flight condition. In every case the test pilot and his family were known to and were on friendly terms with the author's own family.

The author has had close associates and supervisors who have been crippled as test pilots in accidents caused by engineering deficiencies. The author has worked as a designer on aircraft in which test pilots, who were his friends, have been killed because of design engineering faults.

Aerospace companies have tried to minimize some of this personal involvement by hiring test pilots from outside the company, who were unknown to the designers. The author was associated with a project in which such an outside test pilot was killed because of an engineering design deficiency. The fact that the pilot was unknown to the design engineering personnel involved did not lessen the emotional impact of the accident.

As a result of this personal relationship between designer and test pilot, no responsible airframe designer will release an aircraft for flight test that he himself would not fly if he were qualified to do so. No management in the military aircraft industry will override a designer's decision in this area.

The author has investigated hazardous operations in other industries where safety of life is involved; in some cases the safety of large numbers of people has been involved. In no other industry has the author found designers so personally concerned with safety of life. In no other industry is there such a rigorous "formula" for systems assurance. In no other industry does the discipline exist that provides the basis for the strict enforcement of rigorous design control procedures. In all hazardous operations investigated by the author in other industries, safety is maintained primarily by safe operating practices rather than by inherently safe equipment. It has been very difficult for accident investigating boards to recognize and identify design deficiencies, and impossible to trace design deficiencies to responsible designers when such have been identified.

Other industries, which could greatly benefit from the application of aerospace systems development management procedures, must find new ways to motivate their executives and designers to accept a personal sense of commitment and discipline that they have never known before, which provides the necessary environment for the effective application of the systems assurance procedures described in this book.

In an effort to create this sense of responsibility, the hazard accountability system of Section 9.15.3 suggests the recording of the names of the engineers responsible for the elimination of an identified hazard, as well as the name of the representative of the concurring regulatory authority, in a computerized data bank from which they should be easily retrievable at any future date.

12.3.3   Feedback Dynamics

Among the unique features of the formula described above is the capability it provides for utilizing effectively the services of very large numbers of highly trained, creative, technical specialists on a single giant project of staggering proportions. Considering the motivational and morale problems involved in large organizations, this is an achievement unique in the history of the world. The secret of this accomplishment lies in the dynamic feedback loops defining customer-contractor-regulator relationships, which are depicted in many of the figures in this book (such as Figure 6.5).

The overall systems development process is broken down into individual functions, with each function acting as a component of a feedback loop. This results in a complex system of loops operating on various levels, with many lesser loops within the higher-level loops. Deficiencies in the output of any loop are cycled back into the input, and the feedback cycles continue until each deficiency is corrected. In all significant functions, the the feedback loop goes through the customer's and/or regulator's technical monitoring organizations.

In particular, deficiencies observed during demonstration testing or service monitoring are fed back, often through various research cycles, to improve the technical requirements. These in turn control the design, development, and production processes. In this process there is a constant, decentralized improvement cycle taking place in each loop.

Improvements are being made continously in iterative steps, usually in small increments. Spectacular advances such as the introduction of the jet engine or of electronics are rare; nevertheless such advances are much more likely to occur in an environment of continuous progress where everyone is in search of new solutions and where there are few pressures to retain traditional, frozen design concepts. The great overall technical advances in aerospace products rest on the sound base of a steady, continuous effort in an iterative process to achieve small incremental improvements over a wide range of functions.

12.4 THE CUSTOMER-CONTRACTOR-REGULATOR ROLES

Strong professional competence in the customer's and regulator's in-house organization is an essential condition for technical advancement on a broad front. Each feedback loop represents a specialized technical challenge in adjusting the output to correspond with the given input, often of a dynamic or changing nature.

Within these loops, involving the formal organized interaction of customer, contractor, and regulator technical personnel, lies the challenge to the creative technical specialist for advancing the state of the art in small increments. The technical challenges involved in making some of these loops work, within themselves and in conjunction with adjacent and overlapping loops, are sometimes formidable. Men on both sides of the customer-contractor-regulator fences, on all working levels, *do what only they know must be done* to keep the program moving in the right direction. It often requires the assumption of personal risk and considerable personal courage to make the necessary decisions in a timely manner, but this is exactly what happens on any successful aerospace project. Management only enters the process when the lower-level decision making process breaks down; once the design requirements are established, most basic decisions are made by working level engineers.

These numerous, interrelated feedback loops represent a controlled working environment that provides creative technical individuals with opportunities for job identification and for technical achievement and satisfaction in a manner not done any other known way, much less with money alone. These engineers, even at relatively low organizational levels, are more than hired hands; they provide effective technical leadership—sometimes at great personal sacrifice. This is the basic reason that the aerospace industry has been able to attract so many technically outstanding individuals in spite of the basic employment instability in the industry.

12.5 EXTENSIONS IN THE STATE OF THE ART

Estimates on large projects are based on the "success theory," that is, the assumption that every task will be performed in the specified, most efficient manner, and that the performance of the various tasks will follow one another in a timely, orderly, and efficient sequence. In actuality, large projects hardly ever follow this pattern; unforeseen events always occur to interfere with the prescribed scheduling. Since budgets during the proposal stages are always meager, it is impracticable to include funds for contingencies to allow for unforeseen events.

Also, a great deal has been written about the resulting cost overruns of aerospace systems. Cost overruns are not unique, however, to the aerospace industry. Large commercial systems such as construction projects, involving huge sums and no advances in the state of the art, are also characterized by overruns. Even small construction projects such as the author's one-family home involved a painful overrun.

In aerospace projects, in contrast to large construction projects, there is an additional reason for the overruns; namely, the required extensions in the state of the art. As noted in Figure 5.1, costs skyrocket as soon as *any* extension of the state of the art is required in a development program.

To visualize what is happening here, consider the analogy of a championship football game. A top professional football star knows how to handle a ball; under normal circumstances he never fumbles. If he does fumble in a championship game, his failure is caused by actions of equally good professionals on the opposing team. There are three reactions to the fumble: distress on the part of the fumbler; elation on the part of those whose actions caused the fumble; and excitement on the part of spectators. The greater the number of such fumbles during a championship game, the more exciting the event. For each team, the number of fumbles is a measure of the strength of the opposing team.

Similarly, a highly trained team of expert engineers knows how to build an aerospace system to meet technical requirements that are within the state of the art without fumbling; that is, meeting performance requirements without exceeding reasonable, specified weight, schedule, and cost requirements. However, as soon as the application of new technology is involved, and here nature is the "expert" opponent, "fumbles" are to be expected. Design changes, schedule slippages, and weight and cost increases occur as part of the development process; that is, as part of the fight to gain control over yet unknown and undefined forces of nature.

12.6   THE GOAL—SYSTEMS ASSURANCE

Just as the greater strength of the opposing football team causes more fumbles, so do requirements for greater advances in technology cause more design changes to be made to achieve a workable product. Given an expert design and development team, weight and cost increases and schedule slippages provide an index of the technical advances incorporated into the new system. The skilled engineer, in advancing the state of the art, is motivated by much the same excitement as motivates the professional football star. Management questions that arise in this connection are whether the requirements for extending the state of the art are feasible, whether the design team is competent, and whether the problems involved can be solved at all.

12.6.1 The Coutinho Conclusion—The "Least Lousy" Solution

This book outlines the results of the author's experience on procedures for assuring the operational suitability and integrity of big systems. Such new big systems are typically very costly and require a long time for development. They involve the assumption of risks that are so high that they usually can be underwritten only by government agencies. The procedures for the acquisition of such advanced new systems differ basically from conventional practices associated with the procurement of commercial products.

Characteristically, the development of large new advanced systems involves very large organizations, both governmental and commercial. All large human organizations, be they military, civil, governmental, ecclesiastical, academic, industrial, or commercial, are always subject to some extent to inherent inefficiencies, the scope and extent of which are described by Parkinson's Laws, the Peter Principle, and Murphy's Law, as discussed in Section 1.5. In particular, it is difficult in most large organizations to obtain timely decisions; chains of command are long, and there is a seeming lack of appreciation of the high costs associated with procrastination. Great value is laid on not "making waves" or "rocking the boat." There appears to be no ideal, effective format for a very large organization; in any given case, the problem is to devise a workable scheme. The best that normal humans can accomplish is a "least lousy" solution. To obtain a superior solution, in addition to the mastery of systems assurance techniques, the program manager must have a touch of genius. The author feels privileged to have been associated with a few such men in his lifetime; but such men of genius are very scarce, and they have never been program managers.

These are harsh words and will be difficult for many to accept. In the popular image, organizations are often depicted as greater than the men who comprise them. Men allegedly become better by belonging to certain organizations, and they strive mightily to prove themselves worthy of the privilege and responsibility of membership. The quality of the organization itself is generally above question.

The one exception is provided by organizations that deal with the application of science. Natural laws are absolute, democratic, and fair, without exception. If the objective of an engineering organization is to place two men on the moon and return them safely to earth, it becomes the unavoidable responsibility of the organization to do whatever is necessary to accomplish the objective: to appraise the magnitude of the job, to assemble the necessary resources, and so on and on. The organization must accept the full responsibility for success or *failure*. The emphasis is on meeting the objective, not on the quality or "greatness" of the organization. The organization is simply a means to an end.

In contrast, a political organization, for example, need not accept the full responsibility for failure. If it underestimates the magnitude of the job to be done and fails, it can place the blame on the lack of popular support or the unfair acts of domestic or foreign opponents. In the business of saving souls, a religious organization can cite the perversity of the devil, the general moral decay, and similar excuses as the reason for the high crime or suicide rates. Such organizations seldom ascribe their inability to control the external forces that prevent them from fulfilling their objectives to their own internal organizational shortcomings.

To assure success in a systems development program, an organization must accept the full and exclusive responsibility for success or failure, and it must acquire the resources to control all internal and external forces that might prevent it from achieving its goals. On the other hand, the organization need be no better than necessary to achieve success.

References

Books

1. M. J. Peck and F. M. Scherer, *The Weapons Acquisition Process: An Economic Analysis,* Graduate School of Business Administration, Harvard University, Boston, 1962.
2. F. M. Scherer, *The Weapons Acquisition Process: Economic Incentives,* Graduate School of Business Administration, Harvard University, Boston, 1964.
3. H. Kahn and B. Bruce-Briggs, *Things to Come,* Macmillan, New York, 1972.
4. H. Chestnut, *Systems Engineering Methods,* Wiley, New York, 1967.
5. E. B. Flippo and G. M. Munsinger, *Management,* 3rd ed., Allyn & Bacon, Boston, 1975.
6. L. S. Peter and R. Hull, *The Peter Principle,* Morrow, New York, 1969.
7. C. N. Parkinson, *Parkinson's Law,* Ballantine, New York, 1957.
8. H. H. Goode and R. E. Machol, *System Engineering,* McGraw-Hill, New York, 1965.
9. P. F. Drucker, *Management: Tasks, Responsibilities, Practices,* Harper & Row, New York, 1974.
10. B. S. Blanchard and E. E. Lowrey, *Maintainability: Principles and Practices,* McGraw-Hill, New York, 1969.
11. C. E. Cunningham and W. Cox, *Applied Maintainability Engineering,* Wiley, New York, 1972.
12. A. S. Goldman and T. B. Slattery, *Maintainability: A Major Element of System Effectiveness,* Wiley, New York, 1964.
13. B. S. Blanchard, *Logistics Engineering and Management,* Prentice-Hall, Englewood Cliffs, N.J., 1974.
14. W. Hammer, *Handbook of System and Product Safety,* Prentice-Hall, Englewood Cliffs, N.J., 1972.
15. H. W. Heinrich, *Industrial Accident Prevention,* McGraw-Hill, New York, 1959.
16. R. E. Machol, *System Engineering Handbook,* McGraw-Hill, New York, 1965.
17. E. F. Bruhn, *Analysis and Design of Flight Structures,* Tristate Offset, Cincinnati, 1965.
18. A. M. Polovko, *Fundamentals of Reliability Theory,* Academic, New York, 1968.
19. M. L. Shooman, *Probabilistic Reliability: An Engineering Approach,* McGraw-Hill, New York, 1968.
20. C. D. Hodgman, *Mathematical Tables from Handbook of Chemistry and Physics,* 6th ed., Chemical Rubber, Cleveland, 1938.

21. K. D. Lloyd and M. Lipow, *Reliability: Management, Methods, and Mathematics,* Prentice-Hall, Englewood Cliffs, N.J., 1962.

22. R. E. Barlow, J. B. Fussell, and N. D. Singpurwalla, *Reliability and Fault Free Analysis,* Soc. for Industrial and Applied Mathematics, Philadelphia, 1975.

23. H. A. Simon, *Administrative Behavior,* Free Press, New York, 1965.

24. D. McGregor, *The Professional Manager,* McGraw-Hill, New York, 1967.

25. E. B. Roberts, *The Dynamics of Research and Development,* Harper & Row, New York, 1964.

26. R. N. Anthony, *Management Controls in Industrial Research Organizations,* Graduate School of Business Administration, Harvard University, Boston, 1952.

27. A. O. Putnam, E. R. Barlow, and G. N. Stilian, *Unified Operations Management,* McGraw-Hill, New York, 1963.

28. R. F. Harting, *Integrated Design and Analysis of Aerospace Structures,* ASME, New York, 1975.

29. A. Goldsmith, *Guide to System Safety Analysis in the Gas Industry,* Amer. Gas Assn. Cat. No. X51174, Inst. of Gas Technology and IIT Research Inst., Chicago, (undated—published about 1975).

30. S. R. Calabro, *Reliability Principles and Practices,* McGraw-Hill, New York, 1962.

Chapters in Multiauthor Volumes

31. J. Coutinho, "Introduction," in *Transportation: A Service,* New York Academy of Sciences, New York, 1968.

32. J. Coutinho, "The Trouble with Statistics," in *Reliability Control in Aerospace Equipment Development,* TPS Vol. 4, SAE, New York, 1963.

33. J. Coutinho, "Contractual Reliability Acceptance Procedures," in *Structural Fatigue in Aircraft,* STP 404, ASTM, Philadelphia, 1965.

34. J. Coutinho, "Program Management," Part V of *Reliability Control in Aerospace Equipment Development,* TPS Vol. 4, SAE, New York, 1963.

Government Publications

35. *General Specification for Design, Installation, and Test of Aircraft Flight Control Systems,* MIL-F-18372, Para. 3.5.2.3, March 1955.

36. H. J. Grover, *Fatigue of Aircraft Structures,* NAVAIR 01-1A-13, U.S. Govt. Printing Office, Washington, 1966.

37. C. R. Smith, *Tips on Fatigue,* NAVWEPS 00-25-559, U.S. Govt. Printing Office, Washington, 1963.

38. *Configuration Management During the Acquisition Phase,* AFSCM 375-1, Air Force Systems Command, Washington, June, 1964.

39. *Basic Policies for Systems Acquisition,* AR 1000.1, Department of the Army, Washington, December, 1974.

40. *Value Engineering Program,* AMCR 70-8, Department of the Army, Washington, July, 1975.

41. *Selection of Contractual Sources for Major Defense Systems,* DOD Directive 4105.62, Department of Defense, Washington, April, 1965.

42. "Design to Cost," *Defense Management Journal,* Special Issue, U.S. Govt. Printing Office, Washington, September, 1974.

43. *Design to Cost,* DOD Directive 5000.28, Department of Defense, Washington, May 1975.

44. *Life Cycle Cost as a Design Parameter,* Joint Logistics Commanders Guide on Design to Cost, Department of Defense, Washington, October 1975.

45. *Letter of Instructions (LOI) for Implementing the New Materiel Acquisition Guidelines,* LtG. R. R. Williams, DAFD-SDY, Department of the Army, Washington, August 1972.

46. *Joint Design to Cost Guide, A Conceptual Approach for Major Weapon System Acquisition,* AMCP 100-50, Department of the Army, Washington, September 1973.

47. *Guides for Review of Technical Data Packages for Munitions,* PA/HISAR 70-65, Picatinny Arsenal, Dover, N.J., October 1972.

48. *The Army Force Planning Cost Handbook,* Change 1, Comptroller of the Army, Washington, October 1974.

49. *Key Cost Analysis Definitions,* Office of the Comptroller, Army Materiel Command, Alexandria, Va., October 1972.

50. *Weapon/Support Systems Cost Categories and Elements,* AR 37-18, Department of the Army, Washington, October 1971.

51. *Operating and Support Cost Estimates–Aircraft Systems–Cost Development Guide,* Defense Systems Acquisition Review Council, Department of Defense, Washington, May 1974.

52. *Work Breakdown Structures for Defense Materiel Items,* MIL-STD-881A, April 1975.

53. *Unit Work Codes and Maintenance Engineering Analysis Control Numbers for Aeronautical Equipment, Uniform Numbering System,* MIL-STD-780D, November 1967.

54. *Decision Risk Analysis Course,* ALM-63-3463, Army Logistics Management Center, Fort Lee, Va., March 1974.

55. *Reliability Program for Systems and Equipment Development and Production,* MIL-STD-785A, March 1969.

56. *Maintainability Program Requirements for Systems and Equipments,* MIL-STD-470, March 1966.

57. *Maintainability Demonstration,* MIL-STD-471A, March 1973.

58. *Maintainability Prediction,* MIL-HDBK-472, May 1966.

59. *Maintainability Guide for Design,* AMCP 706-134, Department of the Army, Washington, October 1972.

60. *Maintainability Engineering Handbook,* NAVORD OD39223, Naval Ordnance Command. Washington, February 1970.

61. *Maintainability,* Design Handbook 1-0 (DH-1-9), Air Force Systems Command, Washington, December 1973.

62. *Maintainability Program Plan,* DD 1664, DI-R-1740, Department of the Army, Washington, December 1969.

63. *Maintainability Reports,* DD 1664, DI-R-1741, Department of the Army, Washington, December 1969.

64. *Maintainability Mathematical Model(s),* DD 1664, DI-R-1742, Department of the Army, Washington, December 1969.

65. *Maintenance Support Plan,* DD 1664, DI-S-1813, Department of the Army, Washington, December 1969.

66. *Maintenance Engineering Analysis (MEA) Data,* DD 1664, DI-S-6171, Department of the Army, Washington, April 1971.

67. *Integrated Logistic Support Program Requirements,* MIL-STD-1369, March 1971.

68. *Safety Engineering of Systems and Associated Subsystems and Equipment,* MIL-S-38130A, June 1966.

69. *Safety Engineering of Aircraft Systems, Associated Subsystems and Equipment, General Requirements For,* MIL-S-58077, June 1964.

70. *System Safety Program for Systems and Associated Subsystems and Equipment, Requirements For,* MIL-STD-882, July 1969.

71. *System Safety,* Design Handbook 1-6, Air Force Systems Command, Washington, January 1969.

72. *Safety Manual,* NHB 1700.1, NASA, Washington, March 1970.

73. *Quality Program Requirements,* MIL-Q-9858A, December 1963.

74. *TOW Advanced Engineering Test and Test-Limit Analysis,* TOW Report T-68, Revision B, Hughes Aircraft, Culver City, Ca., August 1970.

75. *Airplane Strength and Rigidity,* MIL-A-8860 to 8870, May 1960.

76. *Data and Tests, Engineering Contract Requirements for Aircraft Weapons Systems,* MIL-D-8706B, August 1968.

77. W. H. Miller, *A Review of Bureau of Aeronautics Requirements for the Demonstration of Airplanes,* Department of the Navy, Washington, September 1957.

78. *Department of Defense Authorized Data List: Index of Data Item Descriptions,* TD-3, Department of Defense, Washington, April 1972.

79. *Standard for the Preparation of Component Procurement/Performance Specifications,* KSC-STD-P-0002, NASA Kennedy Space Center, Fl., May 1976.

80. G. S. Peratino, *The Interagency Data Exchange Program,* Defense Industry Bulletin, Office of the Assistant Secretary of Defense (I&L), Washington, November 1967.

81. *Definitions of Effectiveness Terms for Reliability, Maintainability, Human Factors, and Safety,* MIL-STD-721B, March 1970.

82. A. R. Wooten, *Bibliography on Reliability,* AD-773 722/4GI, National Technical Information Service, Springfield, Va., July 1973.

83. *Reliability of Military Electronic Equipment,* Advisory Group on Reliability of Electronic Equipment (AGREE), Office of the Assistant Secretary of Defense (R&E), U.S. Govt. Printing Office, Washington, June 1957.

84. R. Lusser, *A Study of Methods for Achieving Reliability of Guided Missiles,* BuAer TR-75, Naval Air Missile Test Center, Point Mugu, Ca., July 1950.

85. *Reliability,* Command Regulation 705-1, Army Missile Command, Redstone Arsenal, Al., February 1963.

86. W. Yurkowsky, *Data Collection for Non-Electronic Reliability Handbook,* RADC-TR-68-114, Rome Air Development Center, Rome, N.Y., June 1968.

87. *Reliability Stress and Failure Rate Data for Electronic Equipment,* MIL-HDBK-217A, December 1965.

88. *Reliability Prediction,* MIL-STD-756A, May 1963.

89. *Reliability Computation from Reliability Block Diagrams,* Tech Brief B75-10276, NASA Technology Utilization Office KT, Washington, October 1975.

90. C. S. Rank, *Automatic Checkout Systems, A Report Bibliography 1958-1963,* OTS 429929, Department of Commerce, Washington, 1964.

91. *Metallic Materials and Elements for Aeroplane Vehicle Structures*, MIL-HDBK-5B, September 1971.

92. *Reliability Tests, Exponential Distribution*, MIL-STD-781B, November 1967.

93. B. B. Bond, *Spectrometric Oil Analysis*, NARF-P-1, Naval Air Rework Facility, Pensacola, Fl., June 1967.

94. J. M. Ward, "Spectrometric Oil Analysis," in *Approach*, Naval Aviation Safety Center, Norfolk, Va., March 1965.

95. *Elements of Design Review for Space Systems*, NASA SP-6502, Clearinghouse for Federal Scientific and Technical Information, Springfield, Va., 1967.

96. J. C. French and F. J. Bailey Jr., "Reliability and Flight Safety," in *Mercury Project Summary*, NASA SP-45, U.S. Govt. Printing Office, Washington, October 1963.

97. *Cost Information Reports for Aircraft, Missles and Space Systems*, Budget Bureau No. 22-R260, Department of Defense, Washington, April 1966.

98. *PERT/COST Systems Design*, DOD and NASA Guide, U.S. Govt. Printing Office, Washington, June 1962.

99. L. Weglarz, *Special Operational Data Collection Plans*, AD 739438, Army Materiel Command Intern Training Center, Texarkana, Tx., 1971.

100. *Report of the Apollo 204 Review Board*, NASA, U.S. Govt. Printing Office, Washington, 1967.

101. *Apollo Accident*, Hearings Before the Committee on Aeronautical and Space Sciences, U.S. Senate, 90th Congress, U.S. Govt. Printing Office, Washington, 1967.

102. *NASA Safety Management of a Complex R&D Ground Operating System*, TM-X-71697 (N 75-22183), Aerospace Research Applications Center, Indiana University, Bloomington, In., October 1975.

103. *Risk Management Technique for Design and Operation of Liquified Natural Gas Facilities and Equipment*, NASA CR-139183 (N75-19823/LK), National Technical Information Service, Springfield, Va., December 1974.

104. D. Allan, *Technology and Current Practices Transferring and Storing Liquefied Natural Gas*, PB 241048, National Technical Information Service, Springfield, Va., December 1974.

105. *Calibration System Requirements*, MIL-C-45662A, February 1962.

106. *Sampling Procedures and Tables for Inspection by Attributes*, MIL-STD-105D, March 1964.

107. *Multi-Level Continuous Sampling Procedures and Tables for Inspection by Attributes*, MIL-HDBK 106, October 1958.

108. *Inspection and Quality Control, Single-Level Continuous Sampling Procedures and Tables for Inspection by Attributes*, MIL-HDBK 107, April 1959.

109. *Quality Control and Reliability Sampling Procedure and Tables for Life Reliability Testing Based on Exponential Distribution*, MIL-HDBK 108, April 1964.

110. *Sampling Procedures and Tables for Inspection for Variables by Percent Defective*, MIL-STD-414, May 1968.

111. *R&D Test and Evaluation During Development and Acquisition of Materiel*, AR 70-10, Department of the Army, Washington, December 1972.

112. *Reliability Growth Seminar, Supplementary Material*, Army Management Engineering Training Agency, Rock Island, Il., Spring 1972.

113. *Reliability Growth–Methods and Management,* Army Management Engineering Training Agency, Rock Island, Il., April 1973.

114. *Reliability Growth Management,* Army Management Engineering Training Agency, Rock Island, Il., November 1974.

Periodicals

115. J. N. Miller, "The Burning Question of Brown's Ferry," *The Reader's Digest,* April 1976.

116. J. Coutinho, "Failure Effect Analysis," *Transactions,* New York Academy of Sciences, New York, March 1964.

117. J. Coutinho, "Configuration Analysis," *Standards Engineering,* Standards Engineers Society, New Providence, N.J., May and June 1963.

118. M. Mead, "The Nation's Stake in Air and Space," *Astronautics and Aeronautics,* AIAA, New York, January 1976.

119. "Life in the Space Age," *Time,* July 4, 1969.

120. W. Hoffer, "Evolution of an Engineering Manual," *New Engineer,* October 1973.

121. "Guidelines for the Practice of Operations Research," *Journal of the Operations Research Society of America,* Baltimore, Md., September 1971.

122. C. A. Robinson, "Army Nears Armed Helicopter Choice," *Aviation Week and Space Technology,* May 14, 1973.

123. J. Coutinho, "Whither Reliability?" *Journal of Spacecraft and Rockets,* AIAA, New York, January-February 1965.

124. W. A. Stanbury, "QE 2 and the Small Goof," *Product Engineering,* February 24, 1969.

125. I. W. Burr, "Glossary of General Terms Used in Quality Control," *Quality Progress,* ASQC, Milwaukee, Wis., July 1969.

126. R. F. Rolsten, "A Study of the Shock Loading of Materials," *Transactions,* New York Academy of Sciences, New York, May 1974.

127. J. J. Bussolini, "The Application of Overstress Testing-to-Failure to Airborne Electronics," *Transactions on Aerospace and Electronics,* IEEE, New York, March 1968.

128. L. J. Esterby, "Quality Assurance in Design and Development," *Design News,* February 23, 1976.

129. J. Coutinho, "Quality Requirements from the Systems Viewpoint," *Quality Progress,* ASQC, Milwaukee, Wis., October 1969.

130. "Two Year Life Expected for OAO Satellite," *Aviation Week and Space Technology,* May 12, 1969.

131. R. P. Hudock, "The National Scene," *Astronautics and Aeronautics,* AIAA, New York, April 1969.

132. R. Holz, "Apollo 12—New Era for Science," *Aviation Week and Space Technology,* December 8, 1969.

133. E. B. Roberts, "The Myths of Research Management," *Science and Technology,* August 1968.

134. R. A. Frosch, "A New Look at Systems Engineering," *Spectrum,* IEEE, New York, September 1969.

135. E. R. Schubert and L. Calniker, "Synthesizing Aircraft Design," *Space/Aeronautics,* Conover Mast, New York, April 1969.

136. "Condensed Log of 1957–68 Space Projects, and Box Score of U.S. Spacecraft Launches," *Space Log,* TRW Systems Group, Redondo Beach, Ca., Winter 1968–69.

137. S. Levine, "Man-Rating the Gemini Launch Vehicle," *Astronautics and Aeronautics,* AIAA New York, November 1964.

138. "Quality Assurance Program Requirements for Nuclear Power Plants," *ANSI N45.2—1971,* ASME, New York, 1972.

139. J. T. Duane, "Learning Curve Approach to Reliability Monitoring," *Transactions on Aerospace,* IEEE, New York, April 1964.

140. J. Coutinho, "Reliability and Maintainability, the Basic Problem," *Mechanical Engineering,* ASME, New York, February 1966.

141. M. A. Calvert, "Out of One, Many. American Engineering Societies," *Engineer,* Engineers' Joint Council, New York, May/June 1969.

142. J. Coutinho, "The Engineer's Professional Dilemma—Public Interest vs. Private Profit," *Newsletter of the Society for Social Responsibility in Science,* Bala Cynwyd, Pa., May 1967.

143. D. P. Billington, "Engineering Education and the Origins of Modern Structures," *Civil Engineering,* ASCE, New York, January 1969.

Technical Papers

144. E. E. Fawkes, "BIMRAB and Its Influence on the BuWeps System Effectiveness Program," *Annals of Reliability and Maintainability,* Vol. 4, Spartan, Washington, 1965.

145. W. R. Lomas, *Safety Considerations in the Design of Flight Control Systems for Navy Aircraft,* ASME Paper 60-AV-34, 1960.

146. D. A. Guenther, *Product Liability and the Mechanical Engineer,* ASME Paper 75-WA-SAF-1, 1975.

147. A. W. Blackman Jr., *The Role of Technological Forecasting in Analysis and Planning of New Ventures,* ASME Paper 72-DE-26, 1972.

148. J. Coutinho, *Technology and Civilization,* ASME Paper 72-Aero-3, 1972.

149. J. Coutinho, *Systems Engineering—Minimizing Product Defects,* 39th Annual Eastern Regional Safety Convention and Exposition, Greater New York Safety Council, New York, April 1969.

150. J. Coutinho, *Reprioritizing the Weapons Acquisition Process,* Proceedings, Annual Reliability and Maintainability Symposium, IEEE, New York, 1975.

151. D. B. Dickinson and L. Sessen, *Life Cycle Cost Procurement of Systems and Spares,* Proceedings, Annual Reliability and Maintainability Symposium, IEEE, New York, 1976.

152. T. E. Dixon and R. H. Anderson, *Implementation of the Design to Cost Concept,* Proceedings, Annual Reliability and Maintainability Symposium, IEEE, New York, 1976.

153. *Session on Warranties,* Proceedings, Annual Reliability and Maintainability Symposium, IEEE, New York, 1976.

154. D. R. Earles, *Life Cycle Cost-Commercial Application; Ten Years of Life Cycle*

Costing, Proceedings, Annual Reliability and Maintainability Symposium, IEEE, New York, 1975.

155. J. Coutinho, *Vorhersage der Abschussbereitschaft von Raumfahrtsystemen Waehrend der Entwicklungsphase und Abschussvorbereitung,* Fakultaet fuer Maschinenwesen, Technische Universitaet Berlin, Charlottenburg, 1970.

156. L. H. Crow, *On Tracking Reliability Growth,* Proceedings, Annual Reliability and Maintainability Symposium, IEEE, New York, 1975.

157. T. D. Cox and J. Keely, *Reliability Growth of SATCOM Terminals,* Proceedings, Annual Reliability and Maintainability Symposium, IEEE, New York, 1976.

158. T. W. Yellman, *Event-Sequence Analysis,* Proceedings, Annual Reliability and Maintainability Symposium, IEEE, New York, 1975.

159. R. G. Lohmann, *Automatic Checkout Equipment on Military Aircraft,* ASME Paper 60-WA-330, 1960.

160. G. B. Stanton, *The Future of Safety and Health; A Challenge to Engineering Managers,* ASME Paper 75-WA/Mgt-7, 1975.

161. M. A. Schultz and A. Geesy, *A New Approach to Nuclear Safety Shutdown System Design,* ASME Paper 72-Aero-4, 1972.

162. S. Collier, "Reliability and People," *Annals of Reliability and Maintainability,* Vol. 4, Spartan, Washington, 1965.

163. W. Keen, R. Creel, and C. Baum, *The Navy Approach to the Development of Airframe Structural Design Control Requirements,* ASME Paper 60-AV-50, 1960.

164. J. Coutinho, *A History of Selected Reliability and Maintainability Committees and Interested Government Agencies,* AIAA Paper 66-856, 1966.

165. N. Lichter and G. Friedenreich, *Reliability Analysis of Redundancy Mechanism,* Proceedings, Seventh Military-Industry Missile and Space Reliability Symposium, Naval Air Systems Command, Washington, June 1962.

166. R. R. Carhart, *A Survey of the Current Status of the Electronic Reliability Problem,* RAND Corporation, RM-1131, Santa Monica, Ca., August 1953.

167. J. S. Greenberg and G. A. Hazelrigg Jr., *A Reliability, Cost and Risk Analysis of Establishing and Maintaining a Space Communications Satellite Network,* AIAA Paper 73-582, 1973.

168. F. A. Thompson, *Government and Industry Attack the Reliability Problem,* ASME Paper 59-A-326, 1959.

169. L. Bass, H. W. Wynholds, and W. R. Porterfield, *Fault Tree Graphics,* Proceedings, Annual Reliability and Maintainability Symposium, IEEE, New York, 1975.

170. R. L. Eisner, *Fault Tree Analysis to Anticipate Potential Failure,* ASME Paper 72-DE-22, 1972.

171. D. B. Board, *Incipient Failure Detection in CH-47 Helicopter Transmissions,* ASME Paper 75-WA/DE-16, 1975.

172. S. O. Nilsson, *Reliability Data on Automotive Components,* Proceedings, Annual Reliability and Maintainability Symposium, IEEE, New York, 1975.

173. P. H. Wirsching, *On the Behavior of Statistical Models Used in Design,* ASME Paper 75-WA/DE-28, 1975.

174. C. W. Johnson and R. E. Maxwell, *Reliability Analysis of Structures—A New Approach,* Proceedings, Annual Reliability and Maintainability Symposium, IEEE, New York, 1976.

175. D. Kececioglu and L. B. Chester, *Combined Axial Stress Fatigue Reliability for AISI 4130 and 4340 Steels*, ASME Paper 75-WA/DE-17, 1975.

176. M. C. Shaw, *Brittle Fracture Under a Complex State of Stress*, ASME Paper 75-WA/Prod.14, 1975.

177. D. R. Hayhurst, C. J. Morrison, and F. A. Leckies, *The Effect of Stress Concentrations on the Creep Rupture of Tension Panels*, ASME Paper 75-WA/APM-R, 1975.

178. L. A. James, *Some Questions Regarding the Interaction of Creep and Fatigue*, ASME Paper 75-WA/MAT-6, 1975.

179. A. S. Kotayashi, N. Polvanich, A. F. Emery, and W. J. Love, *Corner Crack at the Bore of a Rotating Disk*, ASME Paper 75-WA/GT-18, 1975.

180. R. P. Goel, *On the Creep Rupture of a Tube and a Sphere*, ASME Paper 75-WA/APM-4, 1975.

181. N. Jones, *Plastic Failure of Ductile Beams*, ASME Paper 75-WA/DE-3, 1975.

182. D. W. Parsons, "Reliability Demonstration—Shillelagh Missile Subsystem," *Annals of Reliability and Maintainability*, Vol. 5, AIAA, New York, July 1966.

183. P. J. Bruneau, *Jet-Engine Trend Analysis*, ASME Paper 59-A-297, 1959.

184. K. R. Hamilton and M. H. Chopin, *Diagnostic Engine Monitoring for Military Aircraft*, Proceedings, Annual Reliability and Maintainability Symposium, IEEE, New York, 1975.

185. J. Coutinho, *Contracting for Reliability*, Proceedings, Conference on Advanced Marine Engineering Concepts for Increased Reliability, University of Michigan, Ann Arbor, Mich., February 1963.

186. E. C. Towl, "Fair Price and Fair Play," *Annals of Reliability and Maintainability*, Vol. 4, Spartan, Washington, 1965.

187. A. J. Kullas and W. Bishop, "The Role of Reliability and Quality Assurance in Advanced Systems," *Annals of Assurance Sciences*, ASME, New York, 1968.

188. E. G. Hantz and A. E. Lager, *Configuration Management—Its Role in the Aerospace Industry*, Transactions, Product Assurance Conference, ASQC, IEEE, SES, Hofstra University, Hempstead, N.Y., 1968.

189. W. W. Offner, *Nuclear Power Plant Construction and the Public Safety*, ASME Paper 75-WA/TS-5, 1975.

190. O. H. Fedor, W. N. Parsons, and J. Coutinho, *Risk Management of Liquefied Natural Gas Installations*, Proceedings, Annual Reliability and Maintainability Symposium, IEEE, New York, 1976.

191. G. B. Mumma and W. R. O'Halloran, *Hazard Analysis—Space Applications to Mass Transit*, Proceedings, Annual Reliability and Maintainability Symposium, IEEE, New York, 1976.

192. H. Peter, *Experiences of Receiving and Unloading LNG from an LNG Barge*, Brooklyn Union Gas Co., Brooklyn, N.Y., May 1975.

193. W. H. Friedlander, *Process Control for Reliability*, Proceedings, Ninth National Symposium on Reliability and Quality Control, IEEE, New York, 1963.

Unpublished Work

194. L. Krause, *Industrial Hazards of PVC and TDI*, 96th Winter Annual Meeting, ASME, New York, 1975.

195. R. Barnett, *Relationship of Case Law to Industrial Safety*, 96th Winter Annual Meeting, ASME, New York, 1975.

196. B. Hoyt, Letter, Secretary, ASME Boiler and Pressure Vessel Committee, 345 E 47th St., New York, to J. Coutinho, 31 July 1975.

197. L. I. Medlock, *Quality Control in the Integrated Management System*, Conference on Quality Requirements from the Systems Viewpoint, Engineering Foundation, New York, August 1968.

198. D. S. Feigenbaum, *The Systems Viewpoint and Quality*, Conference on Quality Requirements from the Systems Viewpoint, Engineering Foundation, New York, August 1968.

199. *Report of the Task Group on Transportation to the Committee on Technical Affairs*, ASME, New York, 6 June 1968.

200. Discussion with H. Chanin, Office of the Comptroller, Picatinny Arsenal, Dover, N.J., May 26, 1976.

201. A. A. Weintraub, *Systems Safety Engineering Techniques; Minimizing Human Errors*, 39th Annual Eastern Regional Safety Convention and Exposition, Greater New York Safety Council, New York, April 14–18, 1969.

202. D. S. Bassott and E. C. Corey, *Automatic Checkout for Aerospace Vehicles*, Lectures, University of California, Los Angeles, since 1965.

203. G. D. Breen, *Computing the Incremental Cost of Systems Engineering*, Memorandum, Grumman Aerospace Corporation, Bethpage, N.Y., December 10, 1968.

204. J. Lederer, *Origin of "Risk Management" for Use in the Aerospace Industry*, April 20, 1976.

205. C. D. Edwards, *The Meaning of Quality*, Conference on Quality Requirements from the Systems Viewpoint, Engineering Foundation, New York, August 1968.

206. R. W. Pratt, *Impact of Product Support Engineering on Systems Effectiveness*, Summary Report of the Board Meeting—10 April 1969, Naval Air Systems Effectiveness Advisory Board, Naval Air Systems Command, Washington, 1969.

Index

Acceptance, customer, 49, 257
Acceptance procedure, 315
Acceptance testing, 180
  coordination, 225
  plan, 180
  requirements, 180
  rig (fixture), 334
  specimen, 180
Accept-reject criteria, 180
Accident, definition, 131
  prevention, 131
Accountability, 154
Accounting procedures, required uniform,
  266
Accounting system, 81
  collateral requirements, 272
  standard, 272
Acquisition, *see* Procurement
Action center, 290
Action items, design reviews, 159
Adversary customer-contractor viewpoints,
  160, 253
Aerospace industry objectives, 234
  business, 236
  general, 235
  social, 236
  technical, 235
Aerospace progress, 400
Affordability, 277
Aging, artificial (accelerated), 217
Aircraft, flutter, 356
  flying qualities, simulation of, 176
  program phases, 386
  reliability growth, 388
  system development, 386
  system growth, 388
  weight control, 12, 161
Airplane, impact on history, 4
American Airlines Reservation Center, 19
Analysis, required completeness, 41
  integrity, 41
  purpose, 157
  system, 39
Analyst-designer relationship, 208
Analytical input, timeliness, 162
Apollo fire, 317
Apollo program, 8
Apollo reliability, evaluation, 229
  requirements, 202
Apportionment, 103
  reliability, 202
  requirements, 154
Appropriations, annual, 29
Approval, customer, 150, 155
Approximations, engineering, 12
Artificial aging, 217
As-built records, 48
As-designed records, 48
ASME, *see* Codes
Assembly, automatic, 365
Assembly line, moving, 365
Assessment, technical risk, 40
Assurance, risk management, 321
  sciences, 190
  systems, 6
Attitude, contractor, 329
  management, 267
Automatic assembly, 365

419

Automatic checkout, *see* Test, equipment
Automatic processing, 365
Automobile transportation system, example, 20
Availability, 85
Availability date, 287
Award fees, 61
Award program, safety motivation, 133

Baseline, 313
 allocated, 314
 functional, 314
 government studies, 60
 production, 314
Best effort, 102
Block diagram, functional, 206
Breakdown, bottom-up, 284
 life-cycle cost, 81
 top-down, 284
Break-out, 196, 219
Brochuremanship, 163
Budget, 271
 dollar, 272
 manpower, 272
Business environment, subcontracting, 352

Calibration, 344
Capability, contractor technical, 28, 241
Carpenter shop, example, 242
CCE, *see* Evaluation
Certification, inspection, 359
Challenges, contemporary, 399
Characteristics, scoring, 73
 critical, 74, 337
 qualifying, 74
Chart room, 290
Check-out equipment, *see* Test, equipment
Chits, design review, 159
Codes, 14, 43
 analysis agency, central, 319
 ASME Boiler and Pressure Vessel, 3, 33, 44, 320
 building, 318
 regulatory, 319
Command center, 290
Commonality, 135, 293
Communications, 305
 intra-company, 266
 technical staff, 239
Company standards, 137

Competition, 23, 52, 54, 398
 impact on motivation, 63
 outside, 255
Competitive developments, 53, 63
 contract negotiations, 64
 control, 69
 customer evaluation (CCE), 65, 70, 71
 limitations, 69
 source selection, 68
Competitive factors, scoring, 74, 80, 89
Competitive negotiations, 65
Competitive parallel developments, 53
Competitive points, scoring, 75
Complexity, 288
 equipment, 11
 fixtures, 346
Component, definition, 155
 development, 155
 linear, 171
 non-linear, 171
 design requirements, 172
 requirements, 155, 177
Computer applications, system management, 299
Computer-based systems, 19
Computer record keeping, 299
Computer simulation, 168
Concept formulation, 36
Conditions, controlled, 337
Confidence, engineering, 227
 margins, 174
Configuration, system, 155
 optimum, 154
Configuration analysis, 154, 204
Configuration control, 49, 310
 advanced systems, 311
 limitations, 310
 scope, 312
Configuration identification, 313
Consumerism, 1
Contingency planning, risk management, 322
Contract, 52
 cancellation threat, 63
 changes, 303
 definition, 257
 DTUPC, 62
 milestone reviews, 220
 negotiation, 57
 reference unit, 285

requirements, 26, 101, 327
structure, 303
target price, 118
wording, 58
Contract implementation, adaptive ap-
    proach, 329
legalistic approach, 328
Contracting, competitive, 64
cost-plus, 56
fixed-price, 56, 146
Contractor, attitudes, 329
estimating ability, 84
integrity, 56
motivation, 53, 118, 149
role of, 404
source inspection, 358
strategic advantages, 58
system, 122
Contractor responsibilities, 23, 26, 186
for GFE, 363
for supplier parts, 353
Control, 271
charts, status, 289
configuration, *see* Configuration control
cost, 163
countermeasures, 291
design, 140
by drawing number, 286
engineering documents, 339
hardware design, 12
manufacturing processes, 365, 370
package, 282
project, 289
stockroom material, 366
system engineering process, 401
Controlled conditions, 26, 129, 337, 379
Corrective action, 49, 95, 201
feasibility, 155
gas turbines, 392
Cost, 42
center, functional unit, 249
control, 163
design-to, 52, 59, 64
DTUPC, 52
fixed, 82
impact, of design objectives, 234
life-cycle, 52, 59, 119, 272
maintenance, 59
monitoring, competitive developments, 70
nonrecurring, 82

operating, 60, 83
overruns, 278, 405
of ownership, 52, 59
production, 82
production savings, 119
public scrutiny, 195
recurring, 82
of quality, 225
sunk, 82
support, 59, 83
underestimated, 278
Cost data, quality, 373
Cost estimates, 81
parametric, 276
Cost information report (CIR), 280
Cost management evaluation, 91
Cost-plus, 310
contracting, 56
overruns, 267
Craftsmanship, 268
Creative individuals, management, 239
Creators and evaluators, 261
Credibility, subcontractor monitoring, 295
Criteria, acceptance, 333, 337
competitive evaluation, 77
accept-reject, 101
decision, 73
Critical characteristics, 337
Critical factors, quantification, 205
Critical items, 290
Critical missions, 89
Customer, engineering test, 49, 376
field/service test, 49, 379
in-house capability, 155, 239
monitoring, 56, 149
needs, 23
operational test, 48
priorities, 55
quality system monitoring, 335
relations, 255, 258
requirements, 233
responsibilities, 23, 26, 183
role of, 404
satisfaction, assuring, 330
visibility, competitive development, 70
witness, 46
Customer acceptance, 49, 257
Customer approval, 150, 155
Customer competitive evaluation (CCE),
    65, 70, 71

Customer-contractor interface, 150
Customer-contractor responsibilities, 155
Customer design approval, 318

Data, abstract, 300
  categories, 300
  current-input, 300
  design, *see* Design data
  pick-off, 300
  requirements, CCE, 71
  stored, 300
Data bank, rolling wave, 304
Data collection systems, 299
  duplication, 306
  field troubles, 218
Data item description, 149
  forms, DD 1664, 121
Data package, acceptance, 47
  competitive evaluation, 77
Decision criteria, 73
Decision makers, 41
Decision making, rules, 270
  small organization, 270
Decision points, 45
Decision to proceed, 40, 47
Defect, correction, 146
  definition, 199
  quality, 347
Delivery date, 287
Demonstration, concept formulation, 148
  see also Test, demonstration
Demonstration cycle, system, 35
Deployment, 145
Derating, electronic components, 217
Design, aircraft weight review, 161
  assurance, 188
    effectiveness, 226
  conceptual, 275
  conditions, 157
  customer approval, 158
  decisions, justification records, 154
  discipline, 401
  engineer, characteristics, 140, 251
  engineering assurance techniques, 202
  engineering outlook, 6
  information feedback, 262
  initial maintenance considerations, 123
  limit loads, 212
  objectives, cost impact, 234
  preliminary, 275

professional engineering task, 399
  quality of, 327
  requirements, 141
  reviews, 158, 220
  subsystem, 152
  time, 352
  trade-off, 105
  trade-off tools, 174
  trial and error, 111
  ultimate loads, 212
  verification testing, 47, 165
Design changes, impact on quality, 353
Design control, 45, 47, 140, 157
  analysis, 146
  hardware, 12
  introduction, 15
  safety factors, 213
  subcontractors, 293
  test, 146
  tools, 189
Design data, maintainability, 120
  requirements, CCE, 71
Designer, 140, 251
  versus analyst, 208
Design-to-cost, 52, 59, 64
  priorities, 59
  risks, 60
Design-to-performance, 52, 54
Design-to-unit-production-cost (DTUPC),
    52, 60, 118
  contract, 62
Design units, functional, 243
Deterministic analysis, 190
Development, 39
  advanced systems, 7, 47, 114, 143
  commercial products, 23
  competitive, *see* Competitive develop-
    ments
  component, 155
  engineering, 115
  field/service testing, 380
  hardware, 145
  monitoring, 55
  phases, 278
  postponement, 58
  process monitoring, 262
  prototype, 45
  staff, 233
  test categories, 164
  tools, 174

Development bed, 170
Development cycle, aerospace, 401
  system, 35
Dimensional standards, 347
Discipline, 130
  work, 347
Distractions, development process, 253
Diversification, defense contractors, 238
Document, format, 304
  function, 304
Documentation, engineering, 307
Domino theory, 131
Drawings, quality assurance, 342
  release, 46, 342
  source control, 285
  tree, 284
  work breakdown, 303
Drawing system, need for formal proce-
    dures, 309
Duane diagram, 382
Dynamic characteristics, simulation, 173
Dynamic equivalent, 168

Economic constraints, 398
Economic impact, system development, 28
Effectivity point, drawings and specifica-
    tions, 342
Efficiency, management information sys-
    tems, 302
Element tests, 164
Employment, defense contractors, 237
Emulator, 168
Enforceable contract requirements, 53
Engineer, cognizant, 282
  motivation, 275
  professional orientation, 397
Engineering, change proposals (ECP), 49
  competence, 260
  control, of changes, 342
    of documentation, 339
  critic, 399
  definition, 190
  department organization, 278
  design assurance techniques, 228
  design outlook, 6
  development, 115
  funding, ECPs, 376
  professional orientation, 397
  quality, 307
  required depth, 309

societies, 14
statesmanship, 399
systems staff, 140
Engineering tests, customer, 49, 374, 376
  initial production system, 379, 381
  pre-production model, 379, 380
  qualified component, 379
  qualified subsystem, 379
  system prototype, 379
Environment, controlled working, 2, 337
  system development, 233
Environmental simulation, 172
Equipment, availability, 287
  complexity, 11
  major, 33
  retrofitting, 312
Error, scheduling, 288
Estimate, inflated, 292
Evaluation, customer competitive (CCE),
    65, 70, 71
  cost management, 91
  life-cycle cost, 83
  management responsiveness, 94
  performance growth potential, 97
  program management, 90
  schedule management, 91
  system performance, 87
  technical management, 94
  functional manager leadership, 258
  functional unit, 251
  subcontractor, 251
  system characteristics, 376
Evaluator, government, 259
  professional, 260
Expectation, performance, 199
  growth potential, 97
Experimental method, limitations, 188

Failed component, identification, 209, 219
Fail-safe provisions, 207
Failure, catastrophic, 198
  class, 198, 207
  consequence, 199
  count, 385
  criticality, 207
  data, 198
  definition, 198
  diagnosis, 219
  dynamic factors, in failure definition, 199
  final, 198

gradual, 198
impact, on human performance, 318
independent, 199
instantaneous, 198
intermittant, 198
occurrence, statistical analysis, 194
possible cause, 207
potential, 207
premature, 179, 324
report, 201
Failure effect, and mode analysis, 205
Failure rate, 85
    constant, 111
    generic, 111
Familiarization, project, 278
Fatigue, 15
    analysis, 215
    gas turbines, 391
    life, 215
    strength, 215
Fault tree analysis, 208
Feasibility, 110
    economic, 39
    technical, 45
Feasibility test, 165
Feedback, continuous experience, 401
    cycle, 160
    deployment, 145
    design information, 262
    dynamics, 403
    operational experience, 183
Feedback control loops, 160
    quality data, 335
Fees, 61
    awards, 61
    effectiveness, 61
    incentives, 53, 61, 118, 268
    multiple incentives, 62
    royalties, 53, 61
Field data collection, 218
Field data requirements, maintenance, 121
Field/service tests, customer, 49, 374, 380
    development phase, 380
    initial production system, 381
    pre-production prototype, 380
Financial development risk, 233
Financial incentives, 61
Fitting factor, 212
Fixed price contract, 56, 146
Fixtures (jigs), complexity, 346

Flow diagram, logic, 230
Functional organization, utilization, 249
Functional units, management, 254
    number, 245
    operation, 244
    versus project units, 241
Funding provisions, 29
    contingencies, 42
    downstream pay-offs, 29
    limitations, 28
    structural laboratories, 148
    system integration, 166

Gantt charts, 286
Gas turbines, corrective action, 392
    field monitoring, 391
    product improvement, 390
    reliability, 391
Gate, fault tree analysis, 209
George Washington Bridge, 59
Government furnished equipment (GFE),
    27, 363
Government laboratory, role, 262
Ground support equipment, 124
Growth potential, 61, 97
Guarantee, 146

Handbooks, engineering, 14, 137
    maintainability, 121
    military, 137
Hard tooling, 269
Hardware development, 145
Hazard, analysis, 128
    evaluation, 128
    health, 113
    potential, 322
    reality, 111
Hazard accountability, risk management,
    323, 403
Hazard identification, 128
    risk management, 322
Header unit, 281
Human, in the loop, 173
Human errors, 126
Humanities, 399
Human performance, impact of failure, 318

Ignorance factors, 174
Incentive, financial, 61
Incentive fee, see Fees

Incipient failures, 120
Inherent reliability, 111
In-house technical capability, customer,
    155, 239, 259
Inspected material, separation and protect-
    ion, 359
Inspection, 331, 338, 368
  completed item, 370
  final, 370
  government source, 356
  indirect, 338
  purchased parts, 359
  raw materials, 359
  receiving, 359
  records, 48, 370
  source, 333
  stamps, 361
  status, 361, 370
  structural statistical applications, 361
  tailoring of procedures, 361
Instructions, inspection, 368
  work, 337, 368
Insurance, safety indicator, 192
Integrated logistic support, definition, 191
Integrity, professional personnel, 41, 269
Intelligence operations, example, 243
Intended use, 26
Interactions, component, 166
Interchangeability, 49, 346
Interface, configuration, 313
  customer-contractor, 150
Invention, 106
Investment, 112
Iron monster, 170

Job, definitions, 273, 279
  estimating resource requirements, 301
  satisfactions, 269

Leadership, 256
Lead time, 287
Learning curve, 58
Least lousy solution, 17, 406
Liability, 2
Library, central project documentation, 294
Licensing, mechanics, 334
  personnel, 130
Life, system, 83, 104
Life cycle cost, 52, 59, 119, 272
  breakdown, 81

contractor estimate, 83
customer estimate, 83
data, 62
evaluation, 83
index, 84
scoring, 81
Limit load, 210
Linearity, components and models, 171
Line item, contract, 280
List, qualified products, 350
Loading, dynamic spectra, 215
  limit, 210
  multiple conditions, 215
  ultimate, 212
Logistics and support, 121
Long range planning, 273
Lowest bid, 28

Maintainability, 85, 119
  definition, 191
  design data, 120
  handbooks, 121
  model, 120
  specifications, 121
  warranties, 118
Maintenance, cards, 123
  category, 121
  costs, 59
  depot level, 123
  factory level, 123
  field data requirements, 121
  initial design considerations, 123
  intermediate level, 123
  organization level, 123
  scheduled, 121
  unscheduled, 121
  zones, 122
Maintenance engineering, 119
  definition, 191
Maintenance engineering analysis (MEA),
    121
Make-or-buy, 117, 270
Man, non-linear component, 172
Management, advanced system, 10
  effectiveness, 324
  functional unit, 254
  operating mode, 240
  responsibilities, 265
  responsiveness evaluation, 94
  structures, 17

system development, 265
system engineering, 9
technical staff, 233, 238
visibility, 279
visibility review, 222
Management by crisis, 240
Management by indirection, 240
Management information system, 300
output requirements, 301
program flexibility, 302
special reports, 302
Manager, functional, 258
Mandatory standards, 135
Manpower, budget, 272
loading, 286
Manuals, engineering, 14, 137
policy, 269
Manufacturing, definition, 190
functions, 152
operations, 365
plan, 43, 114, 119
plan review, 226, 344
schedules, 115
Margin of safety, 147, 211
Marketplace, public, 23
Master schedule, 287
Master template, 346
Material, incoming, 333
nonconforming, 368
rejected, disposition, 359
stockroom control, 366
uncontrolled, 369
Material review board (MRB), 334
Mathematical techniques, design assurance, 226
Maturity, 108
Mean-time-between-failures (MTBF), 85
Measuring devices, control, 344
protection, 346
Mechanics, licensing, 334
Milestones, contractual review, 220
Military handbooks, 137
MIL-HDBK-5, 211, 320
Military specifications, 137
Military standards, 137
MIL-STD-721, 190, 198
MIL-STD-780, 82
MIL-STD-781, 53
MIL-STD-881, 81
Missions, critical, 89

Mock-up board, 159
Model, builders, 255
deficiencies, 166
designation, 50
engineering, 11, 14
hardware, 11
limitations, 164
linear, 171
maintainability, 120
predictive, 46, 173
proprietary, 14
reliability systems, 203
software, 16
system, 19, 102
type, 33
Model and test correlation, 179
Modular concepts, 312
Monitoring, competitive development cost, 70
customer, 56, 149, 150
development, 55, 262
dynamic components, 394
process, 338
production, 180
progress, 149
quality cost, 373
quality system, 335
reliability growth, 381
subcontractors/suppliers, 292, 350
technical quality, 292
Monitoring a function, 331
Motivation, 8
awards, 133
competitive, 63
contractor, 53, 118, 149
safety, 133
management and union support, 134
techniques, 347
MTBF, 85
Multiple incentives, 62
Murphy's Law, 17, 130, 131, 406

National Aeronautics and Space Administration (NASA), 9
Need, recognition of, 38
Need date, 287
Negotiation, 57
Next-to-last, race for, 291
Non-linear components, 171
Nonrecurring costs, 82

Notebook, engineer's, 115

Obsolescence, 31
Obsolescent specifications, 329
Operability, 109
  measures, 110
  trade-off, 109
Operating conditions, envelope, 227, 378
Operating costs, 60, 83
Operating mode, alternate, 207
Operating practices, safe, 402
Operational concepts, 39
Operational conditions, combined, 217
Operational integrity, 379
Operational plan, 143, 145
Operational readiness, 109
Operational testing, 48, 374
Operations, continuous flow, manufactur-
    ing, 365
  classification of manufacturing, 365
  discrete-serial manufacturing, 365, 366
Operations research, 139, 143, 189
Operator, skill, 109
  training, 173
Organizations, technical, 406
Output, rejection procedure, 339
Overdesign, 398
Ownership, cost of, 52, 59

Pacing item, 291
Packaging, 368
Paperwork, 53
  delivery time, 306
  distribution, 306
  optimum amount, 305
  quality, 304
  required control, 307
  review of subcontractor, 297
  subcontractor quality control, 351
  volume, 294
Parameter tree, 103, 145, 154
Parametric estimates, cost, 276
Parametric procedures, 162
Parkinson's Law, 17, 406
Performance, 42, 199
  administrative, 289
  design-to, 52
  improvement, 267
  system, 7
  technical, 289

Personality, influence of, 190
Personnel, characteristics, in a large
    organization, 270
  management attitudes towards, 267
  performance, control of, 347
  performance criteria, 130
  turnover, customer, 263
  turnover, large organization, 270
PERT diagrams, 281
Peter Principle, 17, 406
Pilot reaction, evaluation, 175
Plan, 271, 289
  acceptance test, 180
  maintainability program, 121
  manufacturing, 43, 114, 119
  operational, 143, 145
  quality assurance, 328
  reliability program, 114
  safety, 128
  sampling, 362
  test, 40, 46, 158
Planning, 143, 273
  advanced, 119
  amount required, 274
  contingency, 322
  initial quality, 335
  long range, 273
Planning, scheduling, and control, 271
Policy manuals, 269
Potential failure, assumed, 206
  compensating provisions, 207
  consequences, 207
Power, age of, 4
Prediction, expected performance, 157
  technical, 244
Prelaunch operations, space systems, 395
Preliminary design, 275
Premature failure, 179, 324
Price, estimates, 276
  versus quality, 277
Priorities, customer, 55
Probability of success, 195
Process controls, 365
  statistical analysis, 370
Processing, automatic, 365
Process monitoring, 338
Procrastination, cost of management, 406
Procurement, advanced systems, 8, 23
  commercial products, 8
  single source, 23

Producer's risk, 23
Producibility, 114
Product engineering, 150
Product improvement, 316, 374
  aircraft, 386
  gas turbine, 386, 390
  space system, 395
Production, 115
  baseline, 314
  cost, 82
  expansion, 147
  monitoring, 180
  planning, 226
  redesign, 50
  release, 49, 374
  tooling, 344
Productive capability, 28
Product rule, 203
Product support, 125
Professional responsibilities, 397
Profits, company, 257
  versus quality, 251
Profit center, project unit, 249
Program, quality, 182, 335
Program management evaluation, 90
Program phases, 45
  aircraft, 386
Progress, monitoring, 149
  social impact, 5
Project, control, 289
  current, 244
  definition, 275
  versus functional units, 241
  organization, 247
  responsibility, 247
Project engineer, 140
  pre-World War II, 139
Project manager, evolution, 139
  leadership, 256
  orientation, 252
  responsibilities, 265
Proposals, 262, 275
  alternate, 276
  customer rating, 259
  offerer signed, 66
  technical evaluation, 56, 259
Proprietary processes, 26, 137, 197
  customer visibility, 197
  models, 14
Prototype, 115

development, 45
  field test, 380
Provisioning requirements, 125
Public relations, 257
Purchase, control of, 348
Purchase order, 293
  quality requirements, 353
  technical requirements, 353

Qualification, pre-CCE, 77
Qualification test, 33, 47, 160, 177
  applicability, 177
  contractual implications, 179
  plan, 159, 177, 337
  requirements, 177
  specimen, 177
  types, 179
Qualified products list (QPL), 350
Qualifying characteristic, scoring, 74
Quality, cost, 255
  cost monitoring, 370
  defect, 347
  impact of design changes, 353
  initial planning, 335
  monitoring, 292
  policy, 328
  versus price, 251, 277
  program, 182, 335
  purchase order requirements, 353
  records, 338
  responsibility for, 29, 348
  systems, 21
  technical, 290
Quality assurance, 146, 152
  codes, 328
  drawings, 342
  of a function, 339
  historical development, 327
  principles, 327
  program, 328
    definition, 330
    requirements, 331
  role, in design and development, 182, 224
  specifications, 342
Quality control, responsibilities, 191
  subcontractor paperwork, 351
  total, 327
Quality of life, 398
Quality system, customer monitoring,
    335

Railroad system, example, 20
Raw materials inspection, 359
Receiving inspection, 359
Record, as-built, 48
    as-designed, 48
    inspection, 48, 370
    quality, 338
Record keeping, computer, 299
Recurring costs, 82
Regulation, 320
Regulator, 318
    role of, 404
Regulatory codes, 319
Rejected material, disposition, 359
Rejection criteria, 101
Reliability, 85, 191
    aircraft structural, 356
    apportionment, 203
    gas turbines, 391
    improvement, empirical procedures, 391
    incentive warranties (RIW), 60
    inherent, 111
    introduction of the technology, 356
    life testing, 218
    need for control, 193
    probability definition, 198
    program objectives, 194
    program origins, 194
    requirements, 113, 202
    statistical treatment, 201
    warranties, 118
Reliability growth, 85, 114
    aircraft, 388
    application difficulties, 385
    budgeting, 382
    management tool, 384
    monitoring, 381
    cycle, 375
Reports, review procedure, 294
    technical, 154
    test, 46
Requalification, periodic, 316
Request for proposal (RFP), 36, 42
Requirements, acceptance test, 180
    Apollo reliability, example, 202
    apportionment, 103, 154
    component, 155, 177
    conflicting, 41
    contract, 26, 101, 327
    customer, 233

data, CCE, 71
design, 141
design data, 164
diluted, 29
enforceable, contractual, 53
field data, maintenance, 121
formulation, 148
hard, 34, 101
maintainability, 120
non-linear component design, 172
operational, 40
operational test, 375
overlooked categories, 104
performance, 101
proof-of-design, 141
provisioning, 125
purchase order, technical, 353
qualification test, 177
quality assurance program, 331
reliability, 113, 202
safety, 127
soft, 34, 102
systems, 26
technical, 101
versus test results, 31
trade-offs, 118, 214, 358
training, 104
World War II operational, 147
World War II production, 147
Research, free, 22
Research contracts, 258
Residual risk, 111, 321
Resources, available, 273, 388
    required program, 337
Responsibilities, contractor, 23, 186
    contractor-customer, 26, 28, 155
    customer, 23, 183
    division of, 28, 56
    professional, 397
    project manager, 247, 265
    for quality, 29, 348
Retrieval system, management information,
    302
Retrofitting, corrective action, 312
Review, analytical, 223
    contractual milestones, 220
    design, 220
    management visibility, 222
    manufacturing plans, 226, 344
    risk management, 323

safety, 323
subcontractor reports, 295
technical assurance, 222
test plans, 344
Review board, customer competitive
    evaluation (CCE), 77
  material (MRB), 334
RISCA II, 83
Risk, 42
  analysis, 83, 322
  area, competitive evaluation, 73
  customer, 179
  management assumption, 406
  producer, 23
  residual, 111, 321
  supplier, 179
  technical, 45
Risk management, 316
  enforcement, 323
  origin, 317
  scope, 319
Route card, 298
Royalty, 53, 61

Safe operating practices, 402
Safety, 318
  airline, 192
  checks, 323
  criteria, 128
  definition, 191
  factors, 212, 308
  margins, 174, 211
  motivation, 133
  plan, 128
  practices, enforcement, 133
  project reviews, 323
  requirements, 127
Safety of life, 213, 402
Sampling plan, statistical, 362
Scenario, operational, 39, 41, 114
Schedule, 42, 271, 286
  breakdown, 287
  completion, 273
  dominance, 30
  management evaluation, 91
  manufacturing, 115
  slippages, 405
Scheduled maintenance, 122
Scheduling, long range, 271
Scheduling, planning, and

  control system, 271
School health care, example, 242
Scoring, advantages of quantitative, 72
  areas, competitive factors, 89
  life-cycle cost, 81
  methodology, 73
  quantitative techniques, 71
    objections to, 71
  scope, 72
Sequencing diagram, 206
Service organization, 152
Signature, trend analysis, 209
Simulator, 168
  applications, 174
  design tool, 175
  environmental, 172, 176
  justification, 171
  origins, 170
  testing, 168
Simultaneous equations, analogy, 19, 246
Single source, 55
  procurement, 23
  supply, 348
Skills, operator, 109
Slavery, 5
Social values, 399
Software model, 16
Source inspection, 333
  contractor, 358
  government, 356
Source selection, 44, 56, 64
  competitive negotiations, 68
  parallel developments, 66
Space systems, product improvement, 395
Specifications, 285
  formulation, 148
  maintainability, 121
  military, 137
  obsolescent, 329
  pitfalls, 104
  process, 366
  quality assurance of, 342
  release, 342
  role of, 329
  types of, 136
Spectrometric oil analysis, 220
Standardization, 123
Standard parts, use, 358
Standards, 14, 43, 135, 293
  company, 137

dimensional, 347
mandatory, 135
military, 137
voluntary, 135
State of the art, 43, 308
advances, 55
definition, 109
extensions, 106, 189, 404
trade-offs, 105
Static strength, 210
Statistical analysis, process control, 370
Statistical methods, dynamic loadings, 216
Statistical techniques, sampling inspection, 361
Statistical test design, 46
Status, accounting, 315
control charts, 289
inspection, 361
reports, 301
Stockroom material control, 366
Stress analysis, 146
distribution, 148, 212
Subcontracting, 117
Subcontractor, calibration system, 344
capability, contractor support, 355
competitive selection, 352
complimentary skills, 350
cooperativeness, 350
coordination, 355
design control, 293
evaluation, 251
geographical distribution, 293, 350
monitoring, 292, 350
quality programs, 351
selection, 348
shortcomings, organizational, 299
Subsystem, definition, 152
integration test, 166
Success, orientation (philosophy), 17, 55, 257, 404
probability of, 195
Sunk cost, 82
Supplier, geographical distribution, 293, 350
monitoring, 350
qualified, 44
risk, 179
selection, 348
Supplier parts, contractor responsibility, 353

Supply, alternate sources, 196, 350, 351
single source, 348
Support, 121
System, 1
analysis, 39
approach, 5, 25
assurance, 2, 6
goal, 405
procedures, management acceptance, 3
complexity, 288
computer-based, 19
configuration, 155
contractor, 122
definition, 7, 18, 143
demonstration cycle, 35
evaluation, 376
growth, aircraft, 388
hardware, 19
integration, funding, 166
life, 83, 104
military, 7
model, 102, 203
one-of-a-kind, 33, 58
performance, 7
performance evaluation, 87
index, 90
procurement, 18, 23
quality, 21
requirements, 26
size, 288
technical excellence, lack of, 21
testing, 148
System development, 2, 35, 47, 143
aircraft, 386
environment, 233
mainstream functions, 150
space, 395
System engineering, 139
growth, 140
origin, 146
process control, aerospece, 401
scope, 141
staff, 140, 233

Target price contract, 118
Technical adequacy, 251
Technical data package, 315
Technical excellence, 241
Technical integrity, 49, 226, 377, 378

Technical papers, 14, 258
Technical progress, assessment, 274
  formula for, 401
Technical staff, management of, 238
  organization, 233
  responsibility, 241
  stability, 236
Technology, advanced, 1
  hazard of, 1
  impact on society, 3
Technology center, 244, 254
Telephone system, example, 20
Template, master, 346
Test, acceptance, 180, 338
  agency, orientation, 374
  customer engineering, 49, 374, 376,
    379
  customer field/service, 374, 379
  customer operational, 48, 374
  demonstration, 26, 28, 31, 102
  design, statistical, 46
  design verification, 47, 165
  destruction, acceptance, 225
  development, categories, 164
  drop, 215
  dynamic life, 214
  element, 164
  equipment, automatic, 120, 124, 209,
    219
  fatigue, 215
  feasibility, 165
  flight, 212
  laboratory demonstration, 157
  laboratory life, 391
  log, 46
  objectives, 377
  operational, versus qualification, 374
  overstress, 217
  points (test conditions), 227
  proof-load, 334
  qualification, 33, 47, 160, 177
  reliability life, 218
  reports, 46
  results, versus requirements, 31
  simulation, 168
  static, 212
  subsystem integration, 166
  system, 148
Test-fix-test cycle, 382
Test fixtures (rigs), acceptance, 334

  special purpose, 334
Test-model correlation, 179
Test plan, 40, 46, 158
  qualification, 159, 177, 337
  review, 344
Test-to-failure, 147
Theory X, 267, 268
Theory Y, 267, 268
Threats, potential, 235
Time between overhaul (TBO), 210
Tolerances, acceptance tests, 225
  requirements, 102
Tooling, 50, 116
  production, 344
Traceability, 154, 196, 334
Trade-offs, design, 105
  extensions, in the state of the art, 105
  studies, 39
Trailer unit, 281
Training, 130
  operating personnel (risk management),
    322
  professional, 256
  programs, 126
  requirements, 104
Trend analysis, 209, 219
Trouble analysis, 126
  reporting, 126
  shooting, 120, 208

Ultimate load, 210
Uncontrolled materials, 369
Unit, of work, 273
Unit-to-unit variation, 227
Unreliability, technical causes, 195

Value, technical, 290
Value engineering, 53
  royalties, 118
Vertical cut, 295
Visibility, customer requirements, 70
  management, 266, 295
Voluntary standards, 135

Waivers, 47, 77, 103, 334
Warning signals, accident, 128
Warranties, consumer goods, 180
  maintainability, 118
  reliability, 118
  reliability incentive, 60

Wear, gas turbines, 391
Wear-out, prediction, 210
Weight, demonstration, 162
  design review, 161
  prediction, 162
Weight control, aircraft, 12, 161
  effectiveness, 163
  introduction of, 13
Work, breakdown, 280, 303
  breakdown structure, 81, 82,
    266, 278, 280
  discipline, 347
  element, 281
  flow-diagram, 120
  instructions, written, 337
  unit code, 82
Workload, defense contractors, 237
Workmanship, 337
Work order, 366
Work package, 282